Environmental Management in European Companies

Environmental Problems and Social Dynamics

A series of books edited by Peter M. Allen, *Cranfield University, Cranfield, UK* and Sander E. Van der Leeuw, *Université de Paris, Paris, France*

Volume 1
Cities and Regions as Self-Organizing Systems: Models of Complexity
Peter M. Allen

Volume 2
Environmental Management in European Companies: Success Stories and Evaluation
Jobst Conrad

Environmental Management in European Companies
Success Stories and Evaluation

Edited by
Jobst Conrad

Free University of Berlin, Berlin, Germany

Gordon and Breach Science Publishers
Australia • Canada • China • France • Germany • India
Japan • Luxembourg • Malaysia • The Netherlands • Russia
Singapore • Switzerland • Thailand

Amsteldijk 166
1st Floor
1079 LH Amsterdam
The Netherlands

British Library Cataloguing in Publication Data

Environmental management in European companies : success
    stories and evaluation. - (Environmental problems and
    social dynamics ; v. 2)
    1. Industries – Environmental aspects – Europe   2. Industrial
    management – Environmental aspects – Europe
    I. Conrad, J. (Jobst)
    658.4'08'094

    ISBN 90-5699-085-3

CONTENTS

PREFACE

This book is the product of an international collaborative research project investigating exemplary cases of successful environmental management in Denmark, Germany, The Netherlands, Switzerland, Poland and Latvia. The case studies look at the role of company internal and external determinants explaining successful corporate environmental management on the micro-level. As in-depth investigations of social processes leading to such environmental management efforts, these case studies go beyond most corresponding analyses to be found in the upwelling literature on environmental management. Providing the background and context of this research project, presenting the success stories investigated in West European companies, and evaluating them in a comparative empirical as well as theoretical perspective with an environmental policy orientation, this study fits well and extends the tradition of the Environmental Policy Research Unit of outcome-oriented comparative environmental policy analysis of industrialized countries based on environmental performance data.

The research project was prepared and launched in 1991/92, the empirical case studies were performed in 1993/94, the comparative evaluation and writing of the book occurred in 1994/95, and the (publisher's) reviewing process happened in 1995/96. I should like to thank my scientific colleagues who were involved in the research project and contributed to this book, those colleagues who were and are involved in similar research efforts and provided valuable insights and comments for our investigation, the companies and the interviewees, in particular, allowing the inquiry into their environmental management efforts, the collaborating institutions employing the researchers and supporting the project by their infrastructural resources, namely the Environmental Policy Research Unit at the Free University, Berlin (Germany), the Institute of Sociology at the University, Halle (Germany), the Department of Environment, Technology and Social Studies at the Roskilde University (Denmark), the Centre for Clean Technology and Environmental Policy at the University of Twente (The Netherlands), the Center for Environmental Studies at the University of Gdansk (Poland), the Institute of Environmental Engineering at the Technical University of Bialystol (Poland), the Center for Environmental Science and Management Studies at the University of Latvia, Riga (Latvia), and finally, the EU-Commission for funding most of the research undertaken.

Berlin, October 1996   Jobst Conrad

vii

SUMMARY

This study investigates exemplary cases of successful environmental management in West European companies with the aim of discovering the reasons and dynamics underlying them and the role environmental policy did and could play. The nine case studies concern various Danish, Dutch, German and Swiss companies of different sizes and from different industrial branches, and they describe the social processes leading to substantive environmental achievements and the corresponding environmental management systems. They point out the variety of specific formal, organizational, economic and political mechanisms which may lead to such improvements, the interaction among in-house and external determinants, and the general (necessary) characteristics of these development processes, as far as they can be deduced from these and similar case studies. The case studies were carried out in 1993–94 on the basis of a common analytical framework, along with the help of extensive interviews with the main actors involved in the success story, and of related documents and literature. Furthermore, similar case studies were carried out in Poland and Latvia, which point to the enormous importance of the wider social context significantly differing from that of the West European companies investigated. Therefore, these eastern European case studies have not been included in this book, but can be found in the corresponding final report to the European Commission.

Essential background knowledge about successful environmental policy was compiled using 24 cases of (policy-related) successful environmental protection presented at an international conference in the initial phase of the project. The function of this first step was to indicate general conditions and influences relevant for environmental improvements in advanced industrialized countries on the basis of clear concepts evaluating 'success' or 'failure' in environmental policy and management, and to obtain background information about a broad variety of environmental policy and management achievements. These 24 case studies point to the eminent importance of interaction dynamics within the actor constellation, the still major significance of regulatory command-and-control policy, the need for appropriate mixes of policy instruments applied according to structural framework conditions and situational context, and the eminent importance of public awareness or even direct public pressure for successful environmental policy action.

The basic orientation of the comparative research project can be summarized in the following research questions:

1. What are real substantive cases of successful innovative environmental protection?
2. What are the influencing factors and determinants of these innovative success stories?
3. How can environmental policy support these influencing factors?
4. What distinguishes the investigated case studies of innovative environmental management from other (ordinary) cases of successful environmental protection?

These particular case studies address the following processes of improved environmental protection:

1. Optimization of the ecologically and economically favourable reduction of water, energy and chemicals in washing returnable bottles in the medium-sized German Amecke Fruchtsaft fruit beverage producer;

2. The recycling of paint sludge and the reduction and substitution of harmful compounds in the medium-sized German paint manufacturer Diessner Farben und Lacke;

3. The development and introduction of water-based and low-organic solvent paints on big steel constructions in the medium-sized Danish company ABC-Coating, which coats steel constructions;

4. The multiple reduction of environmental burdens in glass wool production and the establishment of an eco-accounting and auditing scheme in the medium-sized glass-wool producing Danish company Glasuld, a subsidiary of the French Saint Gobain Group;

5. The ecological reorientation of the large German retailer Hertie with a corresponding investigation of its range of goods;

6. The establishment of a comprehensive eco-accounting system in the large German textile company Kunert;

7. The striving towards environmentally conscious design in the engineering and design department of the Dutch automobile company NedCar;

8. The installation of a rather comprehensive environmental management system in the large European company General Electric Plastics Europe, which has its head-quarters in The Netherlands and is a subsidiary of the American General Electric Company;

9. Improvements in environmental protection and attributes of the comprehensive environmental management system of the large Swiss chemical company Ciba;

10. Utilization of cleaner technology in the Polish chipboard mill in Grajewo for reducing emissions, sewage treatment and recycling chemicals;

11. Improvements in technology and equipment at the Polish Kujawy vegetable fat factory in Kruszwica in relation to seed extraction and refining;

12. Environmental improvements in the Polish fishery and fishing service company Szkuner in Wladyslawowo relating to processing technology, waste management and energy management;

13. The waste minimization programme of the Latvian chemical company Dautekss;

14. Environmental management efforts by the Latvian pharmaceutical company Grindex;

15. Environmental protection and increased energy efficiency in local heat supply by the new Latvian Malpils boiler house, a gift of the Danish government.

In the comparative evaluation of the nine western European case studies pointing out typical as well as varying actors and structures behind the success stories investigated, an analytical framework connecting five different perspectives is utilized to trace the interaction dynamics leading to successful environmental management. These are an extended two-dimensional model of the sociosphere, a set of various classes of environmental management determinants, five different competitive forces determining industry competition, analytically distinct primary (performative) and secondary (supportive) environmental management activities/fields of a company, and an ecological stress matrix indicating different environmental impacts of the whole cradle-to-grave value chain.

The main results of the (western European) case study investigations can be condensed into the following conclusions:

1. Under circumstances of (radical) change, competent, motivated individuals play a key role in initiating and effecting this change, whereas institutionalized provisions may later replace this precondition.

2. As would be expected, the prominent role of key individuals is especially significant in the smaller companies studied.

3. Typically, the bigger companies having continuous working relations with public authorities emphasize communication policy in environmental affairs.

4. The main actors involved in successful environmental management tend to belong to the upper management levels of the company and of cooperating organizations, such as suppliers or public authorities. Although the shop floor finally has to implement environmental management concepts in everyday working practice, employees on this level at least receive training in environmental management practice, but hardly have any influence on general decision-making in this respect.

5. Cooperation matters in environmental management, but the type of actors collaborating to achieve substantive environmental improvements varies considerably: from essentially in-house actors only (Diessner) via common projects with one or two other companies (Amecke Fruchtsaft, NedCar) to joint ventures with public institutions and environmental consultants, including environmental organizations (Hertie, ABC-Coating, Glasuld).

6. When good profits are being made, financial resources available allow some attempt to sell environmentally friendly products or additional environmental investments which are not profitable and imply economic losses.

7. On a general, abstract level, the initiating events and starting points for improving the company's environmental management were always negative, aiming at avoiding certain effects. But they differed with respect to internal or external causation: sometimes the conditions triggering environmental management efforts are due to competitive strategy considerations (cost reduction, environmentally oriented consumer demand), sometimes they are due to negative consequences and scandals resulting from (illegal) environmental burdens from production or products (ABC-Coating, Glasuld, GEP).

8. Compared with other investigations of environmental management, there are no marked characteristics distinguishing these cases from the others. By the selection criteria chosen, the (western European) case studies are by definition success stories and therefore mark a certain terrain of favourable factor constellations, for instance an environmentally oriented corporate policy and committed management. However, the development path followed by these companies is, in principle, open to most other companies too.

Therefore in general, abstract terms, one may conclude:

1. Overall, successful (innovative) environmental management depends on the positive interaction of company-internal and external determinants leading to the technical, organizational, socio-economic and cultural encroachment of environmental management efforts, thus generating a self-supporting interaction dynamic towards substantive

environmental protection. Therefore, successful innovative environmental management demands a systemic-evolutionary, reticulate perspective, which avoids single unilateral causal models and interpretations.

2. The specific configuration of substantive influencing factors explaining individual success stories is particular for each case, allowing little generalization. Therefore, general explanations should focus on the positive interaction dynamics of determinants and not on specific individual influencing factors of corporate environmental management.

3. Some common ground is given by the preponderance of economically profitable environmental management, by the limited number of individual (10–20) and organizational (1–5) actors usually involved, and, with the exception of the Danish case studies by the minor significance of public environmental policy for innovative environmental management.

4. Apart from agenda setting, environmental policy thus appears to be more important for the subsequent diffusion of examples of innovative environmental management than for their occurrence itself.

5. Furthermore, initial experience of success, the availability of resources, a good working atmosphere, strategic action and cooperation by actors, and only limited (forceful) counteractive influences frequently seem to be a necessary condition for successful environmental management.

6. As is known from general social and organization theory, conceiving a company as an open system is a more challenging, overall more promising but more risky corporate policy than relying on the relatively comfortable and secure scope of corporate action prestructured and delimited by bureaucratic environmental policy. This is because it then has to be innovative to develop a high degree of resilience via flexibility in its situational actions and its adaptation and learning capabilities in order to cope with increasingly varying contextual conditions generating ever more uncertainty. However, taking an innovative stance is not simple and easy, since growing competitive pressures already confront firms with a great deal of uncertainty. To include environmental criteria in strategic considerations adds additional uncertainty and complexity, and firms are thus reluctant to do so. Moreover, some global business trends, such as shorter production life cycles and smaller production series, are not easily compatible with environmental requirements. Therefore, successful environmental management requires rather comprehensive, systematic, resource demanding and long-term concerted corporate action addressing the various levels, functions and groups of a company, and taking into account the whole life-cycle of its products.

Altogether, in spite of the quite interesting characteristics of the case-specific success stories, the case studies showed features of successful innovative environmental management which are to be expected by those familiar with investigations and theories of innovation, organization, political economics, and environmental management.

Whereas companies in western Europe formally have considerable scope for action in their environmental management efforts to go beyond the existing system of public environmental standards and regulations, taking into account their know-how, their resources and the frequently still available potential of economically beneficial environmental improvements, companies in eastern Europe formally have even greater scope for action in successful environmental management because of the lack of environmental policy implementation and worse environmental pollution. At the same time, however, they have a greater dearth of the necessary socio-economic and -cultural resources to act accordingly.

Environmental policy recommendations are concerned essentially with providing adequate framework conditions that facilitate rather than hinder corporate environmental management. They address:

1. the attainment of political and technical in-house capacity and competence in environmental policy programmes, embedding environmental management in a sustainable economy perspective;

2. with an environmental policy orientation towards sustainable development instead of regulations favouring end-of-pipe technologies;

3. internal consistency and reliability in environmental policy, particularly in coordination with other, often opposing policies;

4. the conception of industrial corporations as (in the last resort) cooperative actors with their own legitimate interests when aiming to improve the environment and achieve sustainable development;

5. therefore favouring contextual policy steering by attempting to install ecologically beneficial legal and economic framework conditions;

6. helping to generate actor networks capable of coordinating and organizing relatively comprehensive (cradle-to-grave) environmental management systems;

7. providing valuable information and well-targeted subsidies for environmental management efforts by (small and medium-sized) companies;

8. allowing for the flexible, decentralized organization of concrete environmental management measures and programmes;

9. but combining this cooperative approach to support the diffusion and improvement of (industrial) environmental management with relatively stringent and enforceable environmental regulations as its necessary backbone;

10. supported by the strategic use of prospective intervention, which gives industrial corporations the time to adapt to clearly foreseeable environment-related rules and standards.

If environmental policy follows its own administrative and political requirements, such as the separation of monitoring and administrative action, separation of programme formulation and implementation, intrapolicy cooperation transcending ecological media, interpolicy cooperation networking environmental authorities with other environmentally relevant authorities, regionalized enforcement, personnel education, then it will probably contribute most to improving environmental management and protection in society. In addition, such an environmental policy should be able to improve environmental management within public (political) institutions themselves to which it has easier formal access, thus being able to demonstrate the social (and economic) viability of corresponding environmental protection measures.

EU environmental policy is particularly dependent on such a policy orientation because of the even more limited political resources and power at its disposal (compared to national environmental policies) to make its many directives actually work. So its main efforts to improve environmental management should be oriented towards communication, coordination, networking, besides strengthening uniform environmental rules and standards. This demands competence in intrapolicy cooperation (communication with and coordination of various national environmental policies, as well as different industries in their environmental management practices), and in interpolicy cooperation (internalization of environmental concerns in other policies, especially economic, regional and agricultural policy).

LIST OF CONTRIBUTORS

Jobst Conrad   Freie Universität Berlin, Forschungsstelle für Umweltpolitik, Schwendener Str. 53, D-14195 Berlin, Germany

Jesper Holm   Roskilde University, Dept. of Environment, Technology and Social Studies, Hus 10.1, RUC, Box 260, 4000 Roskilde, Denmark

Joseph Huber   Martin-Luther-Universität, Institut für Soziologie, Abderhalder Str. 7, 06099 Halle (Saale), Germany

Uta Kirschten   Freie Universität Berlin, Institut für Allgemeine Betriebswirtschaftslehre, Patschkauer Weg 38, D 14195 Berlin, Germany

Børge Klemmensen   Roskilde University, Dept. of Environment, Technology and Social Studies, Hus 11.1, RUC, Box 260, 4000 Roskilde, Denmark

Ellen Protzmann   Technische Universität Brandenburg 03013 Cottbus, Germany

Geerten Schrama   University of Twente, Center for Clean Technology and Environmental Policy, PO Box 217, NL 7500 AE Enschede, The Netherlands

Ulrike C. Siegert   Arthur D. Little Consultancy Group, Wiesbaden, Germany

Inger Stauning   Roskilde University, Dept. of Environment, Technology and Social Studies, Hus 11.2, RUC, Box 260, 4000 Roskilde, Denmark

Malti Taneja   Freie Universität Berlin, Forschungsstelle für Umweltpolitik, Schwendener Str. 53, D-14195 Berlin, Germany

Simone Will   University of Potsdam, Faculty of Economic and Social Sciences, Department of Marketing, August-Bebel-Str. 89, D-14482 Potsdam, Germany

I. BACKGROUND AND CONTEXT

1. CONCEPTUAL INTRODUCTION

Jobst Conrad

1 STRUCTURE AND CONTENTS

Over the past decade, the greening of industry has become both an issue in scientific and political debate and a generic substantive development in industry itself (cf. Fischer/Schot, 1993). This development can be seen as part of a gradual, historical and general process of ecologicalization, which will take quite some time, i.e. decades for substantive realization, and which usually proceeds in a hesitant, spasmodic, troublesome and roundabout manner (Conrad, 1995a).

Opportunities for and realization of successful environmental management[1] therefore may well differ in time, location, scope and viability. Favourable conditions should then be able to facilitate and promote implementation. In this context, the purpose of this book is to draw attention to such characteristics of successful environmental management and to possible ways in which (public) environmental policy might enhance corresponding development processes.

What distinguishes this book from many others produced on this topic over the past decade is the combination of:

1. In-depth investigation of the social processes leading to successful innovative environmental management in various industrial companies in different European countries;

2. The comparative evaluation of typical features of successful environmental management in different contexts;

3. The emphasis placed on the (complex) interaction dynamics of various company internal and external determinants on different levels of the sociosphere[2] as an adequate framework for interpreting successful environmental management;

4. The inclusion of the results of related surveys and case studies in the explanation of (successful) environmental management;

5. The normative perspective with respect to starting points of supportive (public) environmental policy and;

---

[1] Throughout this book, environmental management refers to environmentally oriented management of and in (business) organizations in general, and not only to specific substantive environmental practices.

[2] The term sociosphere denotes the domain of the social, as distinguished from the psychosphere (human individuals), both together constituting the noosphere or homosphere, whereas the geosphere (composed of cosmosphere, atmosphere, hydrosphere and lithosphere) and the biosphere refer to the other strata of reality. In colloquial language one speaks of immaterial-spiritual, inanimate and animate nature.

6. The explicit preference for (hermeneutic) understanding within an analytical framework for (causal) explanation of the social evolution of environmental management practices described below.

The empirical basis for the theoretical interpretation of successful environmental management are nine in-depth case studies on Danish, Dutch, German and Swiss companies of different sizes and from different industrial sectors carried out in 1993/94[3], supplemented by the results of other inquiries on environmental management. The investigations deal either with specific substantive cases of successful environmental management and/or with the installation of an environmental management system as such within a company in the course of the past decade. The particular case-studies[4,5] thus address the following processes:

1. Optimization of the ecologically and economically favourable reduction of water, energy and chemicals in washing returnable bottles in the medium-sized German Amecke Fruchtsaft fruit beverage producer;

2. The recycling of paint sludge and the reduction and substitution of harmful compounds in the medium-sized German paint manufacturer Diessner Farben und Lacke;

3. The development and introduction of water-based and low-organic solvent paints on big steel constructions in the medium-sized Danish company ABC Coating, which coats steel constructions;

4. The multiple reduction of environmental burdens in glass wool production and the establishment of an eco-accounting and auditing scheme in the medium-sized producing Danish company Glasuld glass wool, a subsidiary of the French Saint Gobain Group;

5. The ecological reorientation of the large German retailer Hertie with a corresponding investigation check-out of its assortment of goods;

6. The establishment of a comprehensive eco-accounting system in the large German textile company Kunert;

7. The striving towards environmentally conscious design in the engineering and design department of the Dutch automobile company NedCar;

8. The installation of a rather comprehensive environmental management system in the large European company General Electric Plastics Europe, which has its headquarters in The Netherlands, and is a subsidiary of the American General Electric Company;

9. Improvements in environmental protection and attributes of the comprehensive environmental management system of the large Swiss chemical company Ciba, since 1997 part of Novartis.

---

[3] The corresponding research project was funded by the European Commission within its research program on environmental policy instruments, whose support is hereby gratefully acknowledged.

[4] The detailed analyses can be found in the original case study reports: Conrad, 1994; Huber *et al.*, 1994; Holm *et al.*, 1994; Kirschten, 1995; Schrama, 1994, 1995; Taneja, 1994; Will, 1994.

[5] The corresponding eastern European case studies carried out in Poland and Latvia can be found in the corresponding final report to the European Commission (Conrad, 1996).

In order to arrive at an understanding of successful environmental management and to reach conclusions concerning supportive environmental policy, the case studies have to be interpreted in the appropriate context, namely the evolution of environmental management and environmental policy over the past decades. Secondly, they have to be interpreted in a broad analytical framework enabling an understanding of the overall complexity of such social processes of corporate development, if the objective is overall apprehension and not merely the testing of a specific hypothesis. The comparison of different theoretical-analytical[6] perspectives should help to reach more unequivocal conclusions about a possible general pattern of successful environmental management than relying on just one or two approaches.

This line of reasoning underlies the structure of the book. This introductory chapter presents the general methodological and conceptual approach behind the analysis of successful environmental management of industrial companies. The second chapter explicates the context of the project, namely the evolution of environmental policy and management and of the corresponding research, whereas the third chapter describes the design, procedures and development of the project itself.

After having sketched the background and context of the investigation in Part I, the results of the nine case studies undertaken are presented in Chapters 4 to 12, constituting Part II of the book. Whereas roughly the first half traces the social processes leading to specific substantive improvements in environmental protection within certain companies, the second half concentrates on describing the features and evolution of environmental management systems established in the companies investigated, supplemented by references to corresponding substantive environmental protection outcomes. Part III attempts a comparative evaluation of the nine case studies and addresses theoretical and environmental policy perspectives of environmental management. Chapter 13 indicates typical and varying actors and structures underlying the success stories, and Chapter 14 relates the results of these case studies to similar case studies in environmental management. Chapter 15 then attempts to ascertain the major determinants and dynamics of successful environmental management by applying the general analytical framework described below. Chapter 16 provides conclusions on the prospects for environmental management, and in Chapter 17 the key theoretical and political question of whether there is any generalizable pattern of successful environmental management is tackled in view of different theoretical-analytical approaches. Thus, the role of and implications for environmental policy can be elaborated to a certain degree in the final Chapter 18.

2 METHODOLOGICAL REFLECTIONS

Successful environmental management in European companies can be explained by differing methods, juxtaposing the two traditions of (quantitative) analytical

---

[6] This term is used to indicate that the understanding of social phenomena such as successful environmental management, depends on both the analytical categories applied to them and the theoretical hypotheses interpreting them.

explanation and (qualitative) hermeneutic understanding (Haussmann, 1991; Kieser, 1993; von Wright, 1971).

Analytical explanation aims at the (causal) explanation of singular facts by logical deduction from other facts and higher laws (nomological statements), where information content, degree of verification, regularity and degree of completeness are emphasized as evaluation criteria for hypotheses and theories. However, the social sciences recognize only statistical laws and not deterministic ones; hypotheses cannot be directly confronted with reality;[7] judging a hypothesis to be falsified depends finally on the argued consensus of the scientific community; the causal-genetic[8] reconstruction of social processes can never be complete;[8] and mere explanatory sketches usually predominate. Thus analyses oriented on the deductive scheme of explanation at best provide information about regularities limited in time and space, and do so in a way that is always open to criticism. Causal connections which are carried out incompletely, operationalizations whose reliability and validity remain dubious, samples which cannot claim to be representative, levels of significance which are highly dependent on the methods applied – all these have to be accepted if connections discovered by research are to be judged provisionally true (Kieser, 1993: 16).

Hermeneutic understanding underscores that human action is determined by (subjective) meaning and subsequent intention. The social sciences therefore have to reconstruct social processes and events by (also) considering their understanding and interpretation by human beings, even if the consequences need not be the intended ones. Understanding appeals to our experience to infer the meaning of observed action, but does not presuppose the existence of social laws. Without reference to human meaning and intention one cannot really comprehend human action and relations. However, hermeneutic interpretation and rational explanation by intention are not mere (arbitrary) empathizing about the meanings and intentions attributed to social action, but can well be intersubjectively criticized and scrutinized. Again the scientific community has the final say on the adequacy of a specific interpretation of social facts and relations. Furthermore, hermeneutic reconstruction of social action is always concerned, too, with structures and rules constituting and regulating it in the first place. Social meanings and intentions are not arbitrary, but are strongly influenced by these structures and rules. Nevertheless, they matter only if they relate to a certain meaning. Since given intentions need not lead to specific actions, a rational explanation using existing intentions is only a necessary, but not sufficient explanation of human action. Hermeneutic hypotheses can thus be falsified, but need not contain any (social) law, and may well be valid only for an historically singular social fact or event.

Functional analysis and explanation allows certain problems to be removed from the overall context and to compare the prevailing problem solution with other possible, functionally equivalent ones. Such an analysis increases understanding of the functions and modes of operation of social systems, but accomplishes neither

---

[7] The reality of social science analysis is always a constructed one and thus an artificial product.

[8] 'Statistical laws can therefore never be causal laws, since they provide only rational grounds and no causes, let alone demonstrating ineluctable procedures.' (Stegmüller, 1983: 852)

complete causal analysis nor comprehensive hermeneutic understanding, and is at best a partial substitute for these latter procedures.

Analytical explanation and hermeneutic understanding are not mutually exclusive but complementary, since analytical explanations always contain elements of understanding, and since quantitative representative studies indicate regularities in behaviour and structure pointing to potentially typical patterns of action and the corresponding underlying intentions.[9] In the actual practice of social science research, the dispute about explanation and understanding does not play an important role, and the partial compatibility of the two approaches is acknowledged and made use of. The practical difficulties of utilizing theoretical knowledge are not realistically registered by either the concept of critical rationalism or by Habermas' (1981) model of domination-free communicative discourse. Since practicians tend to select and utilize (scientific) theories according to their plausibility and their vagueness allowing for multiple use, their congruence with the practician's convictions, their value for legitimizing his intentions and interests, their agenda-setting power etc. (Lau, 1989), the development of social science concepts and theories should primarily serve to generate scientific knowledge and not straightforward practical implications. Such concepts and theories may provide good reasons for practical programmes and social organization, but they can never justify them; one should therefore be suspicious of the ideological utilization of theories to immunize practical proposals and measures.

Since the focus of the research project was on case studies of innovative environmental management, which by definition depart from established routines, the analysis gives preference to the (hermeneutic) understanding of the reasons leading to each of these individual success stories. On this basis the project attempts to develop an analytical framework to demonstrate regular features of the socio-structural and sociocultural interaction dynamics (causally) explaining the social evolution of environmental management practices.

A similar distinction relates to the two levels of social explanation by structure and by action (cf. Mayntz, 1985; Mayntz/Nedelmann, 1987; Schimank, 1985, 1988; Taylor, 1989; Weyer, 1993). Again, the solution lies in intelligent, productive combination and not in confronting the logic of structure and of action, because on the one hand human intention and action are obviously shaped by existing (perceived) structures, and on the other the genesis of social, and of technological and even natural structures are clearly shaped by human intention and action. The corresponding rationalities of (individual) actors, of systems and of communication relate to different levels of social units, namely individual actors and organizations, sociofunctional systems and social networks, with different time frames for change. These differing rationalities can enter a relationship of tension and inconsistency, experienced by social actors as social coercion. The self-dynamics of

---

[9] 'In brief, the interpretation of quantitative data lives from the qualitative understanding of the social phenomena being examined, and integration of qualitative data lives from knowledge of regular structures within which the individual events being examined belong.' (Wilson, 1982: 501)

'Hermeneutics without latent quantification is just as impossible...as, vice versa, analysis of mass data without hermeneutics.' (Schulze, 1992: 27)

social networks results from the fact that they develop an internal logic of action which can no longer be fully controlled by the participating actors and which is shaped by the principle of communicative agreement (Weyer, 1993).

Consequently, the analytical framework particularly refers to the interaction dynamics between different structural and action levels in explaining the evolution of successful environmental management, such as economic viability, favourable corporate culture, the personal commitment of key individuals, self-dynamics in the development of environmental management efforts due to initial success.

3   ANALYTICAL FRAMEWORK

Compared to earlier models and concepts of social reality, more recent attempts emphasize the need for complex, differentiated (multidimensional and multilevel) explanations of social processes (cf. Bartlett, 1994; Heritiér, 1993; Jänicke/Weidner, 1995; Kieser, 1993; Porter, 1990; Schmidt, 1993). Giving due regard to understanding the specific characteristics of individual social phenomena implies, however, that the regularities and causal connections explaining a specific case may well differ from case to case. As a consequence, the understanding and explanation of certain social facts and processes will tend to be a unique combination of various explanatory modules, which may well claim general validity but need not play a significant role in many cases.

Therefore, the analytical framework used to explain successful environmental management does not refer to a particular social theory, but only provides relevant dimensions and levels for theoretical interpretation, yet without selecting specific theories.

The general analytical framework provides five different perspectives of analysis on which to trace the interaction dynamics leading to successful environmental management. These are an extended two-dimensional model of the sociosphere, developed with reference to Huber (1989); a set of various classes of environmental management determinants, developed by Conrad (1992, 1994); five different competitive forces determining industry competition, taken from Porter (1985, 1990); analytically distinct primary (performative) and secondary (supportive) environmental management activities/fields of a company (see Meffert/Kirchgeorg, 1993; Steger, 1993; Wicke et al., 1992); and different environmental impacts of the whole cradle-to-grave value chain (ecological stress matrix).

These five perspectives offer different valuable insights for the analysis and evaluation of environmental management (development) patterns and activities. However, these perspectives cannot be simply combined to form a multidimensional interpretative superstructure because they represent analytically independent, though partly overlapping, points of reference and delineate conceptually different subdivisions and points of emphasis of the extended sociosphere, comprising aspects of the homosphere, geosphere or biosphere.

The model of the extended sociosphere (see Figure 1) permits verification of whether the social development dynamics of environmental management are dependent on the interaction of its ecological, technological, economic, political,

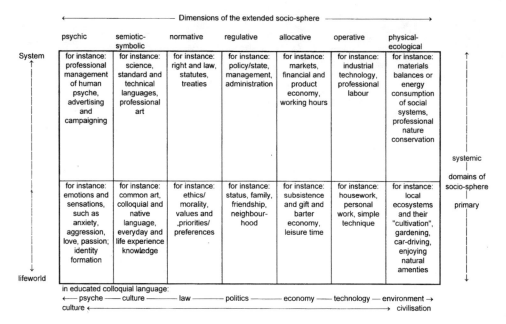

Figure 1 Extended overall system of the socio-sphere (developed from Huber, 1989: 207).

legal, cultural and psychological dimensions[10] as well as its primary and systemic domains.

Similarly, the interplay of significant determinants of successful environmental management can be more comprehensively analysed and understood by distinguishing between substantive (ecological) problem structure, the techno-economic context, the wider social context, the economic boundary conditions, the psychological context, the situational structure, historiographic events and conditions, and the role and personality structure of key individuals (see Figure 2).[11]

The survival of businesses, our crucial unit of analysis, depends on their (long-term) competitive advantage, which is determined by rivalry among existing competitors; the bargaining power of suppliers as well as of buyers; the threat of new entrants; and the threat of substitute products or services (see Figure 3). This may be achieved by different generic company strategies of cost leadership, cost focus, differentiation and focused differentiation. This scheme to locate competitive advantage is able to provide structured information about the economic boundary conditions and the economic dimension of the sociosphere influencing the viability of environmental management.

---

[10] The inclusion of the psychological and the physical-ecological dimension in an *extended* sociosphere addresses non-social, but socially relevant, boundary conditions belonging to other spheres.

[11] So the analytical subdivision of different types of environmental management determinant relates methodologically to another clearly distinct taxonomy than that provided by the two-dimensional matrix of the (extended) sociosphere.

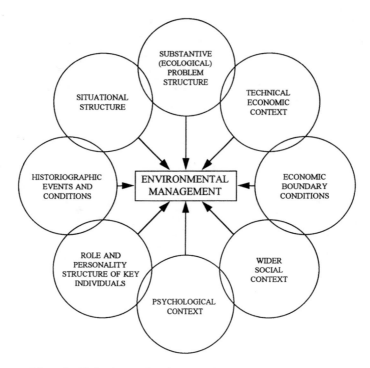

Figure 2   Eight classes of environmental management determinant.

Primary (performative) and secondary (supportive) environmental management activities/fields simply distinguish between different functional and organizational components of a company, such as product development, purchasing/provision of inputs, production, marketing and sales, logistics, waste management; or installations, organization, management systems and instruments, environmental information systems, finance and accounting, personnel, public relations, and corporate culture.[12] This analytical perspective usually refers only to the company-internal division of tasks and allows one to assess the main points of emphasis, development paths and the comprehensiveness of its environmental management activities.

Different environmental impacts of the whole cradle-to-grave value chain (life-cycle assessment) result from the (company-internal and external) subdivision of the total product life-cycle and of the ecological dimension, namely raw materials extraction, transportation and storage, upstream production, mainstream production, distribution, use, maintenance and repair, recycling and waste products, disposal of end products; with the environmental impacts of each life cycle phase relative to the consumption of resources, energy consumption, air, water and soil pollution, noise,

---

[12] In a partly similar way Porter (1985) distinguishes primary and support activities, namely inbound logistics, operations (manufacturing), outbound logistics, marketing and sales, after-sale service, and firm infrastructure (finance, planning), human resource management, technology development, procurement.

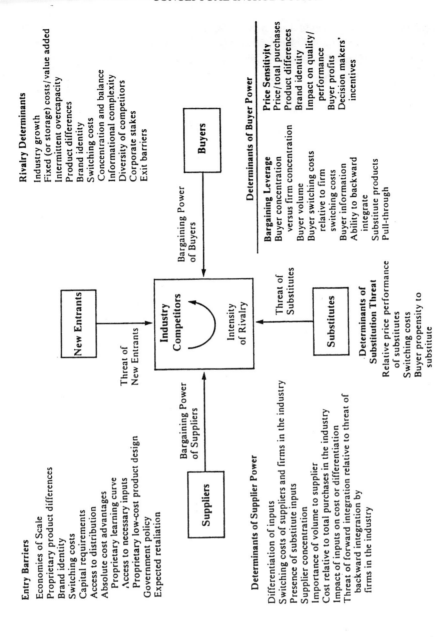

Entry Barriers
Economies of Scale
Proprietary product differences
Brand identity
Switching costs
Capital requirements
Access to distribution
Absolute cost advantages
  Proprietary learning curve
  Access to necessary inputs
  Proprietary low-cost product design
Government policy
Expected retaliation

Rivalry Determinants
Industry growth
Fixed (or storage) costs/value added
Intermittent overcapacity
Product differences
Brand identity
Switching costs
Concentration and balance
Informational complexity
Diversity of competitors
Corporate stakes
Exit barriers

Determinants of Supplier Power
Differentiation of inputs
Switching costs of suppliers and firms in the industry
Presence of substitute inputs
Supplier concentration
Importance of volume to supplier
Cost relative to total purchases in the industry
Impact of inputs on cost or differentiation
Threat of forward integration relative to threat of
  backward integration by
  firms in the industry

Determinants of Substitution Threat
Relative price performance
of substitutes
Switching costs
Buyer propensity to
substitute

Determinants of Buyer Power

Bargaining Leverage
Buyer concentration
  versus firm concentration
Buyer volume
Buyer switching costs
  relative to firm
  switching costs
Buyer information
Ability to backward
  integrate
Substitute products
Pull-through

Price Sensitivity
Price/total purchases
Product differences
Brand identity
Impact on quality/
  performance
Buyer profits
Decision makers'
  incentives

New Entrants

Suppliers

Industry Competitors
Intensity of Rivalry

Buyers

Substitutes

Threat of New Entrants

Bargaining Power of Suppliers

Bargaining Power of Buyers

Threat of Substitutes

Figure 3 Elements of Industry Structure qualifying the five competitive forces (Source: Porter, 1985: 6).

construction, packaging, waste and synergetic impacts. This analytical scheme of an ecological stress matrix (see Figure 4) allows the evaluation of successful environmental management in its ecological substance with respect to its main elements and comprehensiveness. (Dyllick *et al.*, 1994)

4   PROFITABILITY

Since our investigation is concerned with successful environmental management by (European) businesses, which belong primarily to the economic system, their basic boundary condition of survival is that they dispose of sufficient *economic* resources, although this condition holds for non-economic social actors, too, because in the last resort they also depend for their survival on the availability of resources.[13] Therefore industry as a whole finally depends on a return on investments in any case even if some companies may just survive because of the supply of external subsidies and resources.

It is thus no surprise that Winter (1993) distinguishes four levels of environmental management according to their effective costs for a company. These are the adherence to legally prescribed (public) standards and rules on environmental protection, environmental management efforts which are profitable for the company, those management efforts with outcomes that do not affect costs, and those which imply a cost burden to the company. Whereas adherence to legal standards may well require costly investments, this applies for all competing companies subject to and complying with the law, and this adherence can be enforced in principle if and because noncompliance may lead to the shutting-down of production and of the company. Therefore this first level of environmental protection in industry is likely to be realized, at least in advanced industrialized countries, if the average costs of non-compliance are higher than those of compliance (cf. Terhart, 1986), and even if available profitable measures on the second level are not pursued, frequently for reasons of ignorance. Clearly, however, a company will tend to give priority to environmental management measures on the second level over the third level of neutrality with regard to costs, and especially over the fourth level of cost burden. This ranking order thus reflects the key role played by profitability and cost arguments in businesses.[14] Even environmental protection on the fourth level is feasible only as long as these costs are covered by other profits, loans or subsidies.

The criterion of (long-term) economic profitability does not imply an unequivocal course for company strategy and behaviour because there is usually more than one option and various ways to achieve company survival requiring strategic choices to be made. And since alternative (economic) strategies cannot really be compared in

---

[13] If social actors do not earn their own economic income, they depend on the transfer of resources from other actors, at least as long as one has to spend economic resources to earn one's living. Governments cannot exist without taxes or functionally equivalent revenues; children usually live on the economic income of their parents or others.

[14] This is not to deny the possibility of an environmental management pattern not in line with this ranking order, but this will have to be explained by causes outweighing the economic ranking arguments.

product life cycle / environmental impact	raw materials extraction	transportation and storage	upstream production	mainstream production	distribution	use	maintenance and repair	recycling and waste products	disposal of end products
consumption of resources									
energy consumption									
air pollution									
water pollution									
soil pollution									
noise									
construction									
packaging									
waste									
synergetic impacts									

Figure 4 Ecological stress matrix.

view of a highly uncertain future, 'political' compromise between protagonists with diverging 'world views' replaces economic optimization (cf. Cyert/March, 1963; Wiesenthal, 1994). Therefore, economic rationality is an unavoidable *boundary condition* determining success or failure of company strategy, but rarely a *substantive* (*theoretical*) *explanation* of a company's (successful) strategic choices and action (see cf. Nutzinger, 1994).

5   SUBSTANTIVE ENVIRONMENTAL SUCCESS

To judge whether environmental management has proved successful requires specification of the term success. This has to be done in two respects.

First, any social effort towards environmental protection can only be judged successful if it finally leads to a reduced substantive burdening of the environment. However, attempts to define and operationalize environmental compatibility in terms of ecological stability or other general concepts about ecosystems, and to determine critical limits of maximal or minimal population sizes or of the sustainability and recuperative capacity of ecosystems have largely failed. Only the direction of minimizing environmentally burdensome interventions can be indicated by the *idée directrice* of ecological integrity (van den Daele, 1993). So the environmental compatibility and sustainability of human action and systems always partly remains a normative judgement irreducible to ecological criteria.[15]

Since, however, corresponding assessments are not arbitrary, it is possible to specify guiding principles of significant successful environmental management for a company in at least five aspects, although these do not necessarily include other important facets of *societal* environmental compatibility and sustainability, such as sufficiency or consistency (cf. Conrad, 1993, 1995b; Huber, 1995):

1. Substantive, measurable effects of environmental protection may not depend on environmental problem shifting, e.g. by defining dilution as the solution of pollution, although the demarcation of real successes can never be unequivocal due to the persistence of genuine scientific uncertainties.

2. The observable effects of environmental protection need to be more than marginal and lasting ones, which again involves normative demarcation.

3. Measurable effects of environmental protection need not involve the whole life cycle of a product, e.g. rather complete recycling, but may for instance even only refer to a successful installation of only an additive end-of-pipe technology, although typically less innovative conceptually in ecological terms than integrated technologies.

4. Successful environmental management in a specific case does not necessarily require its operation within an extensive concept and practice of environmental management in a business; however, corresponding environmental improvements

---

[15] This reminds us that the social definition of environmental success lastly also depends on what relevant societal groups perceive as constituting such success and how difficult it is in technical as well as in social terms to solve an environmental problem (Jänicke/Weidner, 1995: 16).

may not be a mere unintended bonus effect of other corporate measures and/or policies.

5. Finally, success implies only a relative, but not necessarily absolute, environmental improvement or questioning of whole ecologically-doubtful techno-economic systems such as the automobile, because an enhanced demand for automobiles or higher mileages setting off reduced environmental burden per unit of distance, or an environmentally favourable shift from road to rail as predominant (long-distance) transport technology cannot be simply considered the task and/or responsibility (merely) of an automobile company or even industry alone.

Second, effective environmental protection may be due to direct (intended) or indirect (unintended) impacts of environmental management efforts. The reorganization of an institution or the creation of a new law, may well fail to achieve the intended environmental objectives underlying such procedural arrangements, but lead to other environmental improvements hardly thought of when such measures are decided upon and adopted. So a social arrangement can be a failure with respect to its direct, intended (ecological) outcome but a success with respect to its indirect, hardly intended (ecological) outcomes. Typically, the further impacts of broad social processes and lasting institutional arrangements are of greater (ecological) significance than the direct (intended) ones, because as sociostructural forces they imply a momentum not confined to single (ecological) events.[16] Therefore, it appears

---

[16] Bartlett (1994: 175, 177, 178, 180) elaborates on the example of the US Comprehensive Environmental Response, Compensation, and Liability Act passed in 1980, known as CERCLA or Superfund, in this respect: 'Not surprisingly, evaluations of CERCLA . . . find current policy to be a near complete failure. But each of these assumptions is problematic-politics and policy may only secondarily be concerned with problem solving, policymaking entails a great deal more than decisions, policymaking is partly nonrational and also involves multiple rationalities, and policymaking is concerned with a great deal more than substantive outcomes. The validity and applicability of policy evaluation based on such a model [of outcome evaluation] are more limited than is commonly appreciated. . . . In the real political world, means–ends analysis always gets turned on its head as means and ends are closely intertwined-objectives evolve as agreement is sought on means. . . . Process evaluation is crucial to assessing CERCLA as well. CERCLA profoundly restructured hazardous waste policy processes around liability determination and right-to-know rules. The liability for the costs of hazardous waste site cleanups can no longer be easily evaded by corporate restructuring, poor record keeping, a legal necessity to prove harm, or even bankruptcy or foreclosure. Certain kinds of information about potentially hazardous substances must now be made available by governments and businesses. . . . A blinkered focus on action and outcomes, or for that matter on process, leads policy evaluators to find policy failure everywhere, without any careful analysis of the appropriate criteria for evaluating institutions, much less integrative institutions. If, for example, CERCLA is analyzed in terms of changes it has wrought in rules, routines, and roles, as well as the beliefs, cultures, and knowledge that surround and support those roles and routines, then . . . it can be seen to have been immensely consequential and likely to become more so. Although not mentioned in the legislation, CERCLA can be credited with causing the following institutional transformations, among others, over a very short time: establishment of extensive separation, collection, and recycling programs by state and local governments across the United States; profound changes in commercial lending practices and norms; reversal of longstanding packaging policies by McDonald's and other marketers of fast food; and growth in environmental auditing as an accepted good business practice.'

reasonable to distinguish outcome, process, and institutional evaluation,[17] where the last has the largest potential scope for successful environmental improvements (Bartlett, 1994). Whether this scope is realized over time, however, has to be proved in the last resort by its attributable physical (ecological) outcomes.

Finally, assessing successful environmental management calls for awareness of fundamental limitations to its harmonious development because of its external trade-offs with other company goals and because of its internal trade-offs between different objectives of environmental protection. Environmentally compatible production and products may become important company goals, but they remain always only partial goals supplementary to others like production efficiency, favourable working conditions, meeting consumer demand, which inevitably involve trade-offs. Similarly, environmental improvements in one dimension or medium cannot always avoid (minor) environmental deterioration in another dimension or medium, even if problem-shifting strategies are excluded.[18]

These trade-offs are somewhat neglected in the literature on environmental management, to which we turn in the next chapter.

LITERATURE

Bartlett, R.V. (1994) Evaluating Environmental Policy Success and Failure. In: eds. N.J. Vig and M.E. Kraft. *Environmental Policy in the 1990s*. Washington D.C.: CQ-Press.

Conrad, J. (1992) *Nitratpolitik im internationalen Vergleich*. Berlin: edition sigma.

Conrad, J. (1993) Social Significance, Preconditions and Operationalisation of the Concept Sustainable Development. In: ed. F. Moser *Sustainability—Where Do We Stand*? Technical University Graz.

Conrad, J. (1994) Ökonomischer Wegweiser ins ökologische Mehrweg-Optimum. FFU-Report 94-4, Berlin.

Conrad, J. (1995a) Development and Results of Research on Environmental Management in Germany. *Business Strategy and the Environment* **4**, 51–61.

Conrad, J. (1995b) Grundsätzliche Überlegungen zu einer nachhaltigen Energieversorgung. In: ed. H.G. Nutzinger. *Nachhaltige Wirtschaftsweise und Energieversorgung*. Marburg: Metropolis.

Conrad, J. (ed.) (1996) Successful Environmental Management in European Companies. FFU-Report 96-3, Berlin.

Cyert, R.M. and March, J.G. (1963) *A Behavioral Theory of the Firm*. Englewood Cliffs: Prentice Hall.

Daele, W. van den (1993) Sozialverträglichkeit und Umweltverträglichkeit. Inhaltliche Mindeststandards und Verfahren bei der Beurteilung neuer Technik. *Politische Vierteljahresschrift* **34**, 219–248.

Dyllick, Th. *et al.* (1994) *Ökologischer Wandel in Schweizer Branchen*. Bern: Haupt.

Fischer, K. and Schot, J. (eds.) (1993) *Environmental Strategies for Industry*. Washington D.C.: Island Press.

---

[17] 'Outcomes evaluation is based on the seemingly unassailable assumption that policy is purely instrumental, comprising means for producing results or effects. ... Policy process evaluations are assessments of the merit or worth of policy processes themselves. ... What is central, in this view, are the values of the processes used to define problems, set agendas, formulate alternatives, select actions, and govern implementation. ... Institutional evaluation assesses how processes work and outcomes are produced within a larger institutional framework created in part by policies and within which policies are made and remade. In short, what is evaluated is 'political architecture', architecture that influences outcomes, structures processes, and constructs and elaborates meaning.' (Bartlett, 1994: 170, 176, 179)

[18] For instance, effective protection of ground and surface waters may be locally connected with suboptimal arrangements of nature protection.

Habermas, J. (1981) *Theorie des kommunikativen Handelns*, Volumes I and II. Frankfurt: Suhrkamp.

Haussmann, T. (1991) *Erklären und Verstehen: Zur Theorie und Pragmatik der Geschichtswissenschaft*. Frankfurt: Suhrkamp.

Heritiér, A. (1993) Einleitung. Policy-Analyse. Elemente der Kritik und Perspektiven der Neuorientierung. In: (ed.) A. Heritiér, *Policy-Analyse*. PVS-Sonderheft 24. Opladen: Westdeutscher Verlag.

Holm, J. *et al.* (1994) Two Cases of Environmental Front Runners in Relation to Regulation, Market and Innovation Network. Report, Roskilde University.

Huber, J. (1989) *Herrschen und Sehnen. Kulturdynamik des Westens*. Weinheim: Beltz.

Huber, J. (1995) Nachhaltige Entwicklung durch Suffizienz, Effizienz und Konsistenz, In: (eds.) P. Fritz *et al.*, *Nachhaltigkeit in naturwissenschaftlicher und sozialwissenschaftlicher Perspektive*. Stuttgart: Wissenschaftliche Verlagsgesellschaft.

Huber, J. *et al.* (1994) Fallstudie Ciba AG. Report, University Halle.

Jänicke, M. and Weidner, H. (1995) Successful Environmental Policy: An Introduction. In: (eds.) M. Jänicke and H. Weidner, *Successful Environmental Policy*. Berlin: edition sigma.

Kieser, A. (ed.) (1993) *Organizationstheorien*. Stuttgart: Kohlhammer.

Kirschten, U. (1995) Ökobilanzierung und Aufbau eines Öko-Controlling im Kunert-Konzern. Report, Free University Berlin.

Lau, Ch. (1989) Die Definition gesellschaftlicher Probleme durch die Sozialwissenschaften. In: (eds.) U. Beck and W. Bonß, *Weder Sozialtechnologie noch Aufklärung? Analysen zur Verwendung sozialwissenschaftlichen Wissens*. Frankfurt: Suhrkamp.

Mayntz, R. (1985) Die gesellschaftliche Dynamik als theoretische Herausforderung. In: (ed.) B. Lutz, *Soziologie und gesellschaftliche Entwicklung*. Verhandlungen des 22. Deutschen Soziologentages. Frankfurt: Campus.

Mayntz, R. and Nedelmann, B. (1987) Eigendynamische soziale Prozesse. *Kölner Zeitschrift für Soziologie und Sozialpsychologie.* **39**, 648–668.

Meffert, H. and Kirchgeorg, M. (1993) *Marktorientiertes Umweltmanagement*. Stuttgart: Poeschel.

Nutzinger, H.G. (1994) Unternehmensethik zwischen ökonomischem Imperialismus und diskursiver Überforderung, In: Forum für Philosophie Bad Homburg, (ed.) *Markt und Moral. Die Diskussion um die Unternehmensethik*. Bern: Haupt.

Porter, M.E. (1985) *Competitive Advantage: Creating and Sustaining Superior Performance*. New York: The Free Press.

Porter, M.E. (1990) *The Competitive Advantage of Nations*. London: Macmillan.

Schimank, U. (1985) Der mangelnde Akteurbezug systemtheoretischer Erklärungen gesellschaftlicher Differenzierung. *Zeitschrift für Soziologie.* **14**, 421–434.

Schimank, U. (1988) Gesellschaftliche Teilsysteme als Akteurfiktionen, *Kölner Zeitschrift für Soziologie und Sozialpsychologie.* **40**, 619–639.

Schmidt, M.G. (1993) Theorien in der international vergleichenden Staatstätigkeitsforschung. In: (ed.) A. Heritiér, *Policy-Analyse*. PVS-Sonderheft 24. Opladen: Westdeutscher Verlag.

Schrama, G. (1994) The Internalization of Environmental Management at GE Plastics Europe. Report, University of Twente.

Schrama, G. (1995) The Environmental Factor in the Designing of Passenger Cars by NedCar. Report, University of Twente.

Schulze, G. (1992) *Die Erlebnisgesellschaft*. Frankfurt: Campus.

Steger, U. (1993) *Umweltmanagement*. Wiesbaden: Gabler.

Stegmüller, W. (1983) *Probleme und Resultate der Wissenschaftstheorie und Analytischen Philosophie*, Vol. I: *Erklärung-Begründung-Kausalität*. Berlin: Springer.

Taneja, M. (1994) Erfolgsbedingungen betrieblicher Umweltpolitik am Beispiel der Firma Diessner Farben und Lacke GmbH und Co KG. FFU-Report 94-7, Berlin.

Taylor, M. (1989) Structure, Culture and Action in the Explanation of Social Change. *Politics & Society* **17**, 115–162.

Terhart, K. (1986) *Die Befolgung von Umweltschutzauflagen als betriebswirtschaftliches Problem*. Berlin: Dunker & Humblodt.

Weyer, J. (1993) System und Akteur, *Kölner Zeitschrift für Soziologie und Sozialpsychologie.* **45**, 1–22.

Wicke, L. *et al.* (1992) *Betriebliche Umweltökonomie*. München: Vahlen.

Wiesenthal, H. (1994) Lernchancen der Risikogesellschaft. *Leviathan.* **22**, 135–159.

Will, S. (1994) Hertie und BUND—eine Kooperation im Spannungsfeld zwischen Ökonomie und Ökologie. FFU-Report 94-6, Berlin.

Wilson, Th.P. (1982) Qualitative 'oder' quantitative Methoden in der Sozialforschung. *Kölner Zeitschrift für Soziologie und Sozialpsychologie.* **34**, 469–486.

Winter, G. (1993) *Das umweltbewußte Unternehmen*. München: Beck.

Wright, G.H. von. (1971) *Explanation and Understanding*. Ithaca: Cornell University Press.

2. EVOLUTION OF ENVIRONMENTAL POLICY AND ENVIRONMENTAL MANAGEMENT

Jobst Conrad

This chapter provides an analytical sketch of the multilayered context of the project. Starting from the overall evolution of culture-nature systems of human societies over millenia, the sociostructural conditions of and corresponding requirements for environmental policy development are pointed out for industrialized countries over past and future decades towards ecological sustainability. Then typical features of evolving (corporate) environmental management are summarized, and the need to substantially coordinate environmental policy and environmental management efforts is established. Finally, the contempory development of environmental management research and its conceptual orientation and tools are described.

In a historical human ecology perspective, the interaction between nature (natural eco-systems), human populations (with certain genetically fixed characteristics), and their culture (encompassing all patterns of extra-somatically stored, communicated and handed-down information, including social structure and technology) produces the metabolic-ecological profile of the culture-nature system of human societies and decides their viability and evolution. The Neolithic and the industrial revolutions can be regarded as major watersheds in human cultural evolution, with the transition from hunter-gatherer societies to agricultural societies, and with the delimitation and growth of previously confined physical parameters (population, transportation volumes and speed, energy conversion, per capita consumption of an ever-growing number of substances) via the utilization of fossil energy (Sieferle, 1993). Thus each human society was and is confronted by its specific environmental problems and has developed typical patterns of environmental management.

1 CONTEXT AND REQUIREMENTS OF ENVIRONMENTAL POLICY DEVELOPMENT

Environmental policy as a socially and substantially differentiated policy area has developed (in industrial societies) only since the 1960s, although social concern with nature protection and health policy can look back on a long tradition. Reflected in a temporal learning process, environmental policy has to deal with different types of environmental problems and effects, the related generation of sociostructural problems and impacts, resulting in a range of policy strategies, instruments, and divergent assessment of success or failure. These different features of environmental problems and policy-making can be attributed to different phases of old, modern, and 'post-industrial' society (Wolf, 1992). As summarized in Figure 1, these framework conditions and the corresponding development of environmental policy,

	Old industrial society		Modern industrial society			
Problem	Plants	Existing pollution/ contamination	'Normal accidents	Product risks	Post-industrial society	'Creeping disasters'
Prototype	Steel furnace	Former industrial premises	Chernobyl	PCB asbestos	Traffic	Acid rain, climate
Environmental interference/ burden	Emissions	Soil contamination from emissions and wastes	Highly toxic emissions, radioactivity	Toxicity of products	Emissions, energy consumption, land consumption	Extinction of global resources
Polluter	identifiable	identifiable, but not accountable	identifiable and accountable, but sanctions pointless	only polluter groups localizable	everyone	Accumulation of 1-5
Harmful impact	localizable in space and time	localizable in space and time	not localizable	conditionally localizable	local and global	not localizable
Problem genesis	Joint product of industrial production	In the plant	In the plant	Utilization cycle of products and materials	Everyday behaviour	Accumulation of 1-5
Environmental strategy practised	Emission abatement	Soil decontamina- tion	Risk minimization	Regulation of utilization chains	Local measures	?
Instruments	Adapted regulatory law with technology control by technology	In addition cooperative and corporative strategies	Technology control by technology	Co-operation, infrastructure	Reorganizatio n of infrastructure	?
Success	Visible decoupling effects	Still stage of pilot projects	None	Decoupling effects not yet visible	No or at best local isolated solutions	None
Legal regulation Level of legal protection	High/ functioning neighbour protection	Underde- veloped/ precarious	Increasing legal regulation without individual legal protection	Increasing legal regulation without individual legal protection	Increasing legal regulation with selective legal protection	?
Policy alternatives	Purposive fostering of economic structural change	In addition: modernization, restructuring of old industrial districts	Abandonment	Ecologization of the patenting system, participation in licensing procedure	Modification of use pattern/ raising cost of private transport	Resource conservation

Figure 1　Environmental policy and social change (Source: Wolf, 1992: 372).

though differing from country to country in the industrialized world, may be described in terms of the following characteristics (cf. Brickman *et al.*, 1985; Conrad, 1990, 1991; Enloe, 1975; Glasbergen, 1995; Hajer, 1995; Hartkopf/Bohne, 1986; Héritier *et al.*, 1994; Hey/Brendle, 1994; Jachtenfuchs/Strübel, 1992; Jänicke, 1979, 1986, 1990, 1993, 1996; Jänicke *et al.*, 1992; Jänicke/Weidner, 1995, 1997; Kern/Bratzel, 1994; Kitschelt, 1983; Knopfel, 1993; Knoepfel/Weidner, 1985; Lester, 1989; Mez/Jänicke 1996; Prittwitz, 1993; Ringquist, 1993; Ritter, 1992; Skou-Andersen, 1994; Tsuru/Weidner,

1989; Uebersohn, 1990; Vig/Kraft, 1994; Vogel/Kun, 1987; Wallace, 1995; Weidner, 1996; Weale, 1992; Wolf, 1992):

1. The evolution of environmental politics and policy, where one may distinguish the development phases of latency, thematization, constitution, implementation and conflict, and integration,[1] reflects the growing gradual, inconsistent and erratic penetration and expansion of environmental concerns throughout the overall sociosphere. Corresponding catchwords are the rise of environmental awareness and of environmental movements, the spread of environmental legislation and institutions, environmental monitoring and technology, the development of an environmental industry and of green parties, the propagation of environmental impact assessment and environmental protection programmes and measures, the partial and growing internalization of environmental costs, the slow and subtle ecologicalization of most spheres of social life.

2. As might be expected, the spread of new values and a new sociocultural concern is subject to manifold resistance, stonewalling, detours and reverses due to its multiple (initial) sociostructural weaknesses when confronting conflicting or differing worldviews, interests, and obligations. Environmental policy has thus suffered under existing unfavourable legal, economic, organisational, administrative, manpower and informational conditions and resources, from prevailing patterns of interest and power distribution among the actors involved in the (environmental) policy game and process, from the multiple opportunities to dilute and undermine environmental policy programmes in the course of the policy cycle, and from the collective ecological 'ignorance' and 'organized irresponsibility' attributable to the systemic mechanisms in modern society that prevent due recognition and integration of environmental issues (lacking ecological resonance in the political, economic and other sociofunctional systems; Beck, 1988; Luhmann, 1986; Willke, 1989, 1992).

3. In a way, it is due precisely to the success of modernity, grounded in its structural drive towards technological progress in growth-oriented modern industrial societies, that the environmental impacts resulting from economic growth and technical progress have attained such (global) dimensions that societies can hardly continue to neglect them and in the longer run are obliged to develop effective environmental policies – apart from (unintended) ecologically favourable structural change (cf. Jänicke et al., 1992) – resulting in gradual improvements in environmental management and policy regulation: the internalization of environmental costs, environmental self-regulation replacing bureaucratic

---

[1] These phases, which have occurred in different countries partly at varying periods between around 1960 and 1990, refer to the political treatment of environmental problems as isolated single issues (latency), the separation and restructuring of problems under the common perspective of environmental protection and policy (thematization), the constitution of environmental policy as a genuine policy area by programmes, laws and institutions (constitution), the practising of environmental protection by these newly created institutions and expanding administration, rising criticism of insufficient measures, change in values and counter-concepts for more comprehensive environmental protection (implementation and conflict), and the formal and programmatic integration of environmental protection in all social spheres besides the continued development of environmental degradation structurally entrenched in modern industrial society (integration) (cf. Bechmann et al., 1994).

environmental regulation, environmental (business) ethics and an ecologically oriented corporate culture, slow recognition and testing of ecologically oriented innovations and market opportunities.

4. In more substantive terms, the underlying secular sociostructural causes for the growth of and changes in the environmental problematique can be seen (in technological and in sociostructural terms) in

4.1. the tremendous, still expanding extent of materials extraction, processing, use and wastes;

4.2. the growing prominence of large-scale technological systems;

4.3. the science and technology based transformation of (regenerative) natural resources in artificial substances frequently connected with quantitatively less environmental pollution by processing and by generating waste materials, but qualitatively contributing to new ecological hazards (information and communication technologies, new (raw) materials, genetic engineering);

4.4. extended production–consumption cycles with socially longer and differentiated utilization chains;

4.5. therefore ecological product risks increasingly outweighing those of their production process;

4.6. the growing preponderance of a service-economy (health, education, recreation, culture, information processing) typically consuming less material and energy per unit of value added;

4.7. in parallel the growing (ecological) significance of service and leisure activities, mobility and transportation;

4.8. the social turbulences and ecological consequences resulting from globalizing corporate strategy, action and competition.

5. The corresponding shifts in the structural pattern and complexity of environmental degradation can be systematically conceived as three relatively – not absolutely – new developments: the increasing *extent* of environmental degradation (air pollution, sea pollution, soil erosion), the relative unavoidability of large-scale (global) *environmental disasters* due to avoidable and to normal accidents (Perrow, 1984) (nuclear reactor accidents, tanker accidents, accidents in chemical plants), and the *gradual destruction* of ecological cycles and renewable resources by the *mutual complex interaction* of various individual human activities and pollutants (drying up of Lake Aral, dying forests, desertification), which may well be controllable and less harmful individually: and all this in combination with growing knowledge and political perception thereof.

6. The corresponding coping strategies, typically labelled as environmental protection, minimalization of risk, and sustainable development, respectively, denominate the required line of (secular) development for environmental policy in order to achieve environmental compatibility of an increasingly structurally complex (world) society.

The corresponding requirements of an adequate environmental policy addressing the deeper linkages between ecological risks, socioeconomic change and political regulation therefore typically imply (cf. Böhret, 1990; Jänicke *et al.*, 1995; Wolf, 1992)

in technical terms

1.1. to regulate product risks and product life-cycles;

1.2. to effect the 'muddling out' of high risk technologies;

1.3. (regionally) to limit certain production, consumption, emission volumes in absolute terms;

1.4. to prevent the (uncertain) future, gradually accumulating (synergetic) environmental impacts due to multiple interaction processes between different production, consumption and waste generation processes.

in sociostructural terms

2.1. intelligently to combine sufficiency, efficiency and consistency strategies towards sustainable ecological development (Huber, 1995) in their technical as well as their social dimension, without neglecting existing pollution and contamination accumulated in the past (e.g. consumption of land, soil contamination);

2.2. to be much more concerned with the environmental impacts of services, leisure, mobility and transport;

2.3. to influence (ongoing) structural change of the economy in favour of environmental compatibility.

in sociopolitical terms

3.1. to influence the (conditions of) deeply rooted civilizational patterns (e.g. mobility from private transport, spatial differentiation of housing, workplace, leisure) towards their ecological reconstruction;

3.2. to find viable solutions for the trade-off between ecological and social welfare, when addressing change in ubiquitous civilizational patterns of utilizing the environment, (when e.g. internalizing the environmental costs of (private) motoring by demanding 5,- DM per liter gasoline from every cardriver);

3.3. therefore to transcend genuine domains of environmental policy, to interfere in other (well established and shielded) policy areas and to establish respective interpolicy cooperation (e.g. with policies of transportation, economic and technological development, recreation and tourism, agriculture etc.);

3.4. substantially to internationalize environmental policy to address global environmental problems adequately.

and in political-procedural terms

4.1. to build substantial (environmental policy) capacities actually to realize these demands (at least partly);

4.2. to develop and make use of corresponding environmental indicators as well as policy monitoring and evaluation procedures;

4.3. to develop multiple-self qualities as conditions of successful action under uncertainty, basing interpretations of one's environment and premises for decisions on several disparate world views without being forced to give strong attention to the fact and practice of multiple references, which should improve the chances of ambiguity management and to overcome the limitations of procedural rationality (Wiesenthal, 1990);

4.4. to establish ecologically effective modes of detached cooperation with organized social (industrial) actors polluting the environment, as opposed to traditional approaches of command-and-control;[2]

4.5. to establish strict liability and (partial) reversal of the burden of proof.

It is unlikely, however, for manifold sociostructural reasons (see cf. Böhret, 1990; Hajer, 1994; Jänicke, 1986, 1996; Wiesenthal, 1994; Wolf, 1992) that these requirements will be met in the short or even medium term. So for an (environmental) policy in slow motion against accidents in slow motion (Roqueplo, 1986) time becomes the precarious resource of society.

Nevertheless, against this background, the gradual, by no means warranted, development of environmental policy towards more comprehensiveness, a systemic-evolutionary and multi-actor perspective, linkability to existing structures and interests, social enforceability, extended (political) resources and power basis, becomes plausible and significant.

This process of development in environmental policy (capacity building) and ecological modernization (Huber, 1993; Jänicke, 1993; Zimmermann et al., 1990) can be substantiated (on the social, organisational, and semiotic-symbolic levels) by the following phenomena:

1. development of sociostructural entrenchment from low to considerable, though still quite limited (greening of industry, environmental research and consultancy, export of environmental knowledge, favourable and intense media coverage, National Environmental Plan);

2. development in the institutionalization of environmental concerns and duties from none via separate to integrated;

3. development of (financial, legal, manpower) resources and (political) power from little (mainly symbolic) to middling;

4. development of environmental indicators from scattered data via national emission and solid waste data to national material balances and green GDP figures;

5. development of recognition and significance from local via national to international;

6. from a media and segmented view to a holistic, networked, cradle-to-grave perspective, taking account of the complexity of the environmental problematique, including synergetic effects;

---

[2] These are clearly unsuited to deal with the environmental problems of abandoned waste sites, product risks, synergetic effects, creeping disasters, global environmental (climate) changes which result from the interaction process among many social actors and thus do not permit clear-cut attribution of specified and sanctionable (individual) responsibility.

7. from pollutant concentration and impact-oriented environmental policy programmes via emission-oriented programmes to production and product (risk) oriented programmes;

8. from compensatory and corrective to preventative and integrative environmental protection policies in favour of integrated instead of additive (end-of-pipe) environmental technologies;

9. from a reactive (scandal-triggered) policy to a proactive (preventive) environmental policy;

10. from the technocratic symptom control of mere follow-up problems (Jänicke, 1979) to ecological sustainability as the key problem of modern industrial society;

11. from isolated adversative to multiple cooperative actor constellations and policy programmes;

12. from short-term to (also) medium and long-term policies;

13. from persuasive policy instruments via regulatory, bureaucratic instruments to cooperative, civil law and market-oriented (intelligent) mixes of policy instruments;

14. from segmented, clear-cut information gathering and policy formulation to complex, less unequivocal forms, replacing definite cause-effect relations by multiple cause-effect interactions and uncertain outcomes;

15. from (subordinate) clear-cut environmental protection goals to multi-layered environmental policies, taking into account intra- and interpolicy trade-offs among different ecological goals and among environmental and other socio-political objectives.

These development trends, which often affect society as a whole rather than being specific to the field of environmental policy (see cf. Héritier, 1993), permit the following major policy-strategic conclusions:[3]

1. The in principle limited capacity of the state, or more precisely, of the environmental authorities, to steer and control social development processes in modern societies by means of socially binding decisions and appropriate support and sanctions becomes especially apparent in the field of environmental policy, thus reflecting general characteristics of the modernity project.[4]

2. As a consequence, (successful) environmental policy depends on fruitful cooperation among relevant actors and the strategic utilization and intelligent

---

[3] These conclusions may exhibit some German bias, but should indicate nevertheless generally valid observations of environmental policy development.

[4] The modernity project is usually perceived as the coherent amalgam of rational reason/enlightenment, scientific-technological control and utilization of nature, capitalist industrialism, and individualist democratic society, though some authors point to the hidden agenda of modernity, too (Toulmin, 1990). In more analytical terms, the three dominant sources of the dynamism of modernity can be seen in the separation of time and space, the development of disembedding mechanisms (symbolic tokens and expert systems), and institutional reflexivity with the regularised use of knowledge, and capitalism, industrialism, military power, and surveillance representing its underlying institutional dimensions (Conrad, 1994b; Giddens, 1990, 1991).

combination of those social momenta and forces in different social systems and strata which generate a push-and-pull dynamics in favour of (improved) environmental protection and ecological sustainability (Huber, 1991).

3. This attachment and linkability to existing (predominant) sociostructural modes of action and reasoning imply the recognition of the necessarily limited manageability and long-term horizon of any (effective) environmental policy and the (initial) advisability of a rigorous incrementalism (Maier-Rigaud, 1988), despite the notorious deficiencies of incremental policy-making, such as fragmentation, reactive policy-making, limited alternatives, short-time horizons, pragmatic muddling-through, and selective consideration of policy implications.

4. In addition, the limited capacity of the state to intervene substantially in social process dynamics is demonstrated by frequent tests in the field of environmental policy, where contradiction between the recognition and solution of public sector problems, and effective public policy measures becomes especially obvious (Jänicke, 1993).

5. Furthermore, the results of environmental policy (or the environmental impacts of other (structural) policies) exhibit a high degree of irreversibility, whereas the negative outcomes of other policies, such as social or employment policy, can in principle be corrected later (Maier-Rigaud, 1988).

6. The ecologically necessary long-term time horizon of environmental policy is hardly compatible with the predominant modes of (ordinary) reactive policy-making with the short time horizons of election cycles and learning from past experience; many ecologically detrimental experiences need to be anticipated but not actually made.

7. Environmental policy as a typical cross-sectional policy needs particularly to assert itself successfully against other counteractive public policies and political orientations, and to take into account ecological local–global connexions requiring simultaneous internationalization and decentralization, which imply being confronted with further actors with also diverging intentions and interests.

8. However, environmental policy can neither be supported by a stringent operational goal system (van den Daele, 1993; Maier-Rigaud, 1988) nor be unequivocally related to other socio-economic goals (employment, stability, balance of trade), which justifies pragmatic flexibility and rigorous incrementalism.

9. Nevertheless, the regulatory command-and-control approach that has been followed to date remains an essential, indeed the most effective, environmental policy technique (cf. Jänicke/Weidner, 1995; Huber et al., 1994; Uebersohn, 1990).

10. For all these reasons environmental policy requires an appropriate legal basis to deal with environmental problems within a (comprehensive) environmental law framework in order to prevent arbitrary (discretionary) decision and action finally undermining liberal constitutional democracy (Kloepfer, 1992).

11. The need for ever more comprehensive environmental policy concepts, design and implementation is, however, increasingly recognized by most (environmental policy) actors, as indicated by various types of critique of

environmental policy[5] and the pervasive appreciation of the concept 'sustainable development'.[6]

12. Although induced to a large degree by a broadening social perception of escalating environmental deterioration and disasters (e.g. the greenhouse effect due to various human emissions), the gradual expansion and strengthening of environmental policy over the last decades is based on a both substantive and socio-political logic, tracing out its subsequent course as indicated above and described in similar vein by other authors.

2   DEVELOPMENT OF ENVIRONMENTAL MANAGEMENT

After this sketch of the overall evolution of environmental policy (requirements) the development of environmental management will be summarized in somewhat greater detail.

With the increasing visibility and perceived urgency of substantial remedies and solutions to the environmental crisis, firms need to respond to sociocultural, political and economic pressures in order to ensure continued access to scarce resources, public and political legitimacy, profitability, and financial assurance, although the intensity of these pressures varies by country, industry, sector, and firm. Corporate learning processes – basically not dissimilar to those of other social actors – have modified these responses, evolving from passive via active opposition to adaptation and self-organisation in the effort to gain room for manoeuvre vis-à-vis environmental affairs (Hildebrandt et al., 1994, 1995; Shrivastava, 1992; Ullmann, 1982; Walley/Whitehead, 1994). Two phases in corporate reaction to the public environmental debate, legislation, and scandals can be reasonably distinguished: resistance and reluctant adjustment to external pressures and environmental policy in the period from about 1970 to 1985, and acceptance of such pressures and the incorporation of environmental considerations into corporate policies in a more rigorous way from about 1985 to 1995.

During the first phase, 'environmental protection was seen as an operating constraint that needed to be taken care of because of outside pressures. Some policies were set for the entire corporation, but the implementation of these policies was weak and determined mainly by the local situation. Environmental protection was foremost a local task of facilities, not a corporate task of headquarters.... The dominant pattern...was a lack of willingness to internalize environmental issues. These issues were to some extent accepted as problems that should be managed, but

---

[5] Bechmann et al. (1994) distinguish a cognitive, an ethical, an institutional, a time, and an instrumental dimension of environmental policy criticism.

[6] Social conflicts related to this (new) type of future-oriented thinking and policy-making are typically concerned with wider issues: 'civilization, respect for nature, and the conviction that the natural environment should offer the same opportunities to future generations as it does to the present generation. The resulting social controversies are less isolated, less ad hoc than their predecessors. They involve bringing about ecologically inspired social change in regions of the country, across continents, and even worldwide.' (Glasbergen, 1995: 4)

only in reaction to outside pressures, notably regulation and public pressure.'[7] (Schot/Fischer, 1993: 6, 7)

During the second phase, the gradual evolution of environmental management efforts and strategies on all levels and dimensions of corporate action can be observed, to some degree leading to the development and implementation of environmental monitoring and reporting, the establishment of a considerable number of (specialised) environmental positions, attention to the various dimensions and to upstream and downstream phases in the complete ecological product life-cycle (and the initiation of respective cooperation with other companies), and the inclusion of (offensive, innovative) environmental management in corporate strategy, viewing ecologically oriented production as more of a market opportunity than a (market) constraint. On the average, however, firms hardly develop such environmentally innovative strategies 'but react to outside regulatory, public, and, to an increasing extent, market pressures', and only – depending on the economic situation and the modernization capacity of the company – 'a minority of firms are moving beyond a compliance-oriented approach.' (Schot/Fischer, 1993: 12).

In sum, the debate and research on environmental management[8] point to (Conrad, 1995):

1. The growing significance of environmental problems and concern in the business community, with the unavoidable consequence of having to pay adequate heed to environmental protection in the future;

2. Recognition by business of the (potential and varying) pressure from competitors, customers, employees, and government on companies and industries regarding environmental commitment;

3. The preferability of offensive (innovative) to defensive environmental management, the superiority of exploiting the opportunities offered by an integrative product life-cycle perspective, and of exploiting the integrated technologies this offers to mere (short-term) end-of-pipe technological adaptation to satisfy environmental requirements as they are imposed;

4. The advantages of timely environmental self-regulation by industry over bureaucratic public regulatory regimes for business, government and the environment;

---

[7] 'Just because we are keeping regulations, it doesn't mean that we agree with or like it', one manager put it in Petulla (1987: 171).

[8] Substantive details of analytical schemes and conceptual models of environmental management, as developed for instance in the German literature, can be found in Meffert/Kirchgeorg, 1993; Post/Altman, 1992; Schreiner, 1991; Steger, 1993; Wicke et al., 1992. For more detailed descriptions of the findings of empirical investigations in the field of environmental management in Germany and other countries, the reader is referred to Blansch et al., 1994; Brandt et al., 1988; Coenenberg et al., 1994; Davis, 1991; Dillon/Fischer, 1992; Dyllick, 1991; Dyllick et al., 1994; Fischer/Schot, 1993; Freimann/Hildebrandt, 1995; Gruber et al., 1992; Hildebrandt et al., 1995; Hoffman, 1994; Hoffman et al., 1990; Huisingh et al., 1986; James/Stewart, 1994; Kirchgeorg, 1990; McKinsey & Company, 1991; Meffert/Kirchgeorg, 1993; Morrison, 1991; Nielsen, 1992; Ostmeier, 1990; Piasecki/Asmus, 1990; Prisma-Industrie-Kommunikation, 1992; Rappaport/Flaherty, 1992; Steger, 1992a; UBA, 1991; Vietor/Reinhardt, 1995; Wieselhuber/Stadlbauer, 1992; Ytterhus et al., 1995; Zimpelmann et al., 1992.

5. The actual implementation and necessity of environmental management as an ongoing process at all levels of a company's organization, functions, and production processes, involving many small-scale steps and efforts and representing a gradual ecologicalization of all company activities;

6. Recognition of the necessary interplay and joint operation of many influencing factors and levels in establishing successful environmental management;

7. The need for fairly comprehensive environmental information systems, eco-auditing and eco-controlling to permit substantial environmental management;

8. The significance of learning processes at the level of business organizations and of overall, integrated, corporate environmental-management strategies for the successful introduction of new environmentally sound products and production processes;

9. The growing importance of integrated technological and organizational recycling solutions, especially of waste management, where the producer is required to take back materials, and commerce and the retail trade play a gate-keeper role.

Thus, empirical findings from the 1980s and early 1990s indicate that (in Germany, in particular) environmental protection was considered to be highly important by most company managements, though more due to a long-term risk perspective (securing the continued existence and location of the company) than to opportunities to save costs or increase profits. Frequently, external conditions (market situation, consumer attitudes) were considered less favourable for environmental management than were internal ones, though its actual realization was strongly handicapped by internal factors such as a lack of know-how, staff shortages, or high costs. Actual (perceived) environmental problems, ecological competition intensity, the way of incorporating environmental objectives in a company, and a small number of ecologically oriented and committed individuals appear to be major determinants in favour of installing (offensive) environmental management. Environmental protection was mainly considered to be not so much a genuine company goal as a complementary, incidental condition of the major (economic) goals of a company.

Significantly, measures with positive environmental impact were due mainly to cost-saving measures (reduction of production, energy or material costs). Environmental protection was thus considered important in production or materials management, but less so in marketing and cross-sectional divisions such as organization, personnel, controlling, and public relations. In the past, defensive and repair measures to comply with regulations and to eliminate the environmental burden dominated in a wholly technology-oriented environmental management, whereas innovative and preventive measures were expected to play a more important role in the future. Environmental policy did not appear to enhance offensive (innovative) environmental management but to reinforce (passive) compliance with public regulations. Environmental management was accordingly oriented mainly towards risk prevention and much less towards (strategic) environmental product-innovation and market penetration.

An ecological orientation of companies at the organizational level had yet to be achieved to any significant degree and, in consequence, (legally required) environmental protection officers dominated as the organizational unit in the past with mainly (bureaucratic) control functions and little scope for stimulating environmental integration and innovation. Apart from environmental impact appraisals, to some degree required by law since 1990, traditional planning instruments, such as check lists, preponderated in environmental management, whereas comprehensive life-cycle assessments and eco-auditing (cf. Etterlin *et al.*, 1992; Hallay/Pfriem, 1992; Hopfenbeck/Jasch, 1993; Steger, 1991) were scarcely in evidence, being comparatively expensive, and genuine pollution prevention and thinking in terms of materials cycles had yet to develop. Nevertheless, a steady upward trend in the environmental consulting business can be observed, with the new EU eco-audit regulation 1836/93 fostering diffusion of systematic environmental management and control by legal and political means. Favourable economic conditions and a supportive socio-political climate appear to be necessary conditions for the implementation and diffusion of substantive environmental management. In addition, however, various studies emphasize the appropriate orientation of corporate culture as a necessary precondition for successful environmental management (Dyllick, 1989, 1991; Kirchgeorg, 1990; Nitze, 1991; Rothenberg/Maxwell, 1992; Steger, 1992b; Wild/Held, 1993). Under such circumstances, companies have developed greater openness towards dialogue and even towards cooperation with environmental groups and critics than they had in the past, when they considered them fundamental (irrational) adversaries to be excluded from any substantial moves towards environmental management.

Meanwhile, a considerable number of (descriptive) success stories of environmental management and innovation have been reported, usually combined with the intention of demonstrating their feasibility and advantages for industry as a whole. However, the definition and scope of successful instances of environmental management vary considerably and require critical evaluation. At a concrete level, cases of successful environmental management differ in terms of the actors, concerns, strategies, and instruments involved, or of the role played by public environmental policy; but their direct and indirect results are usually quite similar: the reduction of environmentally detrimental emissions and/or wastes as well as savings of energy and (raw) material inputs, signifying the prevalence of production oriented over product oriented environmental improvements, and the evolution of a more holistic environmental management approach as part of the general ecologicalization process.

The interaction between various classes of environmental management determinants, e.g. between ecology push and ecology pull or between economic and psychographic factors of successful environmental innovation, as substantiated by several investigations (Conrad, 1994a, 1995; Dyllick, 1989; Dyllick *et al.*, 1994; Kirchgeorg, 1990; Ostmeier, 1990; Schmidheiny, 1992; Steger, 1993; Steinle *et al.*, 1994; Vester, 1990), confirms the thesis that successful environmental management requires positive interplay between various levels and reinforcing factors during its social start-up phase.

Altogether, the majority of studies indicate that most companies are aware that environmental management is necessary, and are to a growing extent willing to install it. But the same studies show that only few have already implemented it in practical terms as a necessary medium-term step-by-step process at the various levels and in the different functional spheres of a corporation. Frequently, unsystematic (resource-intensive) eco-activism and pseudo-eco-profile enhancement have substituted comprehensive strategic environmental management (Wieselhuber/ Stadlbauer, 1992). Successful installation of environmental management requires the economic situation of a company to be unproblematic and visible benefits to accrue within a short period.

Relating the results of environmental management research to environmental policy research, one will typically observe quite different modes and operative involvement of the economic and the political systems, of industry and government in tackling environmental problems and effectuating environmental protection, but also a similar variability of case-specific patterns and a similar complex interplay between many influencing factors resulting in successful environmental protection (cf. Jänicke/Weidner, 1995).

Bearing in mind these trends in environmental policy and environmental management, the current state of affairs may reasonably be characterised as a transitional phase, where no clearly predominant, definite pattern in environmental policy and management is to be expected. Specific cases will consequently tend to exhibit a varying overlap of positive and negative explanatory determinants, such as cost saving, image, demand, innovation, diffusion, standards or regulation-driven cases of successful environmental management. And examples of innovative environmental management will still be the exception rather than the rule, although they tend to reflect a general societal (socio-economic) development dynamics.

Furthermore, these development trends indicate both: effective environmental policy must necessarily be conceived as meta-policy setting sociopolitical and socio-economic boundary conditions for (cooperative) self-organized network management within civil society[9] to tackle and solve environmental problems, and environmental policy alone – against the prevailing rules of politics – should not simply be expected to deal effectively with these environmental problems which industry is not expected to solve by itself against the rules of the market.

It is the (strategic) coupling of social forces in different sociofunctional systems and organisations, like industry, government, science etc. – in particular the socio-ecologically productive combination of environmental policy and (industrial) environmental management – which may lead to substantial environmental protection and improvements. And it is the still rather uncommon combination of environmental policy and environmental management research results and concepts which may give rise to conclusions and recommendations on how to orient and organise environmental policy capable of effectively supporting successful environmental management in (European) corporations.

3   ENVIRONMENTAL MANAGEMENT RESEARCH

Since this book focuses on such cases of successful environmental management, it appears appropriate at this stage to summarize the main aspects of (German)

---

[9] See Cohen/Arato, 1992; Walzer, 1992 for an elaboration on civil society.

environmental management research (Conrad, 1995):

1. The (normative) application of micro-economics and innovation theory to environmental management prevails with little scientific controversy and a lack of multidisciplinary research.

2. The gradual development of a genuine, (holistic) environmental management perspective within business firms and the corresponding learning processes are recognised; however, the significance of this historical dimension for the implementation of the environmental management schemes suggested is underestimated.

3. Emphasis is put on offensive and innovative environmental management addressing market opportunities as opposed to defensive environmental management trying only to follow or avoid public regulation and to save environmental costs.

4. Successful environmental management is endorsed by its normative, (conceptual) and empirical description, with an orientation towards practical application, although with a rationalistic-idealistic bias of technocratic models underrating the risks of information overload and of actual actors' behaviour and decision-making processes within business, and lacking a substantive implementation theory of environmental management.

5. Prominent is a pragmatic juxtaposition of company-internal and external determinants of environmental management (for instance environmental concern and policy regulations as well as environmental information systems and an environmental protection officer).

6. A shift can be observed from normative and empirical studies and presentation of the need and advisability of environmental management towards investigations of necessary laborious adaptation of individual company organisation and culture in the event of the consistent, differentiated and multi-level installation of environmental management.

7. For the period up to about 1990, environmental management practice in about 2000 companies in (West) Germany was analysed by detailed questionnaire (Coenenberg et al., 1994; Gruber et al., 1992; Kirchgeorg, 1990; Ostmeier, 1990; UBA, 1991; Wieselhuber/Stadlbauer, 1992).

8. In particular, environmental management success stories were investigated in about 100 companies in Germany by more or less detailed case study (Dyllick, 1991; Longolius, 1993; Meffert/Kirchgeorg, 1993; Oberholz, 1989; Prisma-Industrie-Kommunikation, 1992, 1993; Schmidheiny, 1992; Wild/Held, 1993), whereas practically no case study of environmental management failure can be found in the literature.

9. Altogether, there exist a considerable number of analytically valuable descriptive case studies taking a normative attitude in favour of environmental management, but with an almost complete lack of practically utilizable, theoretically founded reconstructions of the conditions of success or failure of environmental management.

10. This is in parallel to an environmental management research that has rapidly developed micro-economic heuristics on how to use environmental management as an economically profitable option, but which is not yet able theoretically to explain and shape the complex texture of interrelations of a firm and its ambient (ecological) systems beyond rationalistic-idealistic formulae and instruments.

Reflecting the German tendency towards sophisticated model building – as opposed to more pragmatic Anglo-Saxon schemes – the analytical classifications and categories adopted from standard taxonomies and proposed by environmental management research differentiate between, among others:

1. Primary (performative) and secondary (supportive) *environmental management activities/fields*: product development, purchasing/provision of inputs, production, marketing and sales, logistics, waste management; installations, organisation, management systems and instruments, environmental information systems, finance and accounting, personnel, public relations, corporate culture (Meffert/ Kirchgeorg, 1993; Steger, 1993; Wicke *et al.*, 1992);

2. Environmental impacts of the whole cradle-to-grave value chain (*life-cycle assessment*): raw materials extraction, transportation, upstream production, mainstream production, distribution, use, maintenance and repair, recycling and waste products, disposal of end products; with the environmental impacts of each life cycle phase relative to the consumption of resources, energy consumption, air, water and soil pollution, noise, construction, packaging, and synergetic impacts (Beck, 1993; Hopfenbeck, 1990; Meffert/Kirchgeorg, 1993; Oko-Institut *et al.*, 1987; Wicke *et al.*, 1992);

3. *Ecological company goals*: the conservation of resources, the limitation of emissions (prevention, reduction, utilization or controlled disposal of waste and emissions), risk management (with strategies for risk prevention, risk reduction, risk transfer, risk insurance, risk adoption) (Meffert/Kirchgeorg, 1993; Sietz, 1992; Steger, 1993);

4. The inclusion of environmental aspects *at different management levels and in different phases*: normative, strategic, operative, concrete (project) activities; the identification of strategic key factors (ecological portfolios, ecological opportunities and risks, the strengths and weaknesses of a company), environmentally oriented corporate goal-setting, company strategy-formation, environmental-management organisation, enforcement and control arrangements (ecological information systems, controlling and marketing) (Dyllick, 1989, 1991; Hopfenbeck, 1990; Meffert/Kirchgeorg, 1993; Schreiner, 1991; Steger, 1992a, 1993; Wicke *et al.*, 1992);

5. *Strategic environmental management and strategic principles* relating to environmental protection, risk management and competition: the reduction of environmental costs, opening up new business fields, the exploitation of opportunities with new products and production processes, the activation of performance reserves, the minimization of liability risks, the avoidance of confrontation; competence, credibility, commitment, cooperation; cost-leadership, quality-leadership, orientation towards the whole market or towards market niches, timing

strategy; resistance, passivity, retreat, adaptation, selectivity, innovation (Meffert/ Kirchgeorg, 1993; Steger, 1992a, 1993);

6. *Points of departure for offensive environmental management* with respect to production, products, environmental services: environmentally sound purchasing, production, and changes in product range, utilization of environmental counselling and credits; the development and production of environmentally sound(er) products, environmentally-friendly product distribution and pricing, sales promotion pursuant to an environmentally oriented marketing-information policy; the provision of environmental protection goods and services, including overall solutions (Meffert/ Kirchgeorg, 1993; Schreiner, 1991; Steger, 1993; Wicke *et al.*, 1992);

7. *Environmentally relevant actors* internal and external to the business corporation (market or non-market oriented): business units and divisions, subsidiaries, investors, employees, management; competitors, trade, suppliers, customers, service agents, collaborating companies; business associations, public regulatory authorities, political and juridical bodies, scientific and technological institutions, the media, ecologically affected third parties, (environmental) action groups, the public, future generations (Meffert/Kirchgeorg, 1993; Steger, 1993);

8. The *measures* taken by companies *to adapt to the limitations imposed by environmental policy* (and by the social climate) with respect to input, production, and output: variation in the volume and quality of input materials as well as of input factors; changes in or the introduction of new production processes and techniques, the enlargement of production systems, recycling: changing energy and material streams; variation in output volumes and production programmes, waste-management improvement, recycling (Steger, 1993; Wicke, 1982; Wicke *et al.*, 1992);

9. Different *types of environmental technology*, i.e. indirect and direct environmental protection measures: environmental information systems, measurement and control technology; technologies enlarging regeneration capacities, end-of-pipe technologies, recycling technologies, integrated technologies (Hopfenbeck, 1990; Kreikebaum, 1992; Meffert/Kirchgeorg, 1993; Schreiner, 1991; Wicke *et al.*, 1992).

Apart from substantial strategic points of reference in corporate policy (globalization, technology development, ecologicalization, change in values, the information society, growing complexity; see Kreikebaum, 1987; Porter, 1985, 1990; Steger, 1992b), categories originating mainly from sociological discourse (such as (environmental) policy principles, orientations, and instruments, or the levels of system and social integration) are rarely dealt with in an explicit manner in an environmental management research field dominated by economists. Nor do the various publications on environmental management differentiate and combine all these analytical dimensions and distinctions in practice although this is sometimes proclaimed in theory.[10] They usually elaborate on environmental management in the dimensions 1, 2, 4, 5 and 9 listed above.

---

[10] See for instance the 'management cube' in Steinle *et al.*, 1994, containing the six dimensions of corporate policy: basic environmental management strategy, management processes, company functions, personnel selection, management and education, and levels of (corporate) action.

In conclusion, such publications tend to emphasize that it is both advantageous and necessary for a company as a whole to adopt a general environmental-management orientation and commitment rather than undertaking only piecemeal efforts. Similarly, in less strict micro-economic studies, which address the possibilities and limitations of actors' behaviour as well, emphasis is placed on learning processes, reticular entanglements, cybernetic feedback loops, and on a systemic-evolutionary perspective in environmental management (Dyllick, 1989; Huber, 1991; Senn, 1986; Vester, 1992).

LITERATURE

Bechmann, G. *et al.* (1994) *Umweltpolitische Prioritätensetzung.* Stuttgart: Metzler-Poeschel.

Beck, M. (ed.) (1993) *Ökobilanzierung im betrieblichen Management.* Würzburg: Vogel.

Beck, U. (1988) *Gegengifte. Die organisierte Unverantwortlichkeit.* Frankfurt: Suhrkamp.

Blansch, K. Le *et al.* (1994) *Industrial Relations and the Environment.* European Foundation for the Improvement of Living and Working Conditions, Dublin.

Böhret, C. (1990) *Folgen. Entwurf für eine aktive Politik gegen schleichende Katastrophen.* Opladen: Leske + Budrich.

Brandt, A. *et al.* (eds.) (1988) *Ökologisches Marketing.* Frankfurt: Campus.

Brickman, R. *et al.* (1985) *Controlling Chemicals: The Politics of Regulation in Europe and the United States.* Ithaca: Cornell University Press.

Coenenberg, A.G. *et al.* (1994) Unternehmenspolitik und Umweltschutz. *Zeitschrift für betriebswirtschaftliche Forschung.* **46**, 81–100.

Cohen, J.L. and Arato, A. (1992) *Civil Society and Political Theory.* Cambridge: MIT Press.

Conrad, J. (1990) *Nitratdiskussion und Nitratpolitik in der Bundesrepublik Deutschland.* Berlin: edition sigma.

Conrad, J. (1991) Selbstregulierung und Deregulierung im Umweltschutz. Das Beispiel Agrarpolitik. In: (eds.) A. Görlitz and R. Voigt, *Regulative Umweltpolitik. Jahresschrift für Rechtspolitologie 5.* Pfaffenweiler: Centaurus.

Conrad, J. (1994a) Ökonomischer Wegweiser ins ökologische Mehrweg-Optimum. FFU-Report 94-4, Berlin.

Conrad, J. (1994b) Die Entwicklung der Moderne und ihre psychosozialen Folgen. In: (ed.) M. Pieber, *Österreichisches Studienzentrum für Frieden und Konfliktlösung, Europa—Zukunft eines Kontinents.* Münster: agenda Verlag.

Conrad, J. (1995) Development and Results of Research on Environmental Management in Germany. *Business Strategy and the Environment.* **4**, 51–61.

Daele, W. van den (1993) Sozialverträglichkeit und Umweltverträglichkeit. Inhaltliche Mindeststandards und Verfahren bei der Beurteilung neuer Technik. *Politische Vierteljahresschrift.* **34**, 219–248.

Davis, J. (1991) *Greening Business – Managing for Sustainable Development.* Oxford: Basil Blackwell.

Dillon, P.S. and Fischer, K. (1992) *Environmental Management in Corporations: Methods and Motivations.* Medford, Mass.: Tufts University.

Dyllick, Th. (1989) *Management der Umweltbeziehungen. Öffentliche Auseinandersetzungen als Herausforderung.* Wiesbaden: Gabler.

Dyllick, Th. (ed.) (1991) *Ökologische Lernprozesse in Unternehmungen.* Bern: Haupt.

Dyllick, Th. *et al.* (1994) *Ökologischer Wandel in Schweizer Branchen.* Bern: Haupt.

Enloe, C.H. (1975) *The Politics of Pollution in a Comparative Perspective. Ecology and Power in Four Nations.* New York: Taylor Graham.

Etterlin, G. *et al.* (1992) *Ökobilanzen. Ein Leitfaden für die Praxis.* Mannheim: Wissenschaftsverlag.

Fischer, K. and Schot, J. (eds.) (1993) *Environmental Strategies for Industry.* Washington D.C.: Island Press.

Freimann, J. and Hildebrandt, E. (eds.) (1995) *Praxis der betrieblichen Umweltpolitik.* Wiesbaden: Gabler.

Giddens, A. (1990) *The Consequences of Modernity.* Stanford University Press.

Giddens, A. (1991) *Modernity and Self-Identity*. Stanford University Press.

Glasbergen, P. (ed.) (1995) *Managing Environmental Disputes*. Dordrecht: Kluwer.

Gruber, Tietze & Partner. (1992) *Umweltbericht 1992—Stand des Umweltschutzes in führenden deutschen Unternehmen*. Bad Homburg.

Hajer, M.A. (1994) Verinnerlijking: the Limits to a Positive Management Approach. In: (eds.) G. Teubner *et al.*, *Environmental Law and Ecological Responsibility—The Concept and Practice of Ecological Selforganization*. Chichester: John Wiley & Sons.

Hajer, M. (1995) *The Politics of Environmental Discourse. Ecological Modernization and the Policy Process*. Oxford: Clarendon Press.

Hallay, H. and Pfriem, R. (1992) *Öko-Controlling*. Frankfurt: Campus.

Hartkopf, G. and Bohne, E. (1983) *Umweltpolitik 1. Grundlagen, Analysen und Perspektiven*. Opladen: Westdeutscher Verlag.

Héritier, A. (ed.) (1993) Policy-Analyse—Kritik und Neuorientierung. *Politische Vierteljahresschrift*, Special Issue 24. Opladen: Westdeutscher Verlag.

Héritier, A. *et al.* (1994) *Die Veränderung von Staatlichkeit in Europa*. Opladen: Leske + Budrich.

Hey, Ch. and Brendle, U. (1994) *Umweltverbände und EG. Strategien, politische Kulturen und organisationsformen*. Opladen: Westdeutscher Verlag.

Hildebrandt, E. *et al.* (1994) Politisierung und Entgrenzung—am Beispiel ökologisch erweiterter Arbeitspolitik, In: (eds.) N. Beckenbach and W. van Treeck, *Umbrüche gesellschaftlicher Arbeit. Soziale Welt, Sonderband 9*. Göttingen: Otto Schwartz & Co.

Hildebrandt, E. *et al.* (1995) Industrielle Beziehungen und ökologische Unternehmenspolitik. Final Report, Volume 1 and 2, Science Center, Berlin.

Hoffman, A.J. (1994) The Environmental Transformation of American Industry: An Institutional Account of Environmental Strategies in the Chemical and Petroleum Industries. Ms. Sloan School of Management, MIT, Boston.

Hoffman, W.M. *et al.* (eds.) (1990) *Business, Ethics, and the Environment*. New York: Quorum Books.

Hopfenbeck, W. (1990) *Umweltorientiertes Management und Marketing*. Landsberg/Lech: moderne industrie.

Hopfenbeck, W. and Jasch, C. (1993) *Öko-Controlling*. Landsberg/Lech: moderne industrie.

Huber, J. (1991) *Unternehmen Umwelt*. Frankfurt: Fischer.

Huber, J. (1993) Ökologische Modernisierung. *Kölner Zeitschrift für Soziologie und Sozialpsychologie*. **45**, 288–304.

Huber, J. (1995) Nachhaltige Entwicklung durch Suffizienz, Effizienz und Konsistenz. In: (eds.) P. Fritz *et al.*, *Nachhaltigkeit in naturwissenschaftlicher und sozialwissenschaftlicher Perspektive*. Stuttgart: Wissenschaftliche Verlagsgesellschaft.

Huber, J. *et al.* (1994) Fallstudie Ciba AG. Report, University Halle.

Huisingh, D. *et al.* (1986) Proven Profits from Pollution Prevention. Case Studies in Resource Conservation and Waste Reduction Report. Institute for Local Self-Reliance, Washington D.C.

Jachtenfuchs, M. and Strübel, M. (eds.) (1992) *Environmental Policy in Europe*. Baden-Baden: Nomos.

James, P. and Stewart, S. (1994) The European Environmental Executive. Report, Ashridge Management Research Group, Berkhamsted.

Jänicke, M. (1979) *Wie das Industriesystem von seine Mißständen profitiert*. Opladen: Westdeutscher Verlag.

Jänicke, M. (ed.) (1986) *Staatsversagen—Die Ohnmacht der Politik in der Industriegesellschaft*. München: Piper.

Jänicke, M. (1990) Erfolgsbedingungen von Umweltpolitik im internationalen Vergleich. *Zeitschrift für Umweltpolitik und Umweltrecht*. **13**, 213–232.

Jänicke, M. (1993) Über ökologische und politische Modernisierungen. *Zeitschrift für Umweltpolitik und Umweltrecht*. **16**, 159–175.

Jänicke, M. (ed.) (1996) *Umweltpolitik der Industrieländer. Entwicklung—Bilanz—Erfolgsbedingungen*. Berlin: edition sigma.

Jänicke, M. *et al.* (1992) *Umweltentlastung durch industriellen Strukturwandel?* Berlin: edition sigma.

Jänicke, M. and Weidner, H. (eds.) (1995) *Successful Environmental Policy. A Critical Evaluation of 24 Case Studies*. Berlin: edition sigma.

Jänicke, M. and Weidner, H. (eds.) (1997) *National Environmental Policies: A Comparative Study of Capacity Building*. Berlin: Springer.

Kern, K. and Bratzel, S. (1994) Erfolgskriterien und Erfolgsbedingungen von (Umwelt-) Politik im internationalen Vergleich: Eine Literaturstudie. FFU-Report 94-3, Berlin.

Kirchgeorg, M. (1990) *Ökologieorientiertes Unternehmensverhalten*. Wiesbaden: Gabler.

Kitschelt, H. (1983) *Politik und Energie*. Frankfurt: Campus.

Kloepfer (1992) Vom Umweltrecht zum Umweltstaat?. In: (ed.) U. Steger, *Handbuch des Umweltmanagements*. München: Beck.

Knoepfel, P. (1993) Bedingungen einer wirksamen Umsetzung umweltpolitischer Programme—Erfahrungen aus westeuropäischen Staaten. Cahiers de l'IDEAP 108, Lausanne.

Knoepfel, P. and Weidner, H. (1985) *Luftreinhaltepolitik (stationäre Quellen) im internationalen Vergleich*, 6 volumes. Berlin: edition sigma.

Kreikebaum, H. (1987) *Strategische Unternehmensplanung*. Stuttgart: Kohlhammer.

Kreikebaum, H. (1992) *Umweltgerechte Produktion*. Wiesbaden: Deutscher Zeitschriften-Verlag.

Lester, J.P. (ed.) (1989) *Environmental Politics and Policy—Theories and Evidence*. London: Duke University Press.

Longolius, S. (1993) *Eine Branche lernt Umweltschutz. Motive und Verhaltensmuster der deutschen chemischen Industrie*. Berlin: edition sigma.

Luhmann, N. (1986) *Ökologische Kommunikation*. Opladen: Westdeutscher Verlag.

Maier-Rigaud, G. (1988) *Umweltpolitik in der offenen Gesellschaft*. Opladen: Westdeutscher Verlag.

McKinsey & Company (1991) *The Corporate Response to the Environmental Challenge*. Amsterdam: McKinsey and Company.

Meffert, H. and Kirchgeorg, M. (1993) *Marktorientiertes Umweltmanagement*. Stuttgart: Poeschel.

Mez, L. and Jänicke, M. (eds.) (1996) *Sektorale Umweltpolitik*. Berlin: edition sigma.

Morrison, C. (1991) Managing Environmental Affairs. The Conference Board. Report 961, New York.

Nielsen (ed.) (1992) Umweltschutzstrategien im Spannungsfeld zwischen Handel und Hersteller. Report, Frankfurt.

Nitze, A. (1991) *Die organisatorische Umsetzung einer ökologisch bewußten Unternehmensführung*. Bern: Haupt.

Oberholz, A. (1989) *Umweltorientierte Unternehmensführung*. Frankfurt: Frankfurter Allgemeine Zeitung.

Öko-Institut et al. (eds.) (1987) *Produktlinienanalyse: Bedürfnisse, Produkte und ihre Folgen*. Köln: Kölner Volksblattverlag.

Ostmeier, H. (1990) *Ökologieorientierte Produktinnovationen*. Frankfurt: Peter Lang.

Perrow, Ch. (1984) *Normal Accidents*. New York: Basic Books.

Petulla, J.M. (1987) Environmental Management in Industry. *Journal of Professional Issues in Engineering*. 113, 167–183.

Piasecki, B. and Asmus, P. (1990) *In Search of Environmental Excellence—Moving beyond Blame*. New York: Simon & Schuster Inc.

Porter, M.E. (1985) *Competitive Advantage: Creating and Sustaining Superior Performance*. New York: The Free Press.

Porter, M.E. (1990) *The Competitive Advantage of Nations*. London: Macmillan.

Post, J. and Altman, B. (1992) Models of Corporate Greening. In: (ed.) J. Post, *Markets, Politics and Social Performance*, Vol. 13. Preston.

Prisma-Industrie-Kommunikation (ed.) (1992) *Neue Wege im Umweltmanagement* (1. SZ-Umweltsymposium). Dießen.

Prisma-Industrie-Kommunikation (ed.) (1993) *Aufbruch in die Kreislaufwirtschaft* (2. SZ-Umweltsymposium). Dießen.

Prittwitz, V. von (ed.) (1993) *Umweltpolitik als Modernisierungsprozeß*. Opladen: Leske + Budrich.

Rappaport, A. and Flaherty, M.F. (1992) *Corporate Responses to Environmental Challenges*. New York: Quorum Books.

Ringquist, E.J. (1993) *Environmental Protection at the State Level—Policies and Progress in Controlling Pollution*. Armonk/London: Sharpe.

Ritter, E.H. (1992) Von den Schwierigkeiten des Rechts mit der Ökologie. *Die Öffentliche Verwaltung*. 45, 641–649.

Roqueplo, P. (1986) Der saure Regen: ein 'Unfall in Zeitlupe'. Ein Beitrag zur Soziologie des Risikos. *Soziale Welt.* **37**, 402–426.

Rothenberg, S. and Maxwell, J. (1992) Issues in the Implementation of Proactive Environmental Strategies. *Business Strategy and the Environment.* **1(4)**, 1–12.

Schmidheiny, S. (1992) *Changing Course: A Global Business Perspective on Development and the Environment.* Cambridge: MIT Press.

Schot, J. and Fischer, K. (1993) Introduction. The Greening of the Industrial Firm. In: Fischer/Schot, 1993.

Schreiner, O. (1991) *Umweltmanagement in 22 Lektionen.* Wiesbaden: Gabler.

Senn, J.F. (1986) *Ökologieorientierte Unternehmensführung.* Frankfurt: Peter Lang.

Shrivastava, P. (1992) Corporate Self-Greenewal: Strategic Responses to Environmentalism. *Business Strategy and the Environment.* **1(3)**, 9–21.

Sieferle, R.P. (1993) Die Grenzen der Umweltgeschichte, *GAIA* **2**, 8–21.

Sietz, M. (ed.) (1992) *Umweltbewußtes Management.* Taunusstein: Blottner.

Skou-Andersen, M. (1994) *Governance by Green Taxes.* Manchester University Press.

Steger, U. (ed.) (1991) *Umwelt-Auditing: ein neues Instrument der Risikofürsorge.* Frankfurt: Frankfurter Allgemeine Zeitung.

Steger, U. (ed.) (1992a) *Handbuch des Umweltmanagements.* München: Beck.

Steger, U. (1992b) *Future Management.* Frankfurt: Fischer.

Steger, U. (1993) *Umweltmanagement.* Wiesbaden: Gabler.

Steinle, C. *et al.* (1994) Ökologieorientierte Unternehmensführung—Ansätze, Integrationskonzept und Entwicklungsperspektiven. *Zeitschrift für Umweltpolitik und Umweltrecht.* **15(4)**, 409–444.

Toulmin, St. (1990) *Cosmopolis. The Hidden Agenda of Modernity.* New York: The Free Press.

Tsuru, S. and Weidner, H. (1989) *Environmental Policy in Japan.* Berlin: edition sigma.

UBA (ed.) (1991) Umweltorientierte Unternehmensführung. Report 10/91. Berlin: Erich Schmidt.

Uebersohn, G. (1990) *Effektive Umweltpolitik. Folgerungen aus der Implementations- und Evaluationsfor-schung.* Frankfurt: Peter Lang.

Ullmann, A. (1982) *Industrie und Umweltschutz.* Frankfurt: Campus.

Vester, F. (1990) *Ausfahrt Zukunft.* München: Heyne.

Vester, F. (1992) *Leitmotiv vernetztes Denken.* München: Heyne.

Vietor, R. and Reinhardt, F. (1995) *Business Management and the Natural Environment.* Cincinnati: Southwestern Publishing Company.

Vig, N.J. and Kraft, M. (eds.) (1994) *Environmental Policy in the 1990s. Toward a New Agenda.* Washington D.C.: CQ-Press.

Vogel, D. and Kun, V. (1987) The Comparative Study of Environmental Policy: A Review of the Literature. In: (eds.) M. Dierkes *et al.*, *Comparative Policy Research. Learning from Experiences.* Aldershot: Gower.

Wallace, D. (1995) *Environmental Policy and Industrial Innovation. Strategies in Europe, the USA and Japan.* London: Earthscan.

Walley, N. and Whitehead, B. (1994) It's Not Easy Being Green. *Harvard Business Review.* **72(3)**, 46–49.

Walzer, M. (1992) *Zivile Gesellschaft und amerikanische Demokratie.* Berlin: Rotbuch Verlag.

Weale, A. (1992) *The New Politics of Pollution.* Manchester University Press.

Weidner, H. (1996) *Basiselemente eiuner erfolgreichen Umweltpolitik. Eine Analyse und Evaluation der Instrumente der japanischen Umweltpolitik.* Berlin: edition sigma.

Wicke, L. (1982) *Umweltökonomie.* München: Vahlen.

Wicke, L. *et al.* (1992) *Betriebliche Umweltökonomie.* München: Vahlen.

Wieselhuber, N. and Stadlbauer, W.J. (1992) *Ökologie-Management als strategischer Erfolgsfaktor.* Dr. Wieselhuber & Partner Unternehmensberatung, München.

Wiesenthal, H. (1990) Unsicherheit und Multiple-Self-Identität: Eine Spekulation über die Vorausset-zungen strategischen Handelns. MPIFG Discussion Paper 90/2, Köln.

Wiesenthal, H. (1994) Lernchancen der Risikogesellschaft. *Leviathan.* **22**, 145–159.

Wild, W. and Held, M. (eds.) (1993) Umweltorientierte Unternehmenspolitik—Erfahrungen und Perspek-tiven. *Tutzinger Materialie* Nr. 72, Tutzing.

Willke, H. (1989) *Systemtheorie entwickelter Gesellschaften.* Weinheim/München: Juventa.

Willke, H. (1992) *Ironie des Staates. Grundlinien einer Staatstheorie polyzentrischer Gesellschaft.* Frankfurt: Suhrkamp.

Wolf, R. (1992) Sozialer Wandel und Umweltschutz. Ein Typologisierungsversuch. *Soziale Welt.* **43**, 351–376.

Ytterhus, B.E. *et al.* (1995) The Nordic Business Environmental Barometer. Report, Norwegian School of Management, Oslo.

Zimmermann, K. *et al.* (1990) *Ökologische Modernisierung der Produktion.* Berlin: edition sigma.

Zimpelmann, B. *et al.* (1992) *Die neue Umwelt der Betriebe.* Berlin: edition sigma.

3. PROJECT DESIGN, PROCEDURES AND DEVELOPMENT

Jobst Conrad

The contextual background to the project on the evolution of environmental policy and environmental management having been provided, this chapter describes the overall design of the project, its methods of investigation, and its actual development *vis-à-vis* these original premises.[1]

1 PROJECT CONCEPT

The principal objectives of the project are (1) comparative analysis of specified substantive environmental-management success cases in relatively large European enterprises (not yet investigated) and (2) identification on the basis of this comparative analysis of starting points and options for preventive environmental policy favouring offensive environmental management. These objectives are translated in the following four investigative steps:

1. Survey and analysis of current state-of-the-art research, especially empirical case studies on environmental management and the role of environmental policy in this context and respect;
2. Comprehensive reconstruction of selected (prototype) success cases in individual companies taking into consideration all factors and levels of influence on environmental management;
3. The attempt comparatively to reconstruct on a theoretical-analytical level the dynamics of development and of the interplay of influencing factors in the success cases with special regard to points of departure for public environmental policy;
4. Empirical and theoretical outlining of a prevention-oriented environmental policy favouring offensive environmental management.[2]

This basic orientation of the project can be summarized in the following research questions:

1. What are real substantive cases of successful *innovative* environmental protection?

---

[1] This chapter is largely based on Conrad, 1992.
[2] When speaking of environmental policy, I refer to environmental policy of *public (state) organizations* (on federal, state, municipal or EU level), and I refer to *economic enterprises*, when speaking of environmental management in this chapter, although these concepts need not be attributed to specific social actors.

2. What are the influencing factors and determinants of these innovative success stories?

3. How can environmental policy support these influencing factors?

4. What distinguishes the case studies of *innovative* environmental management from other (ordinary) cases of successful environmental protection?

Major implications and delimitations of this envisaged project orientation include the following:

1. The focus on specified innovative success cases of environmental management implies case studies of enterprises that are pioneers in this respect. Neither cases of (successful) environmental management in a whole industry nor cases where companies simply comply with public regulations and environmental law are to be considered. The case studies assume self-initiative on the part of the firms investigated.

2. The case studies are intended to investigate specific substantial cases of environmental management, e.g. the implementation of an environmentally benign integrated technology, and not the successful development of environmental management in a European enterprise in general. The two processes may have mutually reinforced each other, or the case investigated may have played a pathfinder role in this respect, but not necessarily so.

3. The selection of substantial environmental success cases implies their positioning in the (recent) past, probably the 1980s, because more recent (ongoing) probable success cases are not yet able to show substantive environmental-protection effects. As a result, the project should investigate past histories rather than present development trends in (innovative) environmental management.

4. The focus on a business as the key actor in a case of successful environmental management implies the need to distinguish between company-internal and company-external factors influencing the success story. External factors will be relevant only when perceived and processed in accordance with the company's internal codes of reasoning and action. The relative significance of internal versus external determinants of successful environmental management is of special interest for the project (cf. Conrad, 1996b).

5. Environmental policy is only one possible external factor influencing environmental management. As the project is interested in innovative environmental-management pioneers, the role played by environmental policy is an empirically open question, and may well be a marginal or even counterproductive factor in the case studies.[3]

6. As a consequence, the emphasis in analysing the (potential) role of environmental policy is on contextual control and action on the meta-level (procedural arrangements and boundary conditions) and much less on specific policy instruments, the benefit of which is context-dependent.

---

[3] Thus, whereas the project's normative perspective is on options of preventive environmental policy in support of environmental management, the empirical reconstruction of success cases of environmental management in the case studies does not focus on the role of environmental policy (instruments) in this process.

7. The normative orientation of the fourth step of investigation on options of preventive environmental policy favouring innovative environmental management in European businesses implies a concern with ideal-typical modes of environmental policy rather than empirically dominant ones.

8. Since the case studies attempt comprehensively to reconstruct the totality of (positive and negative) factors influencing the particular success story and to reveal the dynamics of their interplay, this implies taking account of the various phases of the case as well as the combination of explanatory factors on the micro, meso and macro-levels of analysis.

9. Comprehensive reconstruction generates problems of overcomplexity. On the one hand, it has to be reduced on substantive and formal grounds. On the other hand, a complex five-dimensional analytical framework provides a frame of reference for interpreting environmental-management success stories and corresponding environmental-policy recommendations, which cannot be used and checked explicitly in detail within the empirical investigations but which allow empirical data to be organized and substantive research hypotheses to be tested.

10. Since the project resources available have limited work on each case study to less than six person-months, reconstruction of the totality of the success cases mostly cannot be fine-grained. The case studies will represent at best good science-based journalistic investigation and will permit only limited generalization.

Main arguments in favour of the above orientation of the project are the following:

1. The research interest of the project is in new constructive ideas for successful environmental protection. It therefore concentrates on case studies of success stories.

2. Innovative substantial environmental protection will usually tend to be located in and performed by businesses; thus the focus on environmental management and the internal company factors of influence.

3. Public policy can support or undermine innovative environmental management, but normally not perform it itself; thus the conceptualization of environmental policy as one external determinant supporting or restricting a company's environmental management.[4]

4. In modern complex and (functionally) differentiated societies, the image of the state as the (potential) central decision-making actor is an illusory one (cf. Willke, 1989). Thus, there can be no one (political) lever for successful environmental protection, but only a mutually supportive interplay of various (equally relevant) actors and determinants. The emphasis is therefore on a holistic perspective reconstructing the development dynamics and interplay of influencing factors.

The project orientation described above supports arguments in favour of the following research foci and strategies:

1. With the focus on successful environmental management of selected businesses, the structures, behaviour, learning processes and development dynamics of *one*

---

[4] Deficits in policy programming and policy implementation are already well known and have to be expected; therefore the relative neglect of these policy processes.

macro-actor are to the fore in each case study. This actor will therefore be investigated in its internal differentiation, whereas other actors tend to be treated as external black-boxes. This perspective presupposes existing margins of action for environmental management in the company selected precluding an interpretation of environmental management as mere 'deterministic' reaction to external, e.g. governmental influence.

2. Due to project interest in the influence of environmental policy as one possible external determinant of environmental management, the role of environmental policy and politics has to be investigated in a relatively differentiated manner, too, which distinguishes different public policy actors, instruments, programmes, fields and modes of regulation and influence.

3. The normative environmental policy perspective of the project requires special emphasis on the analysis of social learning processes, on the self-dynamics of development processes, and on the network-like (cybernetic) entanglements in the system of environmental management.

4. Correspondingly, the investigation of successful realization of environmental management must pay due regard to the consistency of environmental management measures, their linkage capability to, and their compatibility with, existing structures and boundary conditions, and their capacity for accommodating the socially enforceable interests of actors.

5. Similarly, areas of conflict due to the successful case of environmental management and resulting (unintentional) side-effects have to be taken into consideration, because they may (negatively) influence the realization of analogous (future) programmes and possibly even the longer-term success of this programme.

6. Otherwise, (possible) bifurcation points in the reconstructed success story have to be pointed out because these indicate contingencies in its development and perhaps related points of departure for environmental policy.

7. The reconstruction of successful environmental-management cases primarily consists in applying familiar concepts/theories of environmental management, of environmental policy and of the eclectic combination of theoretical perspectives. Thus, special attention will be paid to industrial environmental-management categories and concepts (cf. Meffert/Kirchgeorg, 1993; Steger, 1992, 1993; Wicke et al., 1992).

8. As far as internal factors of influence are concerned, the multidimensional and many-layered sociosphere of the business has to be given due regard, even if profitability has to be considered an inevitable medium and long-term boundary condition in capitalist societies. Existing economy–ecology trade-offs should therefore not be interpreted away by undoubtedly currently available market possibilities of environment-oriented business policy (see cf. Huisingh, 1988 or Porter/van der Linde, 1995 versus Palmer et al., 1995 for opposing emphasis in their viewpoints).

9. It is important to consider the influencing factors not connected with and not (intentionally) targeting the success case and its environment, e.g. currency exchange relations, company mergers, or social policy.

10. The orientation of the project towards a rather comprehensive explanatory framework of successful innovative environmental-management cases demands a systemic-evolutionary, reticulate perspective (cf. Huber, 1991; Vester, 1992), which avoids single unilinear causal models and interpretations.

11. The case studies require competent analysis of the scientific-technical and economic aspects of the success case in order, for instance, to be able to undertake and judge a product life-cycle analysis, which is prerequisite to evaluating substantive success and the actual validity of arguments and justifications offered by interviewees.

The selection of concrete cases of successful environmental management relies on three criteria: empirical measurability, the availability of internal alternative corporate options, and a certain scope.

Empirical measurability requires substantive environmental effects from a case, which tends to limit selection of cases to the past, probably the 1980s.

The availability of internal alternative options requires that some self-initiative by the company played a crucial role, i.e. the possibility that the company could have decided differently, e.g. renounced measures of environmental management. Otherwise one could have expected more or less uniform behaviour throughout the industry concerned. Where there are prescriptive public environmental regulations, a company may in principle still refuse to comply, e.g. in view of a low risk of punishment due to the well known implementation deficit in environmental policy (cf. Mayntz, 1980, 1983; Terhart, 1986). However, mere compliance with environmental law should not readily be taken as environmental management self-initiative on the part of the company.

To demand a certain scope of successful environmental management needs specification in at least five aspects, which have been spelled out in Chapter 1: avoidance of problem-shifting, changes or savings beyond marginal improvements, sufficiency of partial improvements within the whole product life-cycle, no requirement of an underlying environmental management system, sufficiency of relative, as opposed to absolute, environmental improvements.

The cases selected for the reconstruction of social processes leading to successful environmental management hopefully represent prototype examples, but not a (statistically) representative sample in this respect.

The analytical dimensions and taxonomies of environmental management research relevant for the case studies, which have been much better elaborated in the corresponding literature, have been pointed out in Chapter 2. The case studies also refer to general methodological categories, such as the micro, meso, and macro-levels of social explanation, as learning processes, self-dynamics and cybernetic feedback loops in development processes, and as development phases, alternatives and bifurcation points, and are interpreted within the general five-fold analytical framework sketched in Chapter 1.[5]

---

[5] It is this complex analytical framework which the researcher should have in his/her mind but it cannot simply be filled with empirical data during the case studies.

2   PROJECT PROCEDURES AND METHODS

When addressing project procedures and methods, three phases of the project may be distinguished: the phase of knowledge acquisition, clarification and case-study selection, the phase of case-study performance, and that of comparison and theoretical interpretation.

Concerning knowledge acquisition and clarification, a two-step process was followed in the project: first, 24 cases of successful environmental policy and/or environmental management, spread over most Western industrialized countries, were reported and critically evaluated in two international conferences in 1993 (Jänicke/Weidner, 1995); second, 9 in-depth case studies of successful corporate environmental management were carried out in some West European countries, supplemented by further case studies of the same type in Poland and Latvia. This allowed further improved case study design and completion due to the interesting methodological results obtained, to better knowledge about the whole range of influencing factors and the varying role of environmental policy (instruments), and to the personal experiences of most researchers gained by having conducted several of the previous 24 case studies.

With regard to project organization, the individual researchers, and thereafter the national research teams, were to be responsible for the case studies selected. However, these studies were to be carried out in accordance with the common project design to ensure comparability. The project leader was to be responsible for organising joint project consultation, for carrying out the comparative analysis of the cases, and for delivering the final report. Each of the three project phases was to last roughly nine months. Altogether about 6 person-years, plus 3 person-years for the Polish and Latvian case studies, were to be invested in the project.

As far as data and success-story reconstruction is concerned, a combination of the following sources of information appeared to be appropriate: written documents from different (company-internal and external) sources, (twofold) intense interviewing of (company-internal and external) persons, their confrontation with contradictions and counter-evidence, and a problem-centred group discussion with several key individuals about the reliability and validity of the reconstruction of the case and about optimal entry points and options for an ideal-type environmental policy.

Since reconstruction relates to an environmental-management success story, resistance to detailed analysis should be limited. Since the case studies are *ex post facto* analyses, a bias towards rationalized reconstruction and under-exposure of counteractive influencing factors was likely. Longitudinal case studies, however, observing and following the process of environmental management in real-time and thus evading a disadvantage of the historical method, were beyond the scope of this project. So, except for substantive environmental effects, there could be no guarantee for objective reconstruction by the case studies, though they should go beyond subjective self-interpretation by interviewees. The case studies, however, were neither to be limited to (standardized) self-completion questionnaires nor to merely a small sample of interviews, in contrast to many other investigations of environmental management systems and practices (cf. Coenenberg *et al.*, 1994; Gruber *et al.*, 1992; Hopfenbeck/Jasch, 1993; Kirchgeorg, 1990; Mellitzer *et al.*, 1994; Middelhoff, 1991;

Morrison, 1991; Nitze, 1991; Oberholz, 1989; Ostmeier, 1990; Rappaport/Flaherty, 1992; Schmidheiny, 1992; UBA, 1991; Wieselhuber/Stadlbauer, 1992), as the reliability and validity of these research methods had to be considered insufficient for in-depth case-study analysis (cf. Hamel, 1991; Yin, 1994).

The case study inquiry was, apart from an appraisal of the individual interviewed, to address twelve areas of concern, which were usually not all to be dealt with in one interview:

1. Questions concerning the business;

2. Questions concerning the substance of the success case;

3. Questions concerning the history and evolution of the success case;

4. Questions about company-internal factors of influence;

5. Questions about company-external factors of influence;

6. Questions about the role of environmental policy;

7. Questions concerning the interaction of influencing factors;

8. Questions concerning lacking, but potentially relevant, factors of influence;

9. Questions concerning time points of decision-making, critical (bifurcation) points and alternative development possibilities;

10. Questions concerning lines of conflict, consensus building and learning processes, and self-dynamics;

11. Questions concerning side-effects;

12. Supplementary questions.

Carrying out the individual case studies was then to involve more or less the following steps:

1. Study of literature on environmental management and policy with special regard to determinants of successful environmental management;

2. Evaluation of case studies conducted;

3. Selection of success cases with the help of selective screening, special hints, personal contacts, and then more detailed clarification processes relating to the enterprise envisaged for investigation and its environment;

4. Acquisition of sufficient case-specific expertise and hard data outlining and proving the success case;

5. Descriptive reconstruction of the success story on the basis of the written (company-internal and external) documents available and by intense interviewing of different (10 to 20 company-internal and external) persons;

6. Written reconstruction of the success case according to the agreed analytical categories (c. 20 to 30 pages) including an assessment of reliability and validity, prototype characteristics, and the role of environmental policy;

7. Distribution of this paper to interviewees with the request for critical examination and comment, addressing controversial points of view;

8. Supplementary interviews with a few persons already interviewed or with new individuals;

9. Organization of a problem-centred group discussion with 5–10 selected competent (company-internal and external) persons, focusing on controversial perspectives and on the (ideal-typical) role of environmental policy in the case investigated;

10. Analysis of the group discussion and supplementary interviews and preparation of the case-study report;

11. Circulation, critical discussion and final revision of the case study.

3 PROJECT DEVELOPMENT

Turning now to actual project development, the following deviations from the project concept and design described above are to be noted:

1. Whereas the original project proposal (Jänicke *et al.*, 1991) addressed the effectiveness of policy instruments *vis-à-vis* European enterprises as a condition of successful environmental policy, the project design dealt with the social conditions for substantive environmental improvements as exemplary cases of successful environmental management in European companies, coupled with the normative question of how environmental policy could support such favourable conditions (Conrad, 1992).[6] The actual case studies, however, only partly followed this orientation. Whereas three of them (Amecke Fruchtsaft, Diessner, ABC Coating) reconstruct the social processes leading to substantive environmental improvements, three of them (Glasuld, Hertie, Kunert) primarily analyse the social processes leading to the establishment of environmental management systems with corresponding environmental improvements, and three of them (NedCar, GEP Europe, Ciba) describe in detail the environmental management system of and the environmental improvements achieved by the companies. Since the overall eco-performance of a company depends on its general (institutionalized) characteristics of environmental management and not on selected specific cases of individual substantive environmental improvements, the resulting tripartition of the case studies into outcome, process, and institutional analysis and evaluation seems justified although it complicates and partly prevents their comparative evaluation.

2. The screening of success cases of environmental management and the selection of appropriate case studies did not turn out to be a straightforward selection process, which led to delays in the project schedule and even included a change

---

[6] This reorientation was due to the already available knowledge about the dependence on context and specific design of success or failure of policy instruments, and about the limited influence of environmental policy on offensive environmental management, also partly recognized in the critical evaluation of 24 cases of successful environmental policy (Jänicke/Weidner, 1995), which was performed quite independently of the subsequent case study investigations presented in this book.

in companies already under investigation. In the end, companies of quite different sizes and with quite different degrees of environmental management were chosen, and their major efforts in successful environmental management occurred only in the early 1990s. Thus, these companies may still be pioneers in the specific environmental improvements investigated. Frequently, however, they can only partly be considered socio-economic pioneers of successful environmental management in general because after 1990 companies – at least without spectacular environmental improvements – can no longer be simply judged to be front-runners in this respect when corporate environmental management has already gained general recognition in corporate culture and public debate though it is rarely implemented yet in most companies.

3. The case study phase lasted over a year, postponing and curtailing the final phase of comparative evaluation and interpretation. In no case study did the envisaged evaluative group discussions take place because of time constraints, insufficient financial resources, and the unlikelihood of them yielding worthwhile results.

4. Furthermore, the investigation of success cases of environmental management in Poland and of environmental policy and environmental management efforts in Latvia over the previous five years was added to the project in 1993. These case studies address social contexts differing from western European ones, and differ more or less in design from the other case studies. They have therefore not been included in this book, but can be found in the corresponding final report to the European Commission (Conrad, 1996a).

5. Finally, genuine (general) interaction dynamics of company-internal and external determinants beyond the enumeration and classification of, plus the description of, specific interaction between influencing factors have not really been identified. Although the knowledge about such genuine interaction dynamics causing successful environmental management may be improved by systematic comparison of cases of environmental management success as well as failure, this deficit in reconstructing dynamic conditions of successful environmental management is largely due to the fact that social theories aiming at the (simple) explanation of social processes and phenomena refer mainly to concepts relying on social constraints on human action, such as routines, norms, expectations, definitions, and only selectively to variables of individual action, thus underrating processes of normative and cognitive intrusion, breaks in development, institutional invention, and social innovation (Wiesenthal, 1994). No generalizable pattern of interaction dynamics is therefore likely to be discovered in cases of innovative social circumstances such as offensive environmental management.

Altogether, actual project performance only partly accords with the project design and procedures. This makes the answer to the main research questions more difficult, but generally does not prevent them being addressed, as is attempted in Part III after presentation of the individual case studies in Part II of the book.

LITERATURE

Coenenberg, A.G. *et al.* (1994) Unternehmenspolitik und Umweltschutz. *Zeitschrift für betriebswirtschaftliche Forschung.* **46**, 81–100.

Conrad, J. (1990) Do Public Policy and Regulation Still Matter for Environmental Protection in Agriculture? EUI Working Paper EPU 90/6, Florence.

Conrad, J. (1992) Concept, design and methods of the project 'Conditions of success for environmental policy', Ms. Berlin.

Conrad, J. (ed.) (1996a) Successful environmental management in European Companies. FFU-Report 96-3. Berlin.

Conrad, J. (1996b) Unternehmensexterne Determinanten betrieblichen Umweltmanagements. In: (eds.) L. Mez and M. Jänicke, *Sektorale Umweltpolitik.* Berlin: edition sigma.

Gruber, Tietze and Partner (1992) *Umweltbericht 1992—Stand des Umweltschutzes in führenden deutschen Unternehmen.* Bad Homburg.

Hamel, J. (1991) *Case Study Methods.* Newbury Park, CA: Sage.

Hopfenbeck, W. and Jasch, C. (1993) *Öko-Controlling. Umdenken zahlt sich aus!* Landsberg/Lech: moderne industrie.

Huber, J. (1991) *Unternehmen Umwelt.* Frankfurt: Fischer.

Huisingh, D. (1988) Good environmental practices—good business practices. WZB, FS II 88–409, Berlin.

Jänicke, M. *et al.* (1991) Erfolgsbedingungen von Umweltpolitik: die Wirksamkeit von Instrumenten aus der Sicht von europäischen Unternehmen. Project Proposal Berlin.

Jänicke, M. *et al.* (1992) *Umweltentlastung durch industriellen Strukturwandel?* Berlin: edition sigma.

Jänicke, M. and Weidner, H. (eds.) (1995) *Successful Environmental Policy.* A Critical Evaluation of 24 Case Studies. Berlin: edition sigma.

Kirchgeorg, M. (1990) *Ökologieorientiertes Unternehmerverhalten. Typologien und Erklärungsansätze auf empirischer Grundlage.* Wiesbaden: Gabler.

Mayntz, R. (ed.) (1980) Implementation politischer Programme. Empirische Forschungsberichte. Königstein/Ts.: Athenäum.

Mayntz, R. (ed.) (1983) *Implementation politischer Programme II. Ansätze zur Theoriebildung.* Opladen: Westdeutscher Verlag.

Meffert, H. and Kirchgeorg, M. (1993) *Marktorientiertes Umweltmanagement.* Stuttgart: Poeschel.

Mellitzer, J. *et al.* (1994) 'Betriebserfolg—Umweltschutz'. Zusammenhang Wirtschaftsleistung und Umweltleistung. Interim Report, Wien.

Middelhoff, H. (1991) *Die Organization des betrieblichen Umweltschutzes in der schweizerischen und deutschen chemischen Industrie.* Diss. St. Gallen.

Morrison, C. (1991) *Managing Environmental Affairs: Corporate Practices in The U.S., Canada and Europe.* The Conference Board, New York.

Nitze, A. (1991) *Die organisatorische Umsetzung einer ökologisch bewußten Unternehmensführung.* Bern: Haupt.

Oberholz, R. (1989) *Umweltorientierte Unternehmensführung.* Frankfurt: Frankfurter Allgemeine Zeitung.

Ostmeier, H. (1990) *Ökologieorientierte Produktinnovationen.* Frankfurt: Lang.

Palmer, K. *et al.* (1995) Tightening Environmental Standards: The Benefit-Cost or the No-Cost Paradigm? *Journal of Economic Perspectives.* **9(4)**, 119–132.

Porter, M.E. and van der Linde, C. (1995) Green and Competitive: Ending the Stalemate, *Harvard Business Review.* September–October 1995, 120–134.

Rappaport, A. and Flaherty, M.F. (1992) *Corporate Responses to Environmental Challenges.* New York: Quorum Books.

Schmidheiny, St. (1992) *Changing Course: A Global Business Perspective on Development and the Environment.* Cambridge: MIT Press.

Steger, U. (1992) *Future Management.* Frankfurt: Fischer.

Steger, U. (1993) *Umweltmanagement.* Wiesbaden: Gabler.

Terhart, K. (1986) *Die Befolgung von Umweltschutzauflagen als betriebswirtschaftliches Problem.* Berlin: Dunker & Humblodt.

UBA (ed.) (1991) *Umweltorientierte Unternehmensführung. Bericht 11/91.* Berlin: Erich Schmidt.

Vester, F. (1992) *Leitmotiv vernetztes Denken*. München.

Wicke, L. *et al.* (1992) *Betriebliche Umweltökonomie*. München: Vahlen.

Wieselhuber, N. and Stadlbauer, W.J. (1992) Ökologie-Management als strategischer Erfolgsfaktor. Dr. Wieselhuber & Partner Unternehmensberatung. München.

Wiesenthal, H. (1994) Lernchancen der Risikogesellschaft. *Leviathan.* **22**, 135–159.

Willke, H. (1989) *Systemtheorie entwickelter Gesellschaften*. Weinheim.

Yin, R. (1994) *Case Study Research*. Newbury Park, CA: Sage.

II.   SUCCESS STORIES

4. AMECKE FRUCHTSAFT: ECONOMICAL PATHWAY INTO AN ECOLOGICAL OPTIMUM OF RETURNABLE-BOTTLE SYSTEMS

Jobst Conrad

1 INTRODUCTION

This case study[1] investigates the ecological and economic reduction of water, energy, and chemicals consumption in washing returnable bottles by the German medium-sized fruit-beverage producer Amecke Fruchtsaft in Menden, for which the company received an environmental award in 1993. It is a case of successful substantive environmental management because this optimization of the cleansing process led to considerable reduction in input and in waste products relating to water, energy, and chemical resources; it was probably the most advanced system in the world in 1992. In addition, a diffusion process has started with the newly-developed bottle-washing machinery spreading to other fruit-beverage producers, and a move towards further environmental improvements in the company has been triggered by this input minimization process.

The characteristic features of this case are the emphasis on gains in efficiency and economies in input resources for ecological and economic reasons; the focus on individual actors; and the outcome-oriented reconstruction of social processes leading to specific physical environmental improvements.

The case study is based on 18 interviews with 12 persons conducted in two rounds in August and September 1993, 9 of them belonging to Amecke Fruchtsaft and 6 of them being key figures in the success story; a number of documents provided by the company; some literature and reports dealing with the fruit-beverage market and with returnable containers; and the critical comments of the persons interviewed on the draft of the case study.

2 THE ENVIRONMENTAL SUCCESS CASE

In the improving of the environmental compatibility of returnable-bottle systems, reducing inputs and wastes of the returnable-bottle cleansing process is only one important element; others concern the collecting and transportation of used bottles and the circulation rate of returnable bottles.

The ecological optimization of the bottle cleansing system involved considerable efforts of development and of laboratory testing, and is based essentially on the

---

[1] For the complete version of the case study the reader is referred to Conrad, 1994.

following frequently interactive components:

1. The use of softened water;
2. Optimized washing machine for returnable bottles with more splash zones and screen cleaning in the pre-soak phase;
3. Multiple use of water;
4. General use of hot water;
5. Lye storage in insulated batch containers;
6. Lye recovery;
7. Lye filtration;
8. Omission of chemical additives with the exception of an antifoam agent;
9. Label pressing.

Altogether these environmental management efforts had the following direct and indirect impacts:

1. 70% less water demand (only 110 ml/bottle cleaning);
2. 40% less energy consumption;
3. Less lye required;
4. Hardly any chemicals needed;
5. Lower waste-water load;
6. Consequent savings of DM 250.000 or DM 500.000 per year (measured against state-of-the-art technology in 1989 or against the technology used till 1990, respectively);
7. Positive repercussions on in-house environmental management;
8. Influence on other fruit-beverage producers.

3   COMPANY SITUATION AND STRATEGY

The (German) fruit-beverage market[2] is a very competitive one with demand increasing steadily till 1992 (about DM 5 billion in 1992); there are about 450 small German fruit-beverage producers, the 11 largest producers having an approximately 70% share of the market; and there is a growing proportion of returnable bottles (c. 10% in 1980, c. 45% in 1993) being used. The market share of Aldi, the largest retail trader in fruit beverages, particularly fruit drinks, is about 30% (70% for fruit beverages in the state of North Rhine-Westphalia), and it sells only non-returnable bottles and cartons, refusing to order returnable containers.

Amecke Fruchtsaft is part of the family-owned group Amecke-Mönnighof, founded in 1877, and today mainly engaged in selling Coca Cola (as franchisee); in producing and selling fruit beverages; and in supplying and replenishing company

---

[2] Usually, high quality fruit juice (100% fruit content), fruit nectar (30–60% fruit content), and fruit drinks (6–30% fruit content) are distinguished.

catering automates and hot beverages. The company employs about 200 people. Its turnover has expanded more or less continuously (1986: *c*. DM 30 million; 1992: *c*. DM 75 million). From 1986 it expanded, particularly in the fruit-beverage market, from about DM 1 million to about DM 30 million in 1992 (for *c*. 20 million litres of various fruit beverages in returnable bottles only). With an overall market share of 0.6% in Germany, Amecke Fruchtsaft has meanwhile developed into the market leader in returnable-container fruit beverages in North Rhine-Westphalia with over 10% of the market. As an economically sound, medium-sized, owner-operated company, average-range profits were typically reinvested in the company by its directors so that they dispose of sufficient financial resources for strategic investments. Furthermore, the company is characterized by a good working atmosphere, due to its small size, its well-functioning informal organizational structure, and low internal and external manpower fluctuation.

In 1978, Heinrich Amecke-Mönnighof, the key actor in this success story, became director and partner of the Amecke-Mönnighof group after having gained his degree in business management. He initially observed and adapted to the prevailing company habits and practices without trying to enforce his own strategic conception too rapidly. Despite a high degree of personal commitment and dedication, he first took time to form an independent opinion on the existing concrete scope for action allowed by the prevailing circumstances, feeling he could afford to wait for favourable opportunities to show themselves. He concentrated initially on winning personal credibility and reputation with the workforce, a decisive move in implementing projects in the firm and which far exceeded the formal role of managing director. This allowed him to make an essential contribution to preparing for the implementation of his ecologically oriented ideas and programmes, which, launched from about 1985, have hitherto run relatively smoothly. Amecke-Mönnighof himself states that his basic attitude had always been ecologically oriented, seeing an environmentally oriented, but economically profitable corporate strategy as the major opportunity for an individual to produce (exceptional) lasting results.

Unlike incorporated companies, Amecke Fruchtsaft's status as a small and medium-sized enterprise (SME) allowed director Amecke-Mönnighof to design and to realize a credible, homogeneous company strategy. The corporate strategy course set and implemented in the 1980s forms the essential formal and substantive context within which the success story became both possible and understandable. There were essentially five corporate strategic decisions that shaped the present development of the Amecke Mönnighof Group's fruit-beverage business:

1. The decision to rely no longer on the Coca Cola franchise as the only operating area;
2. To concentrate on fruit-beverage production as the principal future field of operation;
3. To sell fruit beverages only in returnable containers;
4. Consistently and professionally to implement the concept developed to expand business with fruit beverages in returnable containers; and,
5. To practise protection of the environment as a deliberate business strategy.

The increasing relevance of environmental protection in company policy resulted from the coincidence of at least three favourable aspects:

1. With rising (public) ecological awareness, superior returnable container systems tending to ecological superiority offer a realistic chance for long-term expansion in a fruit-beverage market dominated by non-returnable packaging. With the need that is gradually becoming apparent of internalizing future environmental costs, environmentally-friendly production and distribution systems will become more competitive.

2. Corresponding innovations in transportation, washing, and re-use rates permit the environmental advantages of returnable packaging to be realized profitably.

3. To a workforce with an growing awareness of environmental issues, the possibility of taking action and gaining a positive image via environmental protection will tend to increase identification with the company and thus create enhanced work motivation and commitment (Amecke-Mönnighof, 1993).

4   THE SUCCESS STORY

Whereas the setting-up of these corporate strategy courses for the future prepared the ground for Amecke Fruchtsaft's successful environmental management efforts, the success story itself developed in three steps, namely the improvement of the old bottle-washing machine, the purchase of a new bottle-washing machine, and the optimization of the latter.

The position of technical manager being vacant, rising (waste) water costs in 1988 caused the managing director, Amecke-Mönnighof, to become concerned about the rising level of water demand of the bottle-washing machine.

After decalcification of the bottle-washing machine had revealed the principal reason for the high water demand and lye carry-over, a softening plant was installed to decalcify extremely hard fresh water obtained from the municipal water supply before it was fed into the bottle-washing machine. This measure, carried out in 1988, which, including the softening plant, cost DM 50,000, proved extraordinarily successful insofar as water demand dropped by more than 40% from c. 800 to c. 500 ml per bottle, so that the investment paid for itself within a few months, without any such economic criterion playing a central role in making the decision to buy it.

The motivational incentive triggered by this success was then decisive in the above-average effort invested in systematically checking out and ecologically optimalizing the new bottle-washing machine. Furthermore, with the rationalization successes, there was greater willingness to invest in a new bottle-washing machine at a time when the old one, little more than ten-years-old, did not have to be replaced yet, and although replacement meant an investment of several million marks, an enormous outlay in the context of this company.

Only a few people were involved in the planning and conceptual implementation of the water-conservation measures relative to the old bottle-washing machine: essentially the managing director, the production manager, and to some extent one foreman and the purchasing manager. Practically no external actors were involved.

The purchase of the new bottle-washing machine had been systematically prepared as from 1988. To this end, the production and purchasing managers, both of whom had been in the firm for many years, inspected bottle-washing machines in some eight companies. Furthermore, the managing director himself, in co-ordination with competent members of the company, developed a comprehensive catalogue of 365 selection criteria and a corresponding computer test programme.

Finally, three offers of around DM 2 million, which were compared in detail with the aid of the test programme, remained in the running. The bottle-washing machine selected was from the firm KHS Maschinen- and Anlagenbau AG at Dortmund, belonging to the Klöckner group, which with a turnover of about DM 4 billion and a 30% share of the market, is the leader on the world market for filling machines, a field predominantly in German hands. The choice was primarily motivated by the fact that, in comparison with the alternatives, this bottle-washing machine offered the greatest potential for subsequent modification and permitted multiple water use on the cascade principle. And the final incentive was the company's willingness to cooperate in subsequent optimization.

After the first contact had been established in 1988 between Amecke-Mönnighof and the KHS sales manager, who lived in Menden and happened to be personally acquainted with Amecke-Mönnighof, the main concern of KHS (apart from sale of the bottle-washing machine itself plus peripheral equipment) in the sales negotiations in the first half of 1989 was the opportunity offered by a neighbouring purchaser to conduct practical laboratory experiments on the technical feasibility and acceptability of the bottle-washing machine with extremely low water-demand which had been specially developed for Amecke Fruchtsaft and was subsequently to be improved in cooperation with the makers. The medium-sized company needed customers for this purpose, since with plants made to order it was unable to carry out such testing itself. The know-how to be acquired at low cost through this development and testing could then be used for other types of machine. In any case, Amecke Fruchtsaft's requirements called for modification of the usual modular system. As long as purchase of the bottle-washing machine was more or less worthwhile, cost considerations were not primordial for Amecke-Mönnighof. The purchase contract for over DM 2.8 million must therefore be seen against the background of profitability calculations submitted by KHS.

While KHS was initially prepared to guarantee a maximum water demand of 400 ml per bottle, under the terms of the purchase contract concluded in June 1989, the company promised to supply a bottle-washing machine using a maximum of 250 ml water per bottle. The terms of the purchase contract also laid down that purchaser and seller were to attempt jointly to reduce water consumption to a minimum. The explanation for the willingness of KHS to cooperate was, apart from the basic interest in the quasi-experimental acquisition of significant technical/ economic know-how, a follow-up sale already under discussion of a bottle conveyor installation, which was then agreed in November 1989 for a sum in excess of DM 1 million. The decisive factor for a later water consumption that was presumably the lowest in the world at only 125 ml per bottle, was the extensible bottle-washing machine conception developed by the head of the KHS construction division, which permitted the later reduction of water consumption from 250 ml to c. 150 ml per bottle to be relatively easily achieved owing to the reserves provided in the machine design. It was only the later reduction of water consumption to a minimum of 100 ml per bottle that demanded additional creative effort on the part of those involved.

After the new bottle-washing machine had been delivered in February 1990, the above-mentioned optimization stages were systematically tackled in cooperation with KHS between 1990 and 1993. Mechanical improvements were principally developed by KHS; the chemical process improvements being largely initiated by Amecke Fruchtsaft.

Furthermore, orders for a further DM 4 million were awarded to other plant manufacturers, mainly concerning juice-processing, packaging, two separate batch tanks, and a bottle inspector.[3] These investments, amounting to altogether DM 7.8 million, were practically equivalent to the complete new installation of the entire filling plant.

The optimization process was by no means a matter of fundamental technical innovation, but rather of the deliberate questioning and rethinking of practically every element of the bottle-washing machine with the aim of creating an awareness of the (ecological and economic) potential for improvement. The optimization process consisted essentially of three components: the experimental alteration of parameters in the cleansing process, the purposive introduction of new process elements, and the systematic recording of changed circumstances in the cleansing process. The relevant learning phases sometimes took months, especially when initially diverging assessments of the feasibility or efficiency of certain measures were checked through by thoroughgoing data collection and (sometimes outside) data analysis.

Overall, the measures were supervised by means of elaborate measurement techniques to monitor experimental results. If one takes into account the additional work involved for both parties in the cooperation, this was, at a cost to each of about DM 300,000, a relatively cost-intensive undertaking.

Plans to use a reverse-osmosis system in bottle-washing for waste-water treatment were not realized because, the water having already been softened, the additional expense was disproportionate to the additional cleansing effects. If technical progress permits, the deployment of such a system at some time in the future would, however, be reconsidered.

The cooperation between Amecke Fruchtsaft and KHS in the period from 1989 to 1993 functioned well on the whole, and was enjoyed by those concerned, even though KHS technicians were to some extent sceptical about the initiatives and extraordinary demands advanced for the main part by Amecke-Mönnighof, sometimes feeling to begin with that too much was being asked of them in the way of technological realization. Without a fixed time schedule, individual improvements to the bottle-washing machine were realized step by step. To this end, the people involved consulted with one another weekly or monthly as called for by the situation, the short distance between Menden and Dortmund facilitating cooperation from both a technical and organizational point of view.

---

[3] This DM 800,000 bottle inspector automatically sorts out technically-unacceptable bottles.

On the part of Amecke Fruchtsaft, those involved in optimization activities and discussions were essentially the managing director and the heads of production and engineering, the latter being later replaced by the head of quality assurance. KHS personnel concerned were the sales manager, the head of the bottle-washing machine construction division, and to some extent the appropriate heads of customer engineering, laboratory or assembly. In discussion of the optimization ideas brought up especially by Amecke-Mönnighof, KHS technicians sometimes felt the need to apply the brakes, being particularly aware of the contradiction between low water consumption and clean bottles, both of which the manufacturer had to guarantee.

At Amecke Fruchtsaft, the pertinent in-house discussions, in which Amecke-Mönnighof as managing director played a decisive part, took place either in pairs or small groups. Apart from the passing on of suggestions, the discussing of ideas, and the planning of the measures to be taken, the emphasis was on weighing up and limiting the scale of certain steps.

Whereas the head of purchasing, because of his earlier experience with fruit-beverage production as head of laboratory, acted as critical audience in these discussions, the technical manager engaged at the beginning of 1990 especially for optimization of the bottle-washing machine, played no significant role. After two years he returned to the technical consulting company he had come from after finding it difficult to get along with the others in the firm involved in the measures, and showing no aptitude for innovation nor willingness to make decisions, his attitude being uneconomic and security-oriented. The head of quality assurance and product development engaged by Amecke Fruchtsaft in the autumn of 1990, as a chemical process engineer equivalent to the KHS engineers, acted as controller whose job it was to evaluate critically the practicability of concrete optimization ideas.

A positive effect within the company was also achieved through the clear delimitation of competencies and responsibilities among the heads of the various divisions (production, procurement, quality assurance, engineering) carried out in 1991, which resulted in a marked decline in ill-feeling and conflict caused by confusion of lines of authority.

The three principal people concerned in Amecke Fruchtsaft, motivated by the economy successes achieved with the old bottle-washing machine, were absolutely committed to optimization of the new plant. Amecke-Mönnighof was the central personality less because of his position as managing director than because of his conceptual leadership and strong ecological ambitions. Without him, all involved agree, this optimization process would not have been realized.

While for Amecke Fruchtsaft cooperation with KHS was hoped to bring the ecological and economic optimization of its bottle cleansing, KHS regarded the process as an opportunity to determine technical and feasibility limits, thus expanding its know-how and consequently its market opportunities. For this reason, KHS also assumed the greater part of the cost of materials accruing in the optimization process. With hindsight, the manufacturer clearly profited from this cooperation, because it now enjoys competitive advantages over other manufacturers in the foreseeable diffusion of this process innovation, and has already been able to sell five bottle-washing machines of the same type up to early 1994.

While there was consensus on objectives, points of dissent arose in the course of cooperation on substantive matters. There was, however, no disagreement on the conception of the bottle-washing machine and hardly any on water consumption.

Optimization of the bottle-washing machine was an ecological/economic learning process through step-by-step learning-by-doing. In unforeseeable, controversial cases, informal agreement was often reached simply to try things out, so that it took a considerable time to collect data. Overall, the cooperation process can be summarized as follows: the managing director of Amecke Fruchtsaft took the main initiative in introducing far-reaching objectives on which he insisted; while each firm made approximately equal contributions, the engineering improvements were more the purview of KHS staff and the process/chemical innovations were predominantly realized by Amecke Fruchtsaft; finally, cooperation proved very successful for both companies.

Apart from the six or more main people involved, no other actors played an essential role in this optimization process. The remaining staff of the production department (20 to 30 persons) had no part in shaping the optimization process with the exception of one foreman, but they were kept informed. Given the nature of their practical activities, this did not present a problem. Nor was the workers' representative council, which was active in safeguarding staff social interests, intimately involved. Similarly, no-one outside the firm was closely concerned in improvement of the bottle-washing machine, with the exception of the few relevant KHS employees. Where external actors were affected, they played a neutral or hesitant role.

Nor did public authorities play an important part. For one thing, the operational emissions subject to the BImSchG (Federal Immission Protection Act) are largely classified as ecologically harmless. In practice, the competent water authority, the State Office for Water and Waste (STAWA), supported the application for a bank loan subsidized by EC funds for the investments undertaken by Amecke Fruchtsaft. For the rest, the interest shown by the authorities in the ecologically-innovative projects of the firm was minimal and formalist. Thus the STAWA required the batch tank, already specially insulated with steel, to be protected by a wall because the lye being stored was a substance hazardous to water. On formal legal grounds such a measure was not deemed necessary for the bottle-washing machine which was much more susceptible to leakage. The matter is still pending and the shield has not yet been built in 1994. It is agreed that the general rise in ecological awareness among the population together with the public debate on the environment had a beneficial impact. However, environmental groups played no part in the success story beyond providing supportive 'background music'.

With the exception of the in-line filtration of the pressed lye and reverse osmosis, all experimental improvements proved to be ecologically beneficial and for the most part economically advantageous as well. Consequent on the positive results of the experiments, the corresponding modifications to the bottle-washing machine were progressively implemented more or less on line, which did not cause any additional downtime. Apart from the usual hardware problems that arise with any similar technical installation, there were no significant readjustment problems with the new machine.

The ecological (and also economically profitable) optimization of the returnable-container system continues on four levels fundamentally encouraged by these successes:

1. Further optimization of the bottle-washing machine, e.g. by experimenting with lower lye concentrations or antifoam-agent dosed according to foam occurrence, waste-heat use, splash-water processing or an additional splash zone as well as reverse osmosis for the purpose of improved wastewater treatment, which has not yet been written off as a possibility;

2. Ideas and measures concerning other components of the returnable-container system, e.g. transport of the beverages in containers by rail and local trucks or also by internal waterway,[4] increasing the return rate of returnable bottles, delivering and collecting bottle tops in containers, wastewater treatment, returnable tops with cap recycling, bottle labels free of heavy metals;

3. Diffusion of the optimized bottle-washing machine among other fruit-beverage producers, so that their ecological advantages can gradually take effect throughout the country;[5]

4. General operational measures to protect the environment, e.g. the installation of catalytic converters in the diesel-driven lift-vehicles (DM 12.000 each); space-saving stacking of board in a container with direct access for the fork-lift trucks; multiple sorting of refuse (paper, board, labels, batteries, plastic cups, fluorescent tubes, white glass and green glass); the propagation generally of waste avoidance and disposal, supported by a specially formed working group; integrated cultivation methods for fruit trees; ideas on autonomous power supplies with total energy units (in cooperation with RWE)[6]; and beyond this, the increasing documentation and stipulation of operating instructions in the ecological field, the preparation of ecological balance sheets and a manual on quality assurance.

On the whole it can be stated that, against the background of an economically successful environmental management, incentives in this respect, and exemplary managerial action, ecological awareness among the workforce was increasingly reflected in appropriate environmental behaviour. The respective measures were for the most part introduced against a background of situational factors and thus far not within the framework of a fully formulated, ecologically-oriented corporate strategy. Upcoming equipment replacements, for example, now always take ecological considerations into account. Management and the members of the workforce already involved make a great effort to exploit the motivational momentum arising

---

[4] This is unlikely to be realized for the foreseeable future, because the costs involved in storage and withdrawal, despite the nearby canal in Dortmund, would increase the final price by about DM 0.30 per bottle. However, with a different basic transport structure, such as the use of pontoons with suitably adapted barges, the building of fruit-beverage bottling plants on waterways, and the internalization of the environmental costs of transport, conditions could perfectly well become profitable. This has encouraged Amecke-Mönnighof to invest further pioneering work in this area to overcome this ecological Achilles' heel of the returnable-container system.

[5] With the renewal of bottle-washing machines or their purchase in the course of conversion to returnable bottles, this diffusion process is a relatively automatic process, insofar as this machine is quite simply the best economic option on the market at the moment because of its low water and energy consumption. Other fruit-beverage producers are likely to prefer it regardless of ecological motives.

[6] The largest German utility, Rheinisch-Westfälische Elektrizitätswerke.

from successful past environmental measures for further ecological improvements, and try not to allow a feeling of being able to rest on laurels already earned. However, whereas the ecological successes obtained to date are essentially based on economically efficient process management and gains in efficiency that are in principle technically easy to achieve, further ecological optimization measures are likely to be less easy to realize, and not so frequently at a profit.

5   ACTOR STRATEGIES AND BEHAVIOUR, AND STRUCTURAL CONDITIONS

The configuration of actors within the framework of this micro-sociological case study is best considered on the level of the individual and not that of the organization, as is usually the case with social science case studies.

The two (leading) members of staff decisively involved in the (ecological/economic) optimization process were grouped about the managing director as principal actor. Beyond this grouping, two (leading) members of the KHS staff, the elderly co-proprietress E. Amecke-Mönnighof, and two further (leading) members of staff at Amecke Fruchtsaft played a certain role via their decisions and activities.

Also to be mentioned are the marginal actors concerned in individual aspects: the officials of the water authority who endorsed the EC-subsidized bank loan and who conditionally approved the batch tanks; the competent bank officials; the institute commissioned to carry out laboratory analyses in addition to in-house laboratory tests; the suppliers of the bottle labels; the suppliers of other equipment components; the VdF[7] presidium; from 1993 the media; as well as members of the firm and KHS members of staff concerned in some way or another with the bottle-washing machine, but who under the given circumstances played no relevant role in determining the direction and course of the success story.

Finally it is important to note that some organizations that were in principle in a position to exercise influence, or representatives of such organizations, did not intervene as actors. Among these were competing companies (except inspection of the new machine on their own behest); approving and supervisory authorities; municipal authorities; bodies promoting environmental policy; environmental groups; consumer associations; as well as (until the award of the Impulse Environment Prize) the media and (potentially) the family of the managing director.

In this success story, unlike in a number of other documented cases of successful innovative environmental management, we are in essence dealing with a limited configuration of actors and a relatively simple and clear structure. Its structure is characterized by a strategically, and to some extent hierarchically, distinctive principal figure; and by cooperation among the main figures, between whom antagonistic interests played practically no role either in relation to particular aspects or in general; and the inclusion of further actors as required, who either supported optimization of the bottle-washing machine *de facto* through their decisions and work

---

[7] Verband deutscher Fruchtsaftindustrie, i.e. the association of German fruit-beverage producers.

or at least did not obstruct it. The configuration of actors as such is thus not a qualifier structurally hindering the course of this optimization process.[8]

Against the background of this configuration of actors favouring the success achieved, it is important to analyse the impacts of the following factors on the development, change, and stabilization of the (environmental) behaviour of actors (see WBGU, 1993: 256):

- The perception and evaluation of environmental conditions;
- Environmentally relevant knowledge and information processing;
- Attitudes and values;
- Incentives for action (motivations, reinforcement);
- Offers and opportunities for action;
- Perceivable consequences of action (feedback).

For the optimization process of the bottle-washing machine, the following can be said in this respect:

1. The environmental problems were initially perceived as useless consumption and consequent potential for economizing, but in the course of optimization were increasingly placed and assessed in a genuinely ecological context by the strongly ecology-oriented managing director. He was relatively successful in conveying this perspective especially to the main parties involved, but also to the less directly concerned actors.

2. In the course of the systematic and comprehensive optimization process, the necessary environmentally relevant knowledge was acquired through an appropriate series of experiments and information processing.

3. Managing director Amecke-Mönnighof is characterized, apart from his marked commercial-efficiency orientation, by unequivocally ecological attitudes and values and his willingness to deploy additional company resources for ecologically motivated measures not only in accordance with profitability criteria. In the course of the optimization process, the other main actors increasingly adopted such ecological attitudes themselves, whereas this cannot be said of the secondary actors. This was not, however, a decisive factor.

4. The vital motivation was triggered by the savings achieved with the old machine and further strengthened by a certain momentum developed in the optimization process in a sort of intoxication with success. Moreover, the environmentally-friendly conservation measures also proved to be economically beneficial. The management supported and did not obstruct staff ecological commitment. Cooperation offered KHS the possibility to strengthen its competitive position without running any risk. And despite her continued commitment to the company, the elderly managing director took the attitude that her nephew as

---

[8] The new technical manager was the only person out of line. During his two years with the firm (1990–91) he made no substantial contribution, despite his function, but who nevertheless did nothing significantly to hinder the optimization process.

representative of the coming generation in the 1980s should now to a large extent determine corporate strategy himself, as long as she did not feel it was foolhardy or too risky.

5. The closer attention paid to the bottle-washing machine gave rise to a series of opportunities to practise environmental protection effectively and with economic advantage to the firm, especially through economy measures.

6. And the results of the optimization process were and are being systematically recorded down to daily production accounting, so that the (ecological) consequences of the measures carried out are clearly apparent.

7. Since the ecological/technical optimization of the bottle-washing machine was a relatively clearly delimited project in structure, there was not too much of a risk that Dörner's (1992) quite plausible logic of failure in complex situations would have a chance. Moreover, the appropriate steps of strategic planning and action in the optimization process had all been explicitly or implicitly taken: the managing director elaborated the goals up to and including exacting optimization standards. Knowledge of the structural and process patterns of the bottle-washing machine became more and more comprehensive and precise as data were collected systematically and in ever greater quantities. Prediction and extrapolation had a role to play, flexibility being respected in individual instances, with correspondingly planned and implemented action. For the purpose of controlling impacts and possibly revising action strategies, especially with respect to controversial estimates, a lot of room was given for systematic data collection and control.

As a whole, the impact analysis of the social-psychologically relevant factors influencing the environmental behaviour of actors provides broad confirmation of this, and thus of the success story.

On the structural level, and thus independent of the actors' (current) articulation of interests and modes of procedure, special account is to be taken of the following relevant structural elements and conditions – from the perspective of ecology (1), economics (2), administrative law and practice (3), corporate organization and culture (4):

1.1. The absence of serious operationally caused (politically or legally relevant) environmental problems;

1.2. The existence of a large potential for efficiency-enhancing, profitable economy measures;

1.3. No significant goal conflicts between environmentally-oriented measures and other corporate goals such as bottle hygiene, production times, production costs (ecological measures well adapted to integration causing little downtime and additional costs);

2.1. A highly competitive market for fruit beverages;

2.2. With, however, relatively separate competitive returnable-container market segments;

2.3. Increasing consumption of fruit beverages;

2.4. Economically sound SME company;

2.5. Reinvestment of its profits for the most part in the company;

2.6. Its strong expansion in the fruit-beverage sector;

3.1. The BImSchG as relevant environmental law framework;

3.2. No critical emission findings;

3.3. No important public authority conditions relating to the successful measure;

3.4. No environmental policy[9] significantly discriminating against certain packaging systems in the past;

4.1. SME, proprietor-managed company with a hierarchical structure, that in practice puts more weight on persuasion than instruction;

4.2. The predominance of informal processes of in-house organization[10] and procedures with correspondingly flexible margins for manoeuvre without great frictional losses and transaction costs;

4.3. Low personnel fluctuation and a good working atmosphere;

4.4. A growing ecological awareness also affecting corporate culture.

These structural conditions had a predominantly supportive or neutral impact on the course of the success story, but had hardly any braking effect. Thus corporate organizational and cultural circumstances enhanced the opportunity for an ecology-minded managing director to tackle and implement relatively rapidly particularly concrete ecology-oriented projects. Delaying bargaining and approval processes involving public authorities were largely avoided owing to the administrative law situation. Despite a highly competitive market situation, the resources available in the company allowed a successful growth strategy in the fruit-beverage sector, and the market structure possibly also permitted environmental protection-oriented measures entailing additional costs. The ecological circumstances were also favourable for the process of optimizing the bottle-washing machine, but did not exert any problem-related pressure prompting action.[11]

To sum up, the structural conditions for action mentioned did not decisively alter the existing pattern of interest, power, and also perception of the pertinent actors, but tended to be to their benefit. Thus, as a result of the also economically advantageous outcome of the process, the positions of the company on the one hand and those of the actors involved in optimization on the other are likely to have been strengthened especially with regard to the increasing ecological demands to be expected in the medium term *vis-à-vis* practically all sectors of society, and their ecological perception is likley to be heightened. These structural conditions (which might well be subject to change in the longer term) tend to support the managing director's hitherto successful profile and reputation strategy of putting his faith in

---

[9] Since 1991, with the advent of the Packaging Ordinance and the DSD (Dual System Germany), this no longer applies in the form then practised.

[10] There is no organizational chart of the firm in the records, nor an environmental protection officer nor even a formal in-house environmental protection organization.

[11] In numerous publicised success stories, by contrast, environmental scandals acted as triggers for correspondingly far-reaching environmental-protection measures (see Hildebrandt *et al.*, 1995).

the returnable-container fruit-beverage market segment and, by doing so, using the potential for savings to make a profit under ecological auspices, thus also serving his interests and power position. Much the same can be reported for the KHS actors. It is also not to be assumed that, outside the structural circumstances described, the actors' pattern of problem, interest and power that determine their action orientations have significantly changed in their basic structure. The outcome was that – in a positive mutual dynamic between the structural and action levels – a beneficial impact in every sense of the word was exerted on the course of the success story by: ecological, economic, political-administrative and company-specific structural conditions; a constellation of actors imposing hardly any restrictions; framework conditions relevant for the environmental behaviour of the actors determined by their interests and perception; a promising corporate strategy in the fruit-beverage sector; and a strategic approach to the optimization process.

6   DETERMINANTS AND DYNAMICS OF THE SUCCESS STORY

If one distinguishes between problem-structural, technical/economic, psychological, situation-structural, historiographical, and societal-context determinants of the success story (see similarly Conrad, 1992), it is possible systematically to list the various explanatory factors, inquire into essential as opposed to non-essential qualifiers, and gain an overall picture of factor interaction and the dynamics of the success story. On the level of specific, concrete factors influencing the success story, the following are to be mentioned:

1. In substance, it was a matter of a relatively clearly determined, well-defined project for the ecological/economic optimization of resource utilization without significant negative side-effects. Because of conceptually clear, thrifty rationalization measures, large, economically beneficial savings had already been achieved. What was essentially envisioned was a systematic improvement of the bottle-washing machine without embedding this in an overall, comprehensive environmental management strategy. All these problem-structural framework conditions permitted relatively smooth and problem-free optimization.

2. The important technical/economic determinants are to be seen first in the successful expansion of the fruit-beverage business targeted by company strategy, with professional marketing oriented on trade and consumer needs, and by no means primarily ecological needs, providing the basis for this expansion; and second in the thoroughly systematic implementation of the new bottle-washing machine project with consistent monitoring and the continued optimization efforts producing significant economic advantages that manifested themselves to some extent only in the aftermath of the measures taken.

3. The managing director as central figure was distinguished by a number of personality traits extraordinarily favourable for the success story. He was in a position to wait with his ecological plans for a favourable situation and a favourable (social) climate while preparing his environment for an opportunity, and was then able to grasp that opportunity when it occurred. His strong personal motivation

to create something lasting through his ecologically-oriented strategies and measures gave essential impetus to the success story. The managing director's work and leadership style was characterized by his commitment, his technical and management skills, and his forcefulness in delegating, convincing, and vesting trust in the skills of his staff.

4. It was also a psychological advantage on the personality level that a benevolent aunt held the position of co-proprietress and managing director, and also the commitment of the other principal members of staff involved and the corporate strategic willingness of KHS staff to cooperate despite initial scepticism.

5. Finally, on the social-psychological level, we must mention the impetus generated by the successful economy measures with the old bottle-washing machine; the psychological momentum provided by the 'intoxication with success'[12]; the in-house 'politicization' of the ecological idea; the recognition and appreciation of Amecke-Mönnighof as a positive central figure; and the cooperative capacity, flexibility and attention to detail on the part of the pertinent actors.[13]

6. As far as the structure of the situation was concerned, it was an advantage that the enterprise was a relatively small one, informal in structure, with a good working atmosphere, few actors, few conflicts of interest, and an exemplary head categorized as an 'ecofreak'. Given smoothly functioning personal cooperation and, in time, clear delimitation of responsibilities, the informal corporate structure permitted effective and uncomplicated operation, since fewer formal rules had to be respected and less paper work was required.

7. Disregarding the process of selecting the new bottle-washing machine, only production and quality assurance, and not other departments such as procurement and sales or administration, were crucially involved in the optimization process. While effective cooperation with KHS (principally on the initiative of Amecke-Mönnighof), for whom the opportunity to acquire know-how and market leadership were situationally decisive, was of central importance for successful optimization, the project was otherwise a purely company-internal affair without additional (disruptive) external actors. The actors involved were not Utopians, but people who took account of the technical and economic possibilities.

8. Finally, the company tradition of ploughing profits back into the firm should be mentioned as an important situational advantage[14] ensuring the availability of adequate financial resources. This also permitted a new bottle-washing and filling plant to be purchased despite significant improvement to, and before depreciation of, the old equipment, a move that was not absolutely necessary but primarily motivated, in the framework of a strategy of expanding in the fruit-beverage sector, by marketing considerations, namely, an enhanced image with (potential) customers.

---

[12] The appreciable improvement potential practically guarantees the success of the measures, so that in the absence of other motives, there is simply no reason to be opposed to them.

[13] Neither the inadequate discharge of his duties by the technical manager nor his insufficient integration had any critical effect.

[14] In the 1980s, they came mainly from Coca Cola franchise income.

9. Important from an historiographical point of view (in the sense of historically and situationally contingent chance qualifiers) were, first, the vacancy of the position of technical manager in 1988, obliging Amecke-Mönnighof to pay closer attention to the question of technical savings potential; secondly, the economic and corporate policy to cease Coca Cola bottling in 1990; and finally the 1985 fundamental, corporate-strategy decision to stake the future on fruit-beverage production in returnable bottles.

10. On the level of societal-context determinants (in the sense of substantively-important general societal framework conditions and development trends), a general increase in ecological awareness constituted an important sociocultural boundary condition.

11. And finally, very important, as summary determinant, was the absence of (contextual) influences typically hindering a success story: no actors and interests seeking to undermine the project; no complex actor-network; no real hindrance by external actors, be it through active opposition or non-participation; no relevant, especially ecological/economic conflicts over goals; no really disruptive structural, institutional, or legal boundary conditions; no distraction from other activities; and no exhaustion through past unsuccessful measures.

This enumeration makes it clear that the success story was a predominantly positive interplay between all levels and factors: this alone allowed it to be realized so rapidly and smoothly, as I have attempted to show in Figure 1 with assumably plausible weighting of the individual qualifiers in the form of concentric circles.[15]

7 CONCLUSION

Looking back dispassionately at the Amecke Fruchtsaft success story, the following characteristics in particular come to mind, permitting it to be classified as strategic, offensive environmental management in environmental strategy terms, a development pattern which hitherto has been little in practical evidence (see Freimann, 1994; Steger, 1993; Wieselhuber/Stadlbauer, 1992), and for the following conclusions to be drawn:

1. Amecke Fruchtsaft is an atypical success story, as practically all influencing variables harmonized to a large extent in its favour, and effective ecological solutions for problems and not solutions for social conflicts were to the fore; whereas an environmental scandal in contrast frequently triggers more consistent environmental management. To this extent, this success story demonstrates the differing patterns, and thus the breadth of range fundamentally possible in society, that successful environmental management can have.

2. It is this positive, dynamic interplay in the pull and push of factors and various classes of determinant that proved decisive in smoothing the comparatively rapid

---

[15] It should be noted that the catalogue of 28 concrete qualifiers, although a sound, relatively comprehensive, and plausible list, is not necessarily a complete one, nor demonstrably unequivocal, and thus scientifically stringent.

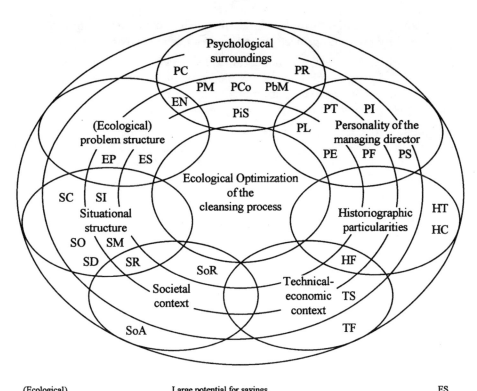

(Ecological) problem structure (E)	Large potential for savings Clearly defined project Hardly any negative-side effects	ES EP EN
Psychological determinants (P) - Environment	Staff commitment Cooperativeness of those involved Respect for the managing director Motivation through initial success Motivational Momentum Benevolent aunt as managing directress	PC PCo PR PiS PM PbM
Psychological determinants (P) - Managing director	Self-realization qua ecological commitment Technical skill Leadership role and qualities Forcefulness Zeal and working intensity Strategic timing	PE PS PL PF PI PT
Historiographic particularities (H)	Returnable packaged fruit beverages as corporate strategy (1985) Lack of technical manager (1988) Cessation of Coca Cola bottling (1980)	HF HT HC
Technical-economic context (T)	Expansion of fruit-beverage business Systematic project realization	TF TS
Societal Context (So)	Growing environmental awareness Hardly any restrictive influencing factors	SoA SoR
Situational Structure (S)	Small Company Favorable operating conditions Informal company organization Only one division essentially involved Financial resources available within the company Additional market opportunities for plant manufacturer	SC SO SI SD SR SM

Figure 1 Classes and significance of determinants of the success story.

course of optimization of the bottle-washing machine into an exemplary case of good, well-functioning, project-related environmental management.

3. Strong stress is placed on psychological explanatory factors in this success story because individuals were leading actors, the company is a small one and environmental management was not institutionalized. The psychological and microsociological explanatory level would otherwise have played a less prominent role.

4. When all factors act in concert, public environmental policy is to a large extent superfluous precisely in such cases where environmental protection measures are generated by companies themselves, and it can at best play a supportive, subsidiary role contributing to diffusion. However, this is both a rather trivial perception and an extremely rare environmental policy constellation.

5. Precisely under such conditions, and at least in smaller companies, environmental protection endeavours and activities by/within an enterprise do not need a more strongly formalized form and structure when those concerned are highly motivated.

6. However, in stable, well-established, corporate contexts, a good working atmosphere is crucial in implementing environment-related changes in the company.

7. Ecologically significant initial successes are especially important from a social-psychological point of view to generate motivational momentum for the effective continuation of ecology-oriented projects.

8. On the other hand, in such 'ideal cases' the risk cannot be excluded that an ecological orientation to corporate activities becomes an end in itself, if high initial success has been achieved with relatively little difficulty, on the euphoric assumption that further environmental measures are possible without a break and without corporate strategy being called into question; but then a severe setback might be suffered under changed general conditions when confronted with technical/economic reality and feasibility.

9. Thus, in reconstructing the success story, some of the people involved evidenced an idealized, positive self-image and a psychologically understandable overestimation of their own contributions.

10. Precisely in situations where actors dispose of a broad margin for decision-making and action, e.g. to take no, few, or far-reaching ecological steps, both the individual psychological dispositions of those concerned and sociocultural conditions and orientations are decisive in determining whether (obvious) environmental protection measures and programmes are tackled at all, as was clearly demonstrated in this case by the pioneering role played by Amecke Fruchtsaft in the fruit-beverage industry.

11. When such steps – in this success story niche-specialization and expansion; exploiting savings potentials; and acquisition of a strong ecological image, all on the basis of a clearly and professionally implemented corporate strategy – result in both demonstrable ecological successes and, on balance, economic benefits, this process is viable. If, on the other hand, no ecological and economic advantages accrue, it is more and more likely to be halted in the longterm.

12. In a broad sense, environmental protection was the essential goal in optimizing the bottle-washing machine. In a narrower sense, however, the intention was more to make the use of material resources such as water and energy more economically efficient, independent of the environmental effects of waste disposal; the issue was thus efficient resource use rather than effective environmental protection. However, such multiple motivation is in the last resort likely to be an advantage rather than a hindrance in getting genuine environmental protection concerns through.

13. The success story I investigated was, on the level of concrete, substantial environmental protection measures, one of a number of important, ecologically-determined processes of restructuring industrial production and distribution on the small-scale level of specific, individual cases. They are accordingly to be classified by environmental policy. One can expect no more of them than this, namely substantial environmental successes on the micro level.

14. Whether such success stories provide broader impetus in the sense of their social (society-wide) diffusion, or in the sense of materially broader ecological progress, i.e. whether, in accordance with the chaos-theory butterfly effect, small events produce big consequences, depends on specific contextual conditions, which in this case favoured diffusion of the resource-conserving bottle-washing machine in the fruit-beverage industry, especially from an economic point of view.

LITERATURE

Amecke-Mönnighof, H. (1993) Ökonomie und Ökologie. Ms. Menden.

Conrad, J. (1992) *Nitratpolitik im internationalen Vergleich*. Berlin: edition sigma.

Conrad, J. (1994) Ökonomischer Wegweiser ins ökologische Mehrweg-Optimum. FFU-Report 94-4, Berlin.

Dörner, D. (1992) *Die Logik des Mißlingens. Strategisches Denken in komplexen Situationen*. Hamburg: Rowohlt.

Freimann, J. (1994) Vom Nutzen des nüchternen Hinsehens. Environmental management in der Abwicklung? In: (eds.) E. Schmidt and S. Spelthahn, *Umweltpolitik in der Defensive—Umweltschutz trotz ökonomischer Krise*. Frankfurt: Fischer.

Hildebrandt, E. *et al.* (1995) Industrielle Beziehungen und ökologische Unternehmenspolitik. Final Report, Volume 1 and 2, Science Center Berlin.

Steger, U. (1993) *Umweltmanagement*. Wiesbaden: Gabler.

WBGU (Federal Government scientific advisory council 'Global Environmental Changes'). (1993) *Welt im Wandel: Grundstruktur globaler Mensch-Umwelt-Beziehungen*. Jahresgutachten 1993. Bonn: Economica.

Wieselhuber, N. and Stadlbauer, W. (1992) Ökologie-Management als strategischer Erfolgsfaktor. Dr. Wieselhuber & Partner Unternehmensberatung. München.

5. DIESSNER FARBEN UND LACKE GmbH UND Co KG: RECYCLING PAINT SLUDGE AND REDUCING POLLUTANTS

Malti Taneja

1 INTRODUCTION

This case study deals with Diessner GmbH und Co KG, which has received the Berlin Environmental Award by the Senate Department for Environmental Protection and Urban Development for the recycling process which the company developed for dealing with paint sludge.[1] This sludge is obtained when manufacturing plants are cleaned out. After suitable treatment, this by-product can be added to a filler. Using this method the Diessner company has reduced the amount of waste it produces by more than 90%. According to the Senate Department, this process could give a lead to the industry as a whole, in view of the fact that no other effective use has previously been found for the paint sludge obtained from the cleaning out of manufacturing plants.

Taking a closer look at this particular case it shows that the awarded process cannot be considered in isolation, because the factors which led to the decision to recycle the material also triggered off other reciprocal processes. Therefore, in dealing with the paint sludge waste, there is a logical reduction in the amount of pollutants, which in turn has a positive influence on the material condition of the paint sludge. For example, a decision was made to replace all the various hazardous substances in use. This has largely been implemented.[2]

At present efforts are being made to find uses for the recycled product outside the company, because the amounts being produced are well in excess of in-house capabilities. From the company's point of view, what was originally a waste by-product has become a secondary product for which a market should be found, and which has to satisfy the quality demands of such buyers. With these aspects in mind even greater emphasis was given to reducing pollutants.

---

[1] These do not consist purely of paint sludges. The company also manufactures small amounts of filler and plaster. Waste water is not separated, and consequently the sludge contains residues from all the various areas of production. However, in chemical terms there are few differences between plaster and filler on the one hand, and dispersion paints on the other, and consequently the properties of the sludges vary only slightly. The majority of the production is concentrated on dispersion paints and varnishes, and for reasons of simplification this study refers only to paint sludges (a similar approach is adopted by the Paint Industry Association).

[2] This case study has been based largely on interviews conducted during the second half of 1993.

Description of the Company

Diessner Farben und Lacke is a medium-sized company with a workforce of approximately 140. Its products range from dispersion paints to varnishes (excluding acrylic varnishes), house paints, plaster and fillers, with dispersion paints accounting for the largest share of its production. No chemical reactions are involved in the manufacture of these products. All products are manufactured solely by mixing various raw materials and primary products.

The characteristic feature of the company is its contract work, i.e. manufacturing is not continuous. Most production is carried out in response to specific orders from customers. The products are bought by professional painters through the wholesale trade. Consequently the product range is aimed at professional users and not at the Do-It-Yourself market.

Within the company a separation is made between the commercial and sales sections on the one hand, and engineering and manufacturing on the other. There are two general managers, whose responsibilities are divided accordingly. The general commercial manager has been with the company for many years, while the general engineering manager is a member of the family that owns the company, and has been in a position of responsibility since 1988.[3] Because he is a partner in the company as well as being a member of the board of management, the management's scope for determining company policy and the company philosophy is wider than that of companies where this is not the case.

Despite being a relatively small company, it has a well structured formal organization, and each department has clearly defined areas of responsibility and a specific remit. However, where environmental protection is concerned, efforts are made to cut through the various hierarchies, to involve staff at all levels. The Environmental Protection Working Group was set up for this purpose in 1990 at the initiative of the general engineering manager, and all the departments involved with environmental issues have to be represented here.[4] Membership of this particular body has been deliberately not restricted to senior management, because it is felt that environmental awareness should also spread through to lower levels in the company. The Working Group has no formal decision-making power, but simply acts in an advisory role. However, since the general engineering manager is a member of the Working Group, and attaches great importance to it, usually the ideas put forward by this Group do become implemented.

Structure of Ecological Problems

The Diessner company obtains the paint sludge when the plants are cleaned out, because the storage containers, pipeline system and mixing containers have to be regularly rinsed out in order to ensure consistent product quality. The pipes connecting the various plants are kept as short as possible in order to keep waste

---

[3] There have always been members of the family in Senior management positions.

[4] The departments involved are those of Research and Development, Materials Management, Sales and Production (Manufacturing). The general technical manager has also taken part in meetings of the working group since it was first set up.

water to a minimum and reduce the intervals at which the equipment must be cleaned. Cleaning is now only necessary when production runs are changed, and not after each batch (see Wesendrup *et al.*: 7).

Nevertheless, no reduction in the amount of waste water can be expected, because changing customer requirements and a growing number of different products have led to the production runs being changed more frequently, restricting the opportunities for achieving economies.

From an ecological viewpoint the best way of dealing with paint sludge would be to introduce a closed circuit system (see Sutter, 1993). However, such a closed circuit requires the availability of a homogenous, non-polluted waste material (see, Sutter, 1988: 64). These conditions are only met by manufacturing plants engaged in mass production. In other words, different products cannot be manufactured using the same plant, and the wastes from individual plants must be collected separately. Such conditions can only be met in large companies, which can utilize all their production capacity without having to switch products. The Diessner company is not able to continuously manufacture the same product in specific plants because it is processing different orders each day.

Initially the Diessner company tried to add the recycled materials to basic products, and although this is the best ecological solution, it is not practicable. According to Sutter the next best solution is to add the recycled material to a filler. However, as the automobile industry has found out in similar experiments, it is difficult to find buyers because of the residual colour in the material. If the difficulties of meeting customers' requirements could be met, annual savings of 40–50 t of filler could be made.[5]

Quantifiable results have already been achieved in reducing the amount of pollutants, and thus can be considered a success. However, unlike recycling, this is not an end-of-pipe measure, but begins at the point where the production process commences. More troublesome materials are not used at all, are therefore not produced as a waste by-product, and consequently do not pose any threat to the environment. The result is an improvement in the composition of the paint sludge and the ensuing recycled material.

2   DETAILS OF THE COMPANY'S SUCCESSES

Preliminary Details

In 1985 the Diessner company decided to specialize in paints and varnishes containing little or no solvents. Following in-depth research, the company was able to launch a complete range of solvent-free paints and varnishes in 1987.

---

[5] No reliable figures are available about the total amount of paint sludge produced in Germany. A study was conducted by the German Research Society for Surface Treatment (DFO) in 1989/90, but this only dealt with solvent-based, mass production paint applications in industry and in the vehicle repair sector. In these areas a volume of 160,000 t of paint sludge was assumed. Given a solid matter content of 45–50%, this results in approximately 85,000 t of solid paint sludge. The Paint Institute assumes that a further 5,000 t of dried paint waste is produced during manufacturing. This total of 90,000 t only represents the waste from around 45% of all manufactured paints. In actual fact this means that the amount of paint sludge produced is far greater.

This decision was not motivated by ecological considerations, but by the need to specialize in a particular market niche. A reduction in the number of new buildings being constructed in the 1980s caused the market to contract, leading to increased competition among the suppliers of paints. This competition could not be countered by introducing innovations, because the paints and varnishes in standard use are already technically advanced. Consequently, the reduced demand forced prices down, and in this situation small- and medium-sized companies such as Diessner were unable to compete with larger firms. Diessner therefore chose to specialize in a particular market niche.

Apart from the pressure of specialization, the company also had to respond to environmental considerations towards the end of the 1980s. In order to survive in this particular niche, Diessner had to continue its activities in this field.

The company's next major environmental efforts were also prompted by external factors: the introduction of the Indirect Discharge Regulation on 31 March 1990 forced Diessner to set up a new waste water treatment plant. This was in the form of a semi-automatic plant,[6] which first filtered the paint sludge, and then air-dried it without requiring the use of any other form of energy.[7]

The new plant commenced operations in October 1989. The waste water now had to be examined by an independent laboratory before it could be discharged. Initial investigations revealed that the waste water failed to comply with the requirements of the Indirect Discharge Regulations. At this point the company had two options. One was to filter out the pollutants in a second treatment plant, and the other was to replace the raw materials that contained the pollutants at the production stage, by introducing new formulations. The general engineering manager consulted with the heads of the departments concerned, and for economic and ecological reasons decided to attempt to alter the formulations.

The criteria which had traditionally been used until then for evaluating the materials used consisted primarily of a technical assessment of various aspects and of the cost of the materials used. They were now extended to include ecological aspects. A critical examination of the lists of raw materials was used to identify any that might cause problems, and in collaboration with the research and development department they were substituted or reduced.

Reducing Pollutants by Changes to the Material Input Side of the Production Process

The Indirect Discharge Regulation focused attention on the material input side of the production process, and this had a positive side-effect, namely that the awareness of those in positions of responsibility in the technical sector was now firmly directed to the composition of their products. Previously the dominant ecological issues within the company concerned the reduction in the amount of solvent, and separate waste collection in the production and administration sectors.

---

[6] See in-depth version of the case study (Taneja, 1994).
[7] The paint sludge is usually dried in waste water treatment plants.

At this point the Indirect Discharge Regulation provided the impetus which effectively influenced attitudes and triggered off a learning process which continues today, and has led to many successes in the substitution of ecologically problematic substances. Ecological compatibility is now taken into account in developing products, along with economic and commercial aspects.

In pursuing the objective of successive substitution or reduction of ecologically and hazardous working materials, the Diessner company made use of two particular instruments. Firstly the research & development department was instructed, in collaboration with the materials management department, to find substitute materials, and secondly, as a result of negotiations with the materials management director, suppliers were required to provide ecologically safer raw materials.

As a result Diessner no longer obtains raw materials that contain heavy metals (which were a common component of the pigment pastes in particular). Since the suppliers of raw materials do not always release all the details about their formulations, they were required to adhere to specific specifications. In many cases a copy of the clearance document was demanded from them when making their deliveries, guaranteeing that their materials did not contain certain undesirable substances.

A further success was the elimination of chlorine from solvent-free dispersions. Although the suppliers of raw materials had been technically capable of doing so for some considerable time, because of the competitive situation they were unwilling to allow other companies to share in this improvement. Only after pressure had been exerted on them by the Diessner company were then persuaded to change their attitude. However, since Diessner is not a large company, and only buys relatively small quantities, its scope for exerting pressure on suppliers is also limited.

Preservatives form another questionable group of chemicals but for the time being they must still be used because they are vital in order to keep the product stable during storage. However some success has already been achieved in reducing the amount of such substances. The preservatives currently in use are less aggressive than those of a few years ago.

Among those products where modern technology still requires the use of solvents, or where solvents are still required because of their specific applications, efforts are being made to use biodegradable solvents such as alcohols, glycols or alcoholesters. Much of the work of the R&D department is involved in reducing, improving and avoiding the use of solvents.

In addition the Environmental Protection Working Group asked for a list of the hazardous working materials used in the company,[8] to which it could refer when discussing ways of dispensing with such substances. It came up with an objective of replacing 50% of the hazardous working materials by the end of 1993, and of eliminating their use entirely by 1994. The 1993 target was met. The Research & Development department, in collaboration with the Materials Management

---

[8] The purpose of the Regulation on Hazardous Substances (GefStoffV) is '... to protect people against health hazards resulting from work process and other causes, and to protect the environment from materially induced damage...' In this respect the regulation deals with the handling of hazardous working materials.

department was appointed to implement this objective by the general engineering manager. The operations manager is responsible for monitoring the use of hazardous working materials and for reducing the amounts required.

The decision to examine the raw materials being processed was not motivated by any particular problems or by governmental pressure. However, one exception applied to those chemicals declared to be hazardous working materials. Regulations stated that these should not be brought into circulation or, if this is unavoidable for technical reasons, personnel should be suitably protected.[9] It is the company's own obligation to ensure that this regulation is observed and that suitable checks are carried out, with the result that it is not automatically applied as strictly as it should be. Companies can often avoid having to comply with this regulation by furnishing a statement that the substances in question are essential. The statutory conditions on storage, handling and compulsory labelling, which accompany the use of hazardous working materials, provide a far greater incentive to find substitutes.[10] We can therefore summarize by saying that the company's actions in this respect have been proactive.

Recycling the Paint Sludge to Produce 'Novoplan' Filler

Simultaneously with the consideration being given to the material input side of production, the Indirect Discharge Regulation was also responsible for initiating another positive ecological process in the company, namely the possibilities of using waste by-products obtained from the manufacturing process, or more precisely, dried paint sludge.

The new waste water plant separates the waste water from the paint sludge, which is stored on the factory site in sacks. Analysis of the waste water revealed that it complied with stricter specified values, leading to a suspicion that, in terms of its composition, the paint sludge need no longer necessarily be classified as a hazardous waste.

Disposing of the paint sludge as hazardous refuse is considerably more expensive than if it is classed as normal waste, and consequently the operations manager and general engineering manager decided to investigate the composition of the paint sludge more closely. During this phase the sludge was no longer disposed of but placed in interim storage.

The analysis conducted by an independent, officially approved laboratory, revealed that the sludge did not contain any pollutants which, had they been present in a sufficiently large quantity, would have meant that this sludge constituted a hazardous waste product. Enquiries with the Berlin city cleaning department (BSR) in 1991 revealed, however, that the BSR would continue to classify the paint sludge

---

[9] The Chemicals Law §19 Section 3.2 states: 'that anyone involved with the manufacture or use of hazardous substances, is required to check whether substances, preparations or products, or manufacturing processes or applications are available that present less risk to human health, and that these should be used or applied wherever possible.'

[10] Further tightening of the Regulation on Hazardous Substances has been announced, but the effects on the Diessner company are as yet unknown.

as hazardous, because, according to its guidelines, sludge from the paint industry always consisted of hazardous waste, irrespective of its composition. There is only one category for paint sludge, based not on the composition, but solely on the type of waste involved.[11]

The response from BSR, taken in conjunction with the Regulation on Residual Waste, led those with the technical responsibility to conduct their own search for possible uses, in order to avoid the expense of disposing of what is actually safe waste as a hazardous material. Their initial motivation was one of cost, because the expense of dealing with hazardous wastes in recent years has risen substantially, and is likely to continue to rise, From an economist's point of view this group of costs is certainly worth consideration.

The operations manager suggested grinding up the dried paint sludge and adding it as an additive to other end-products. The heads of the engineering departments and the general manager welcomed this idea and decided to implement it wherever possible.

Powdered sludge was then sent to the company's own laboratories in order to investigate its usefulness. The experiments that were conducted there revealed that the original idea of adding the raw material as an additive to the end-products was not feasible due to the colouring of the powdered paint. Since the sludge was obtained from the company's general production it was liable to vary in colour, but as a rule tended to be of a greyish hue. In order to obtain a lighter colour, bleaching agents had to be added, the quantity being determined by the colour of the sludge. This was not practical for ecological reasons, and would also have involved enormous technical expense. Therefore the powder could only be used as an additive with one product, the colour of which was irrelevant to the users. The only product made by Diessner that meets these criteria is the filler used to smoothen walls, which are subsequently painted over.

Together with the application systems department, the laboratory was instructed to develop such a filler to which the powdered paint sludge could be added. The resulting product proved qualitatively superior to other fillers made by the company.

The first cost calculations revealed that initial hopes remained unfulfilled, and that the drying and binding of the recycled material was too cost-intensive to be cancelled out by the savings on raw materials. Even including the amount saved by

---

[11] The Federal Hazardous Waste Regulation (Waste Materials Law of 24.5.77 §2 Section 2) states that all paint and varnish sludge, as well as paint products used in industry, come under the heading of hazardous waste. However, this does not mean that all hazardous wastes should be treated in the same way: to aid identification, each different type of residual waste is assigned an identifying number, according to which the preferred type of disposal should be carried out (Himmel, 1990; diagram 8). Dried paint sludge, the contents of which are no longer water soluble, and can therefore be washed out, are not necessarily classed as hazardous waste. However, in Berlin dried paint and varnish sludges are treated as hazardous waste. The situation only altered when changes were made to the Residual Waste Materials Regulation, which was intended to provide a less bureaucratic waste disposal process for water materials from industrial manufacturing, and which do not come under the heading of hazardous wastes. Since then paint and varnish sludges have been classed as domestic refuse in Berlin and are dumped in landfill sites. When the Diessner company was making its enquiries with the BSR this rule had not yet been introduced.

no longer having to dispose of hazardous waste in the calculation did not indicate a positive contribution to the overall costs. From an economic viewpoint, disposal of the waste in the form of hazardous refuse was still the cheaper solution.

Despite the reservations on the commercial side, the general engineering manager decided in favour of the recycling process. His reasons for doing so were an anticipated increase in the future costs of disposing of hazardous waste, the administrative expense associated with waste disposal and the ecological motivation to avoid generating waste materials. The general engineering manager was also initially responsible for putting forward the company's name for the Berlin Environmental Award in 1991.

The company was awarded this prize in June of that year, provoking a number of different responses within the company. Those involved in the development work regarded this as an acknowledgement of their success, and were motivated to give further thought to ecological aspects of production. For them the award confirmed that they had achieved something elective. Sales management staff whose work began at this stage of the process, tended to view the award from the rational aspect of the economist, rather than from an ecological and non-commercial perspective. Lower levels within the company mostly learnt of the award from the company magazine, which featured the news under the heading 'The first recycling product in the sector', and gave an explanation of the process. However, it seems that no far-reaching effects were felt at this level and there was no boost to motivation.

In addition to the normal process of introducing the product to regular customers, the sales department also publicized the environmental award. Articles were written explaining the process and its ecological advantages.

However, within the building trade, there is little scope for advertising filler, since this is one of those product groups which do not enjoy any high regard among users or manufacturers. In chemical terms these are very basic products, sold cheaply because they only involve simple manufacturing processes, with low-cost raw materials and a well developed technology. As a result profit margins are small. The articles printed in trade publications serving the painting and decorating sector met with little response among users.

Unlike the Do-It-Yourself market, normal advertising has little direct commercial impact on professional users, and there is little response to ecological arguments. Diessner reaches its customers mainly through its field staff who promote the products on the basis of price, service and quality.

Success in selling 'Novoplan' therefore depended largely on the work of field staff who are answerable to the head of the sales promotion department. A system of remuneration for field staff selling 'Novoplan' was and is counterproductive, since their commission is based on actual sales figures and on the extent to which the products sold contribute to meeting costs. This cost contribution and the amounts that buyers can absorb are far lower in the case of fillers than they are for dispersion paints, which accounts for the reduced incentive for selling such compounds.

Despite being very easy to work, the filler did not attract a significant response by the market. In addition to the internal difficulties referred to above, the unfamiliar colour and conservative attitudes to buying on the part of the craft trades were the main reasons for a lack of sales success. Even if the colour of the filler is irrelevant for the work for which it is intended, because filled and levelled walls are generally painted over, painters and decorators are usually not prepared to try innovations simply on the basis of ecological arguments. Similar problems were also encountered when trying to introduce solvent-free paints.[12]

One attempted solution to promote 'Novoplan' filler was based on a combination of various factors, intended to reduce the price to a level at which sales of the filler would inevitably increase, if its lack of commercial acceptance really was due to its price. However, this step did not bring any significant increase in sales either. Therefore the unfamiliar colour was identified as the main reason for the difficulties in selling this product. For technical reasons, a bleach would have to be added in order to improve this situation, and therefore it would be difficult to establish this product on the market in the foreseeable future. This will only become possible if customer requirements change. Increased incentives for field staff would probably not provide the necessary breakthrough, although they would no doubt have some positive effects.

In view of the difficulties in selling 'Novoplan', the original intention of re-utilizing all the sludge, and of not disposing of it, could no longer be sustained, because the amount of sludge being produced was well in excess of the amount of 'Novoplan' being sold. There has been no further change in this situation.

The difficulties in achieving sales forced Diessner to reconsider how to deal with the waste by-products from its manufacturing processes. At the present time the simplest and most economical solution would be to allow the BSR to again dispose of the surplus sludge.[13]

The key stake holders involved in the decision-making and implementation process within the company agreed that this would be a backward step ecologically and that other approaches should be tried, even though they might result in increased costs. A resumption of the paint sludge recycling project, despite all the difficulties and the fact that it did not even offer the prospect of medium-term profits, shows that the decision was not taken primarily on economic grounds. There are good reasons for believing that one of the decisive factors was the awareness of the ecological factors, particularly on the part of the general engineering manager. Therefore funding for environmental activities was not entirely dependent on considerations of short-term profitability.

Prospects: Reaction to the Poor Commercial Response to 'Novoplan' Filler

Because Diessner did not want to dispose of the sludge itself, nor did it have the means of processing it all, the company was obliged to seek elsewhere for buyers.

---

[12] The conservative attitude of buyers in the painting and decorating trade with regard to more environmentally friendly products has been confirmed by the Association of the Paints Industry (see information brochure from the Deutsches Lackinstitut No. 30/May 1993: 10). Although there has been an increased demand for low emission paints, this is attributable to the popularity of these products among DIY enthusiasts (annual report of the Verband der Lackindustrie, 1992: 4).

[13] Particularly since the dried sludge now probably has to be disposed of by the BSR as domestic refuse and not as hazardous waste.

Contacts were established with companies which might be interested in this product.[14]

The efforts to sell the paint sludge led to an altered perception of the product. What was originally seen as a waste by-product of the manufacturing process came to be viewed within the company as a product which had to satisfy the same quality criteria as those of a raw material. The standards it had to meet were not those of state-imposed conditions and obligations as assumed by the company itself, but the requirements of the market. However, the decision to sell rather than dispose of this product meant a greater expense, with little prospect of worthwhile returns.

As a production output, the paint sludge is dependent on the input, which is the point at which action could be taken to influence the situation. It was this realization that led to a further critical scrutiny of the various substances making up the raw materials used, and once again a number of substances were replaced.

Despite ecological improvements, the two companies which might possibly have used the sludge decided not to. Having made the decision not to dispose of the material through the BSR, the company was determined to find alternative uses.

By changing the formulation it was possible to double the proportion of paint sludge in the filler. At the same time it was decided to alter the marketing strategy, since it was obvious that references to the environmental award and the proportion of the product derived from recycled materials did not promote sales, but may even have acted to inhibit them. The decision was therefore taken to avoid any references to the ecological benefits. The emphasis is now being placed on the excellent workability of the product. For example, the labelling points out how economical the filler is, and that it can be easily rubbed down.

Moreover the field staff were instructed to intensify their efforts to sell the filler by focusing on new selling points.

Figure 1 summarizes both processes of Diessner's significant environmental improvements.

3   ANALYSIS OF THE KEY STAKE HOLDERS

Analysis of the Key Stake Holders 1: Recycling the
Paint Sludge to Produce 'Novoplan' Filler

Only four key stake holders, all from the technical sector, were involved in the process from the initial idea to the decision to recycle the paint sludge. It was initiated by the general engineering manager and the operations manager, who together questioned the existing practice of disposing of the sludge, and considered various possible applications. The other two departmental heads in the engineering division, whose areas of competency were affected by the project, were subsequently asked for their own opinions, and about various aspects of technical feasibility.

---

[14] Experiments have been carried out in cooperation with two companies: one wanted to use the powdered paint sludge as a filler for compound concrete blocks used as the paving for cycle paths, while the other planned to use it as a filler for injection-moulded plastics for the manufacture of buckets and barrels.

Description of the process

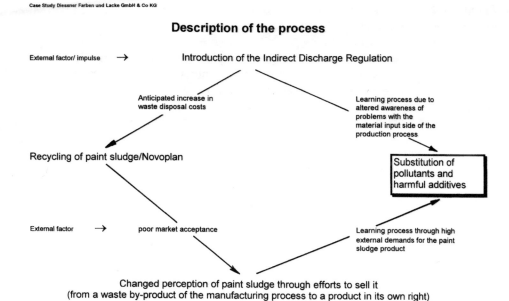

External factor/ impulse   →   Introduction of the Indirect Discharge Regulation

Anticipated increase in waste disposal costs

Learning process due to altered awareness of problems with the material input side of the production process

Recycling of paint sludge/Novoplan

Substitution of pollutants and harmful additives

External factor   →   poor market acceptance

Learning process through high external demands for the paint sludge product

Changed perception of paint sludge through efforts to sell it
(from a waste by-product of the manufacturing process to a product in its own right)

Figure 1   Summary of processes of Diessner's environmental improvements.

After implementation of the requirements imposed by the Indirect Discharge Regulation and the associated ecological improvements (water savings, reduction in the amount of pollutants) the experience of actually being able to achieve something elective provided a high level of motivation to tackle other ecological problems in the company.

A final solution to the waste disposal problem would have met with approval in other sections of the company too, because it had become steadily more cost-intensive, with a corresponding increase in bureaucracy as well. On first considera-tion it seemed that the additional load imposed on individual departments was fairly insignificant compared with the considerable advantages that would ensue. Conse-quently the project met with no resistance at this time.

Due to the relatively small size of the company and the clear allocation of corporate responsibilities, the project only affected three sectors, whose depart-mental heads did not have divergent interests. As a result they were able to come to an agreement with one another and hold discussions without difficulty. Another favourable factor was that the project had been approved by the general engineering manager, thus ensuring that sufficient time and financial capacities would be earmarked for it. Those involved had the confidence of knowing that, should it reach a stage at which it could be practically implemented, the project would not simply be left to gather dust, and this increased the motivation to devote more time to it. During the decision-making process the project was not assessed according to economic aspects, and not immediately abandoned when it became apparent that it would fail to meet these criteria in the short term. The room for manoeuver that was available to the participating heads of department in implementing the project was certainly a motivating factor.

Research and development personnel below departmental management level were not involved in the decision-making process, since this is not the normal practice in this company. The same applied to production staff engaged in the manufacture of the end products. As far as staff in the laboratory and in applied technology were concerned, the work of developing the filler was seen simply as a somewhat difficult task confronting them. Those staff who were involved did assess the project in a positive light, which eventually meant that they were prepared to pursue the project, but they were mainly concerned with fulfilling their own specific tasks effectively. When the product was finally developed, its manufacturing process was no different from that of other projects as far as the staff engaged in manufacturing and packing were concerned, and did not require any additional effort.

The sales department was still not involved in the decision-making process, because neither the disposal of production waste nor the development of new products came within their remit. Sales staff were only concerned with the launch of the product onto the market. Despite receiving the Environmental award, 'Novoplan' was of far less significance to sales staff than it was to those in the technical sector. Although it might be assumed that the environmental award would help sales of the product, when compared with ordinary fillers, this would not have resulted in any significant returns for the company because, as previously mentioned, such compounds are cheap products anyway, and are not required in vast quantities. The success of the sales department is measured in terms of turnover and the position achieved on the market, and not by ecological or technological refinements, and therefore attitudes there differ from those in the technical sector. Consequently there was less euphoria among sales staff about this new, ecologically praiseworthy product.

Field staff whose earnings depend on actual sales and the contribution they make towards meeting costs, did not have any incentive to step up their efforts to promote this product among their customers. They are not interested in the waste disposal problems or in the ecological successes of the company; their earnings are their primary consideration. Consequently it is the company's task, especially the sales department's, to introduce special incentives into the remuneration structure in order to influence the action of staff in the field. However, for the sales department too, the success of 'Novoplan' is only of incidental importance, and therefore no alterations to the remuneration structure were made specifically for this product.

The Senate Department for Environmental Protection, which learnt of the process when the company entered the competition, briefly became one of the stake holders through the award of the environmental prize. The Senate Department itself stated that one of its reasons for providing this award was to initiate a process of diffusion, which they regarded as desirable, but no further efforts were made apart from the award itself. As far as the market was concerned the award on its own was not adequate because, as has already been explained, the market does not respond to ecological advertising. Therefore the award failed to achieve its intended effect, and did not exert an influence.

Indirectly, by imposing stricter waste water conditions (Indirect Discharge Regulation), the state had created the conditions for the development of this process, because the new waste water treatment plant established the technical prerequisites, and this initiated certain learning processes. The motivation to solve the problem was further encouraged by constantly increasing costs and by the greater bureaucratic efforts required for the disposal of waste materials.

It is a feature of this case that it was influenced by only a few key stake holders. Initially the process benefited from the fact that its implementation only depended on the actions of a few, because this meant that it could be pursued without interference during this phase. Until the product was actually launced the interaction between the stake holders and those with a particular interest in it was extremely favourable, and this explains why the project did not immediately collapse once the initial unfavourable results became known. Once the product was ready for the market a number of different interests converged, which had not been considered beforehand by the initiators, and could in fact not have been predicted in some cases. Too little attention was paid to divergent internal interests within the company, and the special features of the professional market were underestimated.

Buyers did not play a serious part as stake holders. There was very little feedback, in contrast to the introduction of most new products, and as a result, competitors were not obliged to react and launch similar products. Moreover disposal of the waste material is still cheaper than recycling, and therefore there was no particular reason to follow suit.

Analysis of Key Stake Holders 2: Reduction in Pollutants by Changes in the Material Input Side of the Production Process

It proved more difficult to study the stake holders, their interests and learning curves in the process of reducing pollutant levels, because this did not follow a strictly defined time scale, but consisted of continual improvements.

As previously mentioned, the process was initiated by the successes achieved in implementing the Indirect Discharge Regulation (saving water, reducing pollutants, using energy required). At this point the award became aware of the potential for improvements if the input was altered. One advantage was that the substitution of questionable materials did not depend on decisions by numerous different parties, and neither was it necessary to make major alterations to work processes. Moreover, for each project to replace a particular substance a separate cost analysis had to be drawn up, in order to provide the basis for the decision, i.e. decisions on specific substitutions did not have far-reaching, unpredictable financial consequences. Therefore this changeover attracted little attention, and this explains why no significant resistance was encountered.

Within the company, various interests favoured the replacement of questionable materials, because this sometimes offered advantages for individual departments, or at the least, there was no conflict of interests.

The general engineering manager introduced the idea of revising the input for ecological reasons to the Environmental Working Group. After the substances involved were listed, this group then suggested which of them should be replaced.

Once again he was motivated by personal ecological values and by the company's motto 'paints for people and the environment', which the company was expected to uphold. With these factors in mind he provided the necessary financial resources.

The operations manager, who took over the new waste water plant, and who is responsible for dealing with the waste materials from the production process, is also responsible for observing that the conditions of the Hazardous Materials Regulations are observed. Therefore he must ensure that the processing of hazardous substances and other questionable raw materials is kept to a minimum, since he has to deal with the bureaucracy involved and with the disposal of the waste materials from the manufacturing processes (including waste water). Consequently he pursues his role actively, exerting a continuous influence on the departmental heads of the laboratory and the material management division, to ensure that they do not introduce any new hazardous substances, and to replace any existing ones.[15]

The replacement of substances causes more work and expense for the materials management department and the R&D department, because many of the hazardous materials have been in regular use for some time. However, they are able to integrate this work into their routine processes. It is also an advantage for these departments if they can have their concepts prepared in advance of statutory regulations. Although resistance is more likely from this quarter, they are aware that they also have to comply with the general trend towards a reduction in pollutants.

With its capacity for introducing regulations in this sector, the state is always a potential stake-holder, because the regulations applying to individual chemicals and to waste disposal tend to change more frequently, forcing companies to respond accordingly.

4   DETERMINING FACTORS

Encouraging Factors and Economic Capacities

The decision in favour of recycling the sludge and to reducing the level of pollutants took place during a period of expansion in this industry and within the company itself, because of the markets that were being developed in the new federal states. A substantial pent-up demand existed there because of the poor state of building fabric there. As a rule the willingness to fund expenditure on projects offering certain ecological advantages, but without the expectation that they would pay their way in the short term, met with little resistance within the company, provided that the financial capabilities were available and could be expected to continue.

The decision to invest in a new waste water plant was closely linked with the Indirect Discharge Regulation, which prescribed stricter limiting values for waste water. Since the existing plant was already quite old, it was not replaced primarily for ecological reasons. In fact its replacement had become a matter of some necessity to the company, and therefore there was no internal resistance to the construction

---

[15] Since the heads of department were all on an equal level in the company hierarchy, the operations manager was unable to exercise sanctions.

of a new plant. The accompanying ecological improvement was simply a fortuitous side-affect.

The recycling project required hardly any further investment since it was able to capitalize on the possibilities provided by the new waste water plant. It is doubtful whether Diessner would have constructed this plant without the threat of stricter waste water regulations, because it would have been difficult to justify such expenditure within the company simply for environmental reasons.

Thus the paint sludge recycling project did not give rise to any initial investment costs, but only generated continuous operating costs, which were within the company's capabilities. Neither was it necessary to change any of the company's best-selling products, which would have presented a risk if sales had declined as a result of the changes. All that occurred was the introduction of a new product, whose success was by no means a vital factor for the continued economic health of the company.

As previously mentioned, the reduction in pollutants is an on-going process, the costs of which have to be continuously determined, along with assessments of the likely returns. The substitution of individual raw materials, or the decision to dispense with them altogether, can be more easily implemented than in the case of a major project, which leads to greater investment and for which the decision-making process can be vastly more complex and difficult.

Corporate structure: As a result of the clear separation of tasks between the general managers and the departments, the decision-making structure was clearly defined. Due to the hierarchical structure, each individual field of competence only had a limited number of decision-makers, which facilitated the coordination process. The clear separation between the levels responsible for decision-making and those engaged in carrying out the work meant that the lower level did not encounter any obstacles, because the projects consisted only of orders, and part of their work consisted of effectively implementing such orders.

The attitude within the company with regard to environmental protection: With the appointment of the general engineering manager in 1987, a number of projects were initiated to achieve working processes with a more ecologically acceptable structure. Some resistance was encountered from the workforce while these projects were being implemented,[16] because it was felt that some of these new developments constituted an unwelcome additional burden. Failure to comply with requirements led to constant discussions, so that staff gradually became accustomed to complying with environmental considerations. Another factor determining the attitude within the company with regard to environmental protection was the inclusion of environmental obligations in the company guidelines, the introduction of a Working Group for Environmental Protection, and efforts to reduce the amount of solvents used. As a result, environmental protection as a corporate task was no longer questioned.

---

[16] For company employees the separate disposal of waste in different containers created more work, but this was accompanied by a reduction in paperwork in the administrative departments.

Structure of the industry: The paint and varnish industry[17] is dominated by powerful international companies such as ICI, BASF, AKZO and Hoechst, as well as several medium-sized companies. However, there are also many smaller companies, which have mainly specialized in specific areas. The dominant position of the large companies, which also cooperate with one another, exerts pressure on smaller companies to be more innovative and to identify niche markets in order to survive. The major players are, of course, able to discover and capture such lucrative markets, with the result that smaller firms cannot simply remain in one particular niche but are constantly having to tackle new challenges. In contrast to other, less innovative areas, changes are not automatically questioned, and potentially this makes it easier to introduce innovations for environmental purposes. This means that, because work is always going on to produce new formulations, it is not necessary to convince people to face the issues of change as such. They simply have to accept a wider scope, that includes ecological criteria.

The paint and varnish industry forms part of the chemicals sector and is therefore closely scrutinized by the state and by environmentalists, because critics have traditionally regarded the chemical industry as a source of pollution. In view of this fact, companies are well aware of the advisability of taking pre-emptive action, because any discussion about possible regulations is bound to involve the chemical industry. Thus many of the substances used by the paints industry are already prohibited, or else restrictions have been imposed on their use. Those companies which foresaw developments and adapted their production now enjoy a competitive advantage.

Legal conditions and regulations imposed by the authorities: Compared with large concerns, a relatively small company with limited personnel resources faces various problems in complying with laws and regulations imposed by the authorities. This is not only due to financial reasons, but to the time factor as well. Obtaining details about new regulations is time-consuming and effectively prevents staff from carrying out their normal daily work. Moreover, registration requirements, for example, impose a regular burden because of the paperwork involved.

Therefore the company seeks to keep such efforts and expense to a minimum, structuring its production methods and input of raw materials in such a way that no additional regulations need to be observed. For example, hazardous waste is much more difficult to dispose of than normal materials, and this is a further financial incentive not to produce hazardous wastes.

The use of hazardous materials requires compliance with appropriate regulations, and various safety data sheets, storage regulations and work safety regulations have to be prepared and monitored. A company that no longer uses such substances enjoys a number of advantages: its output contains less pollutants, its production is more ecologically acceptable, its paperwork is reduced, and if it does not deal in statutorily-prohibited substances, the company is not forced to take any action.

---

[17] The paint and varnish sector is jointly represented in the Verband der Lackindustrie (Association of the Paints Industry). This association in turn is part of the Deutsches Lackinstitut (German Paint Institute), which has taken over publications and PR work. Where publications report on varnish sludges, this also includes all paint sludges.

Apart from these advantages there is also a downside, in particular the research facilities required by such substitution, and the expense involved in the case of several of these substances.

The waste legislation that has been in force so far, consisting of the Duales System Deutschland (the 'Green Dot') and the possibilities for using specialist companies to dispose of hazardous wastes, are all measures that have already become expensive, but have not required any major ecological changes. However, companies are likely to find that legislation will become stricter in the future. The paints industry is aware of this and has advised its members of the expected changes in an information sheet published in September 1993: 'The influence of environmental legislation on the paint industry will continue to grow. This applies to the entire life cycle of its products. The implications are that companies should not only ensure that manufacture is ecologically safe, which includes the raw materials, but should also engage in far-reaching recycling and correct disposal of products after use'. The subject of paint sludge in particular has attracted greater interest, and any company able to solve these problems will gain a competitive advantage.[18]

By reducing the amount of pollutants and introducing recycling, Diessner was quick to recognize future problems, which, if solved before the introduction of the relevant regulations, would offer competitive advantages.

The Indirect Discharge Regulation referred to several times before regulates the quality of waste water, and plays a decisive role in this case history, especially in the preliminary period, because it was responsible for the introduction of a number of learning processes.

Inhibiting Factors

Scant environmental awareness among the professionals: The inadequate commercial acceptance of 'Novoplan' meant that no significant quantities of the paint sludge could be recycled. For the same reason the benefits of recycling were not passed on through the industry. The company's actions will only be imitated by others when decisions by buyers in this market are also influenced by environmental issues. Currently, products with little or no solvent content are gaining a larger share of the market, although they are only slowly gaining ground in the professional sector. Therefore we can expect recycled products to enjoy greater commercial acceptability in the future too.

---

[18] The Federal Environmental Office has been commissioned by the Federal Environmental Ministry to conduct a survey of paint and varnish sludges (including sludge from dispersion paints) and has prepared evaluation models. At the members' meeting of the Association of the Paints Industry the paint sludge expert from the Federal Environmental Office, Dr. Sutter, presented this concept by the paints industry, supporting the retention of thermal utilization, because the industry was not yet in a position technically to carry out effective recycling, and because this appeared the most effective ecological solution. This report by the Federal Environmental Office appears to have alarmed the paints industry, because in the most recent edition of the information brochure considerable space was devoted to the subject, and the Paints Association published a brochure on the problems of recycling in 1993. At the time Diessner decided in favour of recycling, the study had not yet been published and neither had there been the ensuing pressure from the governmental side.

Less opportunities for exercising pressure on suppliers: For a relatively small company like Diessner it is much more difficult to persuade suppliers to provide ecologically acceptable materials than it is for large concerns. The threat of withdrawing one's custom has less effect on a supplier if it is made by a smaller company, and consequently they find it hard to assume a pioneering role in this field.

Limited possibilities for obtaining information: Small and medium-sized companies are often unable to delegate staff to carry out such tasks, because they cannot spare anyone. Moreover, compared with major international companies, which maintain entire departments for dealing with these aspects, they as a rule have fewer contacts for example with scientists and with associations.

Summary of the Determining Factors

This case[19] would be inconceivable without the interaction of various factors and forces. The various determining factors are identified below, and considered in terms of their significance and general applicability.

It was definitely an advantage in this case that the projects involved clearly definable tasks, which had no far-reaching consequences for the company, and in which only a few key stake-holders and decision-makers were involved to any extent. If one compares the two projects, it is evident that this limited number of stake-holders was a definite advantage. This structure is primarily found in smaller companies, where the division of labour is not so strictly differentiated. The relative lack of a division of labour enables new projects to be more easily implemented, since the discussion and coordination processes involve fewer people and are therefore less complex.

Another factor favouring the projects was the view that not only could environmental improvements be made, but internal aspects could be simplified as well, leading to a greater freedom from the threat of state-imposed regulations.

The motivations and attitudes of individual people, especially management staff are more important in small and medium-sized companies, because their scope for action and influence is often considerably greater. In this case study we can see that the ecological attitudes of the general engineering manager were a very significant, possibly even vital factor, but it was not until interaction between other factors occurred that he was provided with adequate scope for action, and this interaction also triggered off learning processes in the company.

The path was made considerably easier by various aspects such as the structure of this sector, where the pressure to innovate is quite powerful, the specialization on low-solvent products, the favourable economic situation at the time the project was first initiated, the general trend towards reducing pollutants, and efforts to deal with the problem of waste.

The main impetus came from the introduction of the Indirect Discharge Regulation because, when it was implemented it created the technical prerequisites for the recycling process and attracted attention to ecological issues. Although this

---

[19] The reduction in pollutants and the recycling process were considered together here.

particular influencing factor is an important one, in order for it to be elective it required the other factors and the intention of solving the problems in an ecologically effective way. It would have been equally possible simply to meet the minimum requirements of this regulation.

LITERATURE

Deutsches Lackinstitut. (1993) Lack im Gespräch, Information brochures from the Deutsches Lackinstitut, various publications.

Diessner intern. (1989–1992) Information for staff.

Gewerbeordnung. (1992) Beck texts from im dtv-Verlag, 25. Edition, Munich.

Himmel, Joachim. (1990) Das Abfallrecht in 23 Schaubildern: eine Gesamtdarstellung des Abfallrechts des Bundes und der Länder, Frankfurt am Main.

Sutter, Hans. (1988) Vermeidung und Verwertung von Sonderabfällen, Berlin.

Sutter, Hans. (1993) Abfallwirtschaft in Forschung und Praxis, Stoffökologische Perspektiven der Abfallwirtschaft, Berlin.

Taneja, Malti. (1994) Erfolgsbedingungen betrieblicher Umweltpolitik am Beispiel der Firma Diessner Farben und Lacke GmbH und Co KG, Recycling von Farbschlämmen und die Reduzierung von Schadstoffen, Forschungsstelle für Umweltpolitik der FU-Berlin, FFU-Report 94-7.

Wesendrup, Meister et al. (WS 92/93) Paper for the seminar 'Abfallvermeidung und Sekundärrohstoffwirtschaft II', under Prof. Dr. Fleischer WS 92/93, TU-Berlin; Subject: Lack und Farbenherstellung Firma Diessner GmbH & Co KG.

6. ABC-COATING: DEVELOPING AND DIFFUSING WATER-BORNE COATING SYSTEMS

Jesper Holm and Inger Stauning

INTRODUCTION

This case-study deals with a transition towards the use of cleaner technology at a small Danish company, ABC-Coating. The company has integrated environmental improvements in their corporate strategy by innovative means. During the period 1990–1992 the company developed a procedure for the treatment of steel surfaces with a coating system containing very low amounts of organic solvents compared to traditional coatings. ABC-Coating was the first company in the European trade to use water-borne and low-organic solvent paints on big steel constructions. The system was developed as an experimental project in close cooperation with the central environmental authorities, a technological service institute and the management of ABC-Coating.

The use of this system resulted in substantial reduction in emissions of organic solvents and reduction in health hazards in the working environment due to the water-borne topcoat. The new type of coated surface lasts longer than usual and the surface may be repaired with water-borne paint.[1] The reduction in use of organic solvents is thus amplified when viewed from a life-time perspective. The management states that the system has been used for about 2/3 of the total coating treatment in 1993. This implied a reduction in the use of organic solvents by 89%, compared to traditional systems.

The main environmental problem of the new system is that it is based on Epoxy, which is allergic. Furthermore, small amounts of organic solvents and other toxic and ecotoxic agents are still present in the paints. Thus, it is not the cleanest possible technology, but a best available technology representing strategic balances between economy and ecology. Therefore, the question of the possibilities for a further diffusion and improvement of the technology must be discussed.

In order to analyse and discuss the perspectives in this story about the preventive environmental policy, it is necessary to assess the environmental impacts given the choices of technology and strategy. Fundamentally, the environmental problems of the production processes and ways to solve them cannot be attacked from the viewpoint of the firm, but have to be put in a broader analytical framework of both ecological considerations and societal regulation of the production.

The article is structured in three main parts: *1. Presentation* of the production plant and company, the environmental problems and the regulation; *2. The story of the transition* in 4 phases: The change in strategy, the development project, the running of the system and the diffusion of cleaner technology; *3. Factors determining*

---

[1] Corrosion Protection of Structural Steel. Danish Technological Institute, Taastrup, 1993.

the success. What made the firm make this transition, what made the development and diffusion of cleaner technology a success, what were the barriers.

1   THE COMPANY, THE TECHNOLOGY AND THE ENVIRONMENTAL REGULATION

1   ABC-Coating and the Relations of Production

ABC-Coating is a small production division situated in a small Danish town. The division is involved in sandblasting, painting, mounting and repairing of big steel constructions: buildings, power stations, windmill towers, off-shore installations, containers, train wagons and the like. It is an independent division of the company ABC Steel and Coating, with a total of 15% of the orders coming from the parent company. The company employs 40–70 people, depending on orders and the season. The turnover has risen from 12.5 mill. kr. in 1988 to almost 20 mill. in 1992.

ABC-Coating is one of the biggest plants in the Danish trade. There is a total of approximately 100 companies doing surface treatment of steel constructions in Denmark. 32 coating firms are organized in 'Sandblæse- og maleentreprenø-rforeningen' ('Trade organization of sandblasting and paint contractors'). It is estimated that these firms cover *c.* 60% of the market for coating steel products.[2] The rest of the market is covered by small firms. Apart from coating done in the specialized firms, there is a great amount of coating done at the steel-manufacturing firm itself. Factories with production of scale of steel products can make investments in a special workshop, where the painting is done. Small factories often choose to do the surface treatment at their own factory without ever telling the authorities and without investing in the equipment necessary to follow the environmental regulation, as this is a cheap solution.[3]

ABC-Coating faces a strong competition. It produces to order and normally the customers invite tenders for the jobs, and the profits are low. As a consequence, investments in new equipment are seldom made, and even investments in the equipment necessary to comply with the law are often not carried out. Development of new technology seldom happens in the firms, but at the paint suppliers or in research institutions. The paint suppliers are, on the other hand, dependent on the products delivered from the chemical firms.[4]

2   Technology and Environmental Problems of Corrosion Protection

The technology used in corrosion protection of steel constructions has, up to now, almost exclusively been based on the use of paints with a high content of organic solvents.[5] It is assumed that the annual consumption of paint used in the corrosion

---

[2] Interview with Ralph Nielsen, Head of the Trade Organization, 23.12.1993.
[3] All the interviewed persons presuppose that a certain amount of the corrosion protection is done this way, but nobody can give an estimate of the size of this amount.
[4] Interview at Hempel; Danish Technological Institute and ABC-Coating.
[5] Miljøprojekt nr. 140 (1990): Vandige malematerialer til korrosionbeskyttelse. Miljøstyrelsen.

Table 1 Methods of corrosion protection of structural steel. Overview based on DIF (1982); EP 140 (1990); Hempel (1994).

Corrosion protection of steel is normally achieved by applying either a metallic coating (zinc or aluminium) by melting or galvanic processes, or a painted coating. The paints are applied either as a fluid substance, which dry up or harden after some time, or as a powder, which is electrically attracted to the steel and hardens after a heating to around 400 degrees. The powder paints do not contain solvents. Of these methods, only the painting with liquid paint is in practice used on the big constructions from technical reasons, among others because the other methods need excessive use of energy to the heating, melting or galvanic processes. The degree of protection depends on the thickness of the paint-layers and the type of the pigments.

The main contents of the paints are the binder and the solvent. Different types of paints need different amounts and types of organic solvents (VOCs = Volatile Organic Compounds), dependent on the bindertype:

In the <u>organic solvent-borne</u> paints the binder (alkyd, chlorinated rubber, coal tar etc.) is dissolved in more than 40 % VOC, all of which evaporates to the air. In the <u>water-borne</u> paints the binder (alkyds, acrylics etc.) is dispersed in water, which evaporates, and then the protecting pigments, which are dissolved in a little amount of VOC (4-5 %), melt together and harden, as the VOCs slowly evaporate. The <u>High-Solids</u> or <u>two-component paints (e.g. epoxy or polyurethane)</u> hardens when a hardener is applied/activated. These paints can be powders or solvent- or waterborne. If low-molecular forms are used, they need only a smaller amount of VOC to be fluid (often 10-20 %). Paints with less than 20 % organic solvents are called Low Organic Solvent paints.

Apart from the binder and the solvent, the paint consist of <u>pigments/-metals</u>. The pigments often contain metals, either to yield the colour (cadmium, chromium, titaniumdioxide etc.) or to yield high corrosion protection (e.g. lead-oxide, zinc-chromates, zinc-powder, zinc-silicate). Finally the paints contain a lot of socalled <u>additives</u>, small amounts of chemical substances, that are added to give various functions, for example fungicides, dispergents etc.

Before the paints can be applied, the surface must be cleaned by <u>sandblasting</u> or other methods, as grinding or high-pressure water-blasting.

Table 2 Environmental problems from surface treatment of steel. Overview based on EP 140 (1990); EP 126 (1990); EP 72 (1986); AMU (1980); OECD (1982); WHO (1985); Sørensen *et al.* (1992).

Some of the most serious environmental problems from the entire coating process are:

VOC-emission. All VOCs evaporate to the atmosphere during or after the paint process. VOCs are known to give harmful effects on the central nervous system and on the breathing organs etc, both acute effects (dizziness, irritations of skin and organs) and chronic effects (brain damage). Some VOCs are especially harmful to the health with carcinogenic, reprotoxic or allergenic risks (xylene, formalin among others). VOCs also contribute to the formation of ozone and smog in the atmosphere. Some VOCs are especially harmful to the environment (some of the chlorinated types, causing ozone layer depletion, pollution of water resources).

Epoxy- and polyurethane products. Epoxy is highly allergenic by contact with skin and mucous membranes. In the hardening process of epoxy is used amine, which is toxic, and for polyurethane is used isocyanate, which is highly toxic. When hardened, the particles emitted from abrasive work on epoxy may cause health risks. They are not suspected to cause any harm to the natural environment.

Heavy metals as lead, cadmium, tin, and chromium and chromates were earlier used in large scale for corrosion protection. Heavy metals in the pigments are still contained in some types of paint. They cause environmental problems for health and/or nature in all processes of coating, cleaning and reusing of the steel, as well as the flake off and the scrapping of the steel.

Various other chemicals in the paints and in the cleaning agents used to clean tools etc. may cause health risks and environmental problems in the working processes and the waste handling.

Particles from sandblasting may cause severe lung illness, and irritate breathing passages, skin, eyes and mucous membranes. In the environment, the particles and the old paint and iron surface cause problems, if they contain heavy metals or other harmful parts.

Grinding, repair, burning off of the surface may cause environmental problems from the reasons mentioned above.

---

protection of steel constructions in Denmark is *c.* 7,500,000 kg. Half of this amount is estimated to be organic solvents that are evaporated into the atmosphere. In 1988 only 5 out of 15 suppliers of paint in Denmark were able to provide water-borne paint systems, and its sale was almost non-existent; in 1990 it was estimated to be less than 2%.[6]

Technically, the paints based on solvents have many properties that help minimize costs and yield high quality. The coating can be done without big investment in equipment and buildings – except the investment in ventilation and equipment prescribed by the authorities. ABC-Coating was the first firm in the trade to change strategy and introduce a Low Organic Solvent-system for corrosion protection of big steel constructions.

The choice of paint system influences several stages in the life-cycle of steel products: The treatment of the old surface, the painting process, the cleaning of tools and handling of waste, the repair after mounting of the construction, the preservation, the removal of the paint when the steel is to be re-used, and the final deposit of the scrap. Thus the technology used at ABC-Coating affects the subsequent processes on both economic matters (it might be cheaper to preserve and re-use the steel, if the coating is easy to remove) as to their environmental problems.

An overview of the environmental problems makes two points clear: the assessment of the environmental impacts of the surface treatment technologies is very complex, and the health hazards and the impacts on the natural environment have to be assessed together. The environmental assessment of technology has to comprise the whole life-cycle[7]: Cleaner technology in surface treatment of steel must maximize lifetime and reusability of the steel, while minimizing the environmental loading and health hazards from the entire life-cycle of the steel and the surface technology itself.

3   Regulatory Demands Concerning the Corrosion Protection Process

In Denmark, there have been great debates, union activities, and research about the health hazards, especially concerning the brain-damaging effect of solvents in the '70s and '80s. These activities have resulted in a comprehensive regulation of paint work. All paint products have to be registered and marked with code numbers to indicate their toxicity. According to the code, special rules of ventilation and safeguards have to be followed. Furthermore, regulation of the concentration of solvents by ventilation in the working environment is done on the basis of limit values. Since 1985 it has been mandatory to reduce the concentration as much as possible without regard to the limit value. In the course of time, the limits of organic solvents have been downwardly adjusted, and recently a new statutory order sets high demands on the amount of vertical ventilation in rooms were organic solvents are used.

---

[6] Miljøprojekt nr. 140 (1990): Vandige malematerialer til Korrosionsbeskyttlse. Miljøstyrelsen. pp. 9–11, 26.

[7] 'Cleaner technology' is defined as changes in the production process leading to reductions in the environmental loading from the total life-cycle of the production in question in the Danish Programme for Cleaner Technologies, EPA 1992.

From the early '80s, it has been stated that substitutions for organic solvents should be used as far as possible. There are several rules for specific products and working conditions, for example for epoxy and polyurethane products. All these rules are controlled by the local Labour Inspectorates, who give advice on the choice of paints and how to comply with the rules, etc.

Since 1974 there have been efforts to control air pollution from organic solvents. The main focus has been on reducing the risks to the nearest environment by filtration and dilution. The requirements have been and are still aimed at: *emission limitation* (burn-off, dilution) down to a fixed value, if the total amount of emission exceeds a limit value, and *immission limitation* (diffusion, raising the chimney height) if the concentration of the substance in the environment exceeds a certain limit (fixed either from toxicity or smell). In 1990 a new – and for organic solvents sharper – rule concerning air pollution control came into force. It is the local authorities who express the specific conditions in the environmental approval. It is necessary to obtain an approval if a company wishes to modify or establish a plant. The authority may also issue orders for working plants, if there are complaints or other reasons for this.

Internationally, there is a growing pressure for the reduction of the VOC-emissions. In the US limits are set for maximum VOC-contents of paint materials, and yearly quota of VOC-emissions are set for painting shops. In Germany, limits are set to solvent emission rates and maximum concentrations. In the Netherlands the paints industry is committed to the so-called 'KWS 2000 Project', an agreement of reduction of solvent emission. There is a Directive on VOC-reduction for paint-users on the way in the European Union. In 1990 an ECE-agreement was entered into, which aimed at a 30% reduction of VOC-emissions by the year 2000.

Within the last ten years the attention of the National Environmental Protection Agency has been drawn towards the possibilities of substitution and use of cleaner technology in working with solvents. In the '80s in Denmark, The National Environmental Protection Agency and others subventioned research and development to substitute the solvent-based paints in all areas. Substitutions were found in most areas of surface treatment of metal and wood, based on either water-borne paints or powder paint methods. But in the area of surface treatment of big steel constructions, demanding a good corrosion protection, cleaner technology had not been introduced.

What was needed, was an industrial partner, who was willing to make experiments and to develop a system, that could be used in practice. At this moment, ABC-Coating entered the stage with a need for a system with less VOC-emissions than the traditional systems.

2   THE STORY OF THE TRANSITION

In this section we will describe the interplay between the company, the environmental authorities and the market in the historical progress during which the company goes through a transition towards an environmental strategy. In this reconstruction of the story of the transition we build upon interviews with the involved actors who from their respective standpoints have contributed to the understanding of the course of events and the motives and strategies of the parties involved.

As we have seen in the preceding section, many parties influence the development of technology in this area. On the one hand, we can point to the economic relations in the trade, the *product chain* and the *competitors*. We have interviewed the *paint supplier* and the *head of the trade organization* to get a better understanding of these relations. On the other hand, we have the authorities and institutions trying to influence the firm to achieve environmental improvements. The *local authorities* are the Local Labour Inspectorate in Aabenraa, and the environmental authorities at the municipality of Røde Kro. The *central authorities* of the natural environment is the National Environmental Protection Agency, where we made an interview. Furthermore, we have interviewed the persons on the *technical service institutions* that were involved in this case: The Danish Technological Institute and The Industrial Health Service. Finally, we have interviewed the *local manager* and *two paint workers* at the firm.[8]

1   The Background for the Choice of a New Strategy at ABC-Coating

The pre-history of ABC-Coating up to 1987/88 is characterized by small investments, a low degree of expansion and almost no restrictions from the environmental authorities to the plant neighbouring a residential area. But The Labour Inspectorate intervened regularly in the paint work with demands concerning ventilation and safeguards.

From the mid-'60s up till 1974 sand-blasting and coating of the ABC products was carried out in a separate room of the workshop. In 1974 a special surface treatment plant was established with separate sand-blasting and coating halls both equipped with exhaustion and ventilation.[9] Environmental approval was *not* required for plants established before 1974 and no demands from the authorities were imposed till 1986 where a new sand blasting hall was built. In 1986, an environmental approval demanding filtering of emissions was given to this hall.

In 1986 temporary halls were established for spray-coating to keep up with increasing orders. But as the plant was situated near a residential area and caused much noise, the local council had to ban continuous production in the temporary halls in 1987 after several claims from neighbours.

---

[8] The quoted actors are: *Oluf Lauridsen*, division commander at ABC-Coating. *Kurt Schau Christensen og Ove Nielsen*, painters at ABC-Coating. *Ole Møller Nielsen*, representative from the local environmental authorities in Røde Kro. *Niels Lund Jensen*, The Danish Technological Institute, Surface Coating. Technologies. *Jørn L. Hansen*, The National Environmental Pro-organization for sand-blasters and paint-coaters. *C. Bjerge Petersen & Søren Nysteen*, Hempel's Marine Paint a/s. *Mogens Kragh Hansen*, Service of Industrial Health in Painting.

[9] The sand-blasting and coating processes were and still are manual; only cranes and trucks are used to handle the objects. After cleaning and sand-blasting by spray guns in special rooms, the objects are stored on the floor, suspended in conveyers or placed on carts. The succeeding spray-coating is primarily done with spray guns. All kinds of jobs involving sand-blasting, painting with epoxy products and traditional coating products require fresh-air respiratory protection and personal safeguards. The work is organized in two-shifts.

In 1987 the plant was handed over to new owners, Danish Steel Construction. The new owners took an interest in profiling the firm as technically advanced contractors and had the necessary investment capital at their disposal. Three months after the take-over, the planning of a new building for sand-blasting and spray-coating was initiated and the building started in 1988. The environmental authorities granted an environmental approval of this plant as if it were the only production in the area; they failed to take account of the total noise and VOC-emissions from the old and new production units.

Besides requirements concerning noise reduction and working hours, the environmental approval of 1988 included requirements concerning the chimney height to reduce odour nuisances from the coating plant. The local council allowed the company to take up the entire regional immission quota of solvents permitted in the air in spite of the fact that other car and industrial coating firms with considerable emission of solvents were situated close-by. Thus, the approval overlooked the fact that the factory already had reached the ceiling of legal emissions.

In the course of events, the old (not-approved) plant burned down in 1988. The company applied for a quick approval of the establishment of a new plant ready for use in 1989, largely identical to the newly-built one. When the new coating plant was being built, the company chose to cooperate with The Labour Inspectorate concerning questions of dimension and organization of the new hall. This choice was motivated by the previous good experience in improving the working conditions together with The Labour Inspectorate.

As with the previous hall, the company wanted to size the ventilation plant and the hall so that only the technical requirements of fire safety by air exchange were observed. However, The Labour Inspectorate demanded observance of the limit values of occupational health calling for more high-powered ventilation. After some discussion, an agreement was made on the ventilation, which included a division of the hall into different work zones and a special work zone for small objects with more strict ventilation protection. On the basis of this cooperation with The Labour Inspectorate the company was granted an approval of the hall-dimensions and process organization of the hall, and they were subject to special terms of running and construction.

Having been given the dimension of the hall and the ventilation requirements, the company applied for environmental approval. But with the recent calculations and approval in mind, the local authorities now felt obliged to consider the pollution from the *whole* factory. As the first plant had already reached the limit of the total immission concentration of solvents in the area, the authorities required a massive dispersal from the new plant realized by a new 40 metre-high chimney. Calculations were made to find the number of spray-paint workers who might work at the same time in the old and the new halls in order to keep emissions low to ensure that the immission contribution would not be unacceptably high. Finally, the total noise from the company was taken into consideration and led to requirements implying that due to noise from the ventilation they could not operate two coating plants during the night.

The company chose to build a hall with a plant of very high technical standards in order to class themselves as suppliers of high quality. With a large amount of invested capital compared to competitors very low costs, they chose to take a different road than the usual. Normally costs to halls and equipment are kept low and competition is based on wages, and the prices are dumped or the production is partly closed down in hard times. But ABC-Coating needed a large hall close to the other one. And the buildings had to be that large in order to have space for the objects produced on ABC Steel. The new owners wanted growth and investment – the insurance paid for the burnt hall also left capital for investment.

Obviously, the company now had a problem in expanding production within these regulatory frameworks. To make the two halls profitable with the ventilation and the chimneys and to meet the customers' demand during a massive expansion the company had to aim at a maximum production utilization, a yearly production of a minimum 150,000 m² treated surface. To make this possible night shift was necessary, but the limits of noise from the ventilation plants laid down in the environmental approval was a hinderance. Add to this the requirements from both the environmental authorities and The Labour Inspectorate as to the number of spray-paint workers who were allowed to work at the same time; this restrained the capacity utilization. Finally, the new air control guidelines (valid from 1990) would in reality place the running emissions far above the emission limit values, if full capacity was reached. Massive ventilation, maximum use of the floorage and large-scale operation would altogether cause large emissions of solvents to the environment.

2   What were the Possible Strategic Options for the Firm?

Had the company chosen to build without consulting The Labour Inspectorate they probably would not have built a plant with such a high air renewal. But as the company had already consulted The Labour Inspectorate during the planning and, thereby, reached a juridically-binding decision it was not possible to evade the requirements of the working environments.

As to the emission of organic solvents, Oluf Lauridsen considered it possible to use the old method of organizing production on the new plant. Above all it would have been difficult to estimate the actual consumption of solvents both from the information normally given and from measurements. The local council had not so far controlled the emission and had no intention of doing so during the period of operation. The local council based their calculations on the number of work places operating and the number of square metres times the thickness of coating. They derived the emission of solvents using data sheets stating the contents of solvents in the paints. From these facts a *theoretical* emission was calculated, which would render a continuous emission within the environmental limits possible. But *realities* are something quite different from theoretical calculations. 'Data sheets specify what is called a theoretical disposal; but the disposal is more than twice the amount of the theoretical one. It has something to do with – and I dare say this now because we *have* solved the problem – extra sprays on knots and difficult corners, losses, and the like.' (Oluf Lauridsen)

The company considered another possibility: What could be done to extract or burn-off the dissolution vapours so that the plant would observe the emission requirements? After a dialogue with the local council they investigated the expenses of a plant that would observe the emission requirements; it came up to about 20 mill. kr. for a plant that could handle the necessary quantities of air, and in addition to this were the running expenses. 'We might as well stop here,' stated Oluf Lauridsen.

According to Oluf Lauridsen the background for the choice was first and foremost the attitude of the management which gave Oluf Lauridsen free scope for his creativity: 'Our concern manager wanted to see some development and he would definitely like to see this company as a leading company within many areas, and it had to be a company that stood for good quality and was really environmentally-acceptable.' 'The way my former boss looked at it, he would never dream of doing it the way it is to day. It is very personal. He did not see the possibilities in it, he did not see it as a necessity to make enormous environmental investment. We would have patched it up as long as possible with botched work to meet the noise nuisances we had. For if the noise nuisances had been solved, I believe it might have been kept running on really low investment. In that case the environmental problems would have continued unobserved.'

It was a new situation for Oluf Lauridsen, 'but it fitted in nicely with the fact that when I realized we would have problems then I was free to think of something new because I had a boss who stood behind me – that's when you find greater interest in your everyday life.'

Perhaps the most important factors behind the choice of a new strategy were the economic considerations. The new halls and plants for sand-blasting and ventilation had cost 26 mill. kr. 'We would make calculations backwards and say that if it took 36 hours to paint a surface, it would be too expensive. If you must make 26 mill. kr. pay, then you may in principle calculate an interest per square metre surface. If a surface treatment takes 48 hours, then it is twice as expensive if it takes 24 hours. To pay for the plant the painted surface had to run through in 24 hours.' This calculation required from the paint system, that it should be possible to apply and dry all the necessary layers (traditionally 2–5 layers) in 24 hours. This was not possible with traditional paints, if they complied with the constraints in night work due to the noise, and the constraints in the number of workers and simultaneous work, due to both the approval from the Labour Inspectorate and the local authorities. Even if these constraints were ignored, many of the traditional paint systems could not be applied in 24 hours.

It is an important factor in the turn of strategic thinking that the management actively began to take the environmental thinking of the authorities and the customers into account. The company had already learned something from the benefit of the cooperation with The Labour Inspectorate in the form of organization of work and the dimensions of the last hall built. They had achieved a good knowledge of their own emissions as well as of the restrictions that could be expected on the emission of VOCs in the future. By aiming at a strategy of complying to future regulation they hoped to gain competitive advantage when the regulations imposed new conditions of production for everybody in the trade. However, they overestimated the restrictions to come in the new guidelines from 1990. Letters from The Environmental Protection Agency stated that the forthcoming guidance concerning air pollution would contain requirements of total outlet limits. The first draft of the new guidelines for reduction of air pollution included setting limits to the mass-flow of the solvents – but in the final version this principle of regulation was removed after pressure from industrialists opposed to this rule!

Further, the managers from ABC-Coating had realized that environmental concern had to be seen as a future parameter for development. They had tried to find coating systems which did not need sand-blasting at preservation and repair and understood this as an important parameter for the customers. That would mean an environmental and a working health advantage as well as profit for the customers. 'Our aim is to be five years ahead of the other companies in our sector and right now we are taking a jump into the next generation' (Oluf Lauridsen). When Oluf Lauridsen seriously started to think about the development of alternative solutions based on coating systems with a lower content of solvents it was out of a requirement of an overall solution. 'We began to put together the five things that normally do not agree: effectiveness, environment, working environment, higher quality and future maintenance – and then it must not cost more.' The investment was to be paid by higher productivity.

3   The Cooperation and the Considerations Behind the Development Project

Oluf Lauridsen was inspired to enter the development project through his contact with The Danish Technological Institute, Surface Coating Technologies. This institute is a state-supported service institute supplying technological advice and development, mainly directed towards the needs of small- and medium-sized firms in Denmark. It was something new in the trade to enter development projects and to co-operate with The Technological Institute: 'We were the first in the trade whole-heartedly to enter the cooperation. We wanted to use them – the others saw them more or less as representatives of the building owner, which gives an opposite relation.' (Oluf Lauridsen). Normally The Technological Institute controls the quality of the surface treatment on behalf of the building owner.

A purely water-borne coating system did not satisfy Oluf Lauridsen's demands to secure the return of investments in halls and other plants. The painting process demands just as long a time or even longer as for the organic solvent-borne process, the system is more expensive and calls for more repairs and maintenance, and it is new to the customers. Therefore, Oluf Lauridsen investigated other possibilities, including contacting a paint supplier, Hempel's Marine Paints a/s. The idea had occurred to combine an epoxy primer with a water-borne topcoat. The epoxy primer has been used and developed for 35 years, and Oluf Lauridsen had had good experiences using it.

An epoxy primer with 14% solvents and a water-borne topcoat with 4% solvents (slowly vapouring) were combined. For low demands on corrosion protection a water-borne primer and topcoat is sufficient. For very heavy demands on protection a primer with a solvent-based paint and zinc powder (51% organic solvents) is necessary before coating with the epoxy coating and the water-borne topcoat. This combined system was named the Hybrid System.

This combination offers several technical advantages: Mechanical strength and long durability in all the corrosion classes connected with an environmentally-acceptable laying and maintenance of topcoating. Only 2, or 3 at the most, layers are necessary against the usual 2–5 layers, which means savings in working hours and time of process flow. Only one layer, two at the most, consists of solvent-based epoxy paint. Thus it seemed that ABC-Coating would be able to obtain the production capacity necessary to make the new halls pay without reaching the limits of emissions previously described. But the epoxy primer had never before been used in combination with a water-borne paint. To The Technological Institute it sounded unthinkable: 'An experiment comes in when we mix the traditional epoxy with a water-borne paint. That would immediately make one shout, and one would consider the chances of making them merge to be very small...' (Niels Lund Jensen). The paint suppliers were able to deliver the requested paint types, as they had started experimenting with water-based paints already in 1982, but 'we had never had the idea of combining an epoxy primer with a water-borne topcoat, this was solely the idea of Oluf Lauridsen' (C. Bjerge Jørgensen & Søren Nysteen).

The Danish Technological Institute made an application for a grant to a cleaner technology project from The National Environmental Protection Agency. They attached great importance to the fact that The Labour Inspectorate, the trade organization and The Danish Federation of Semi-skilled Workers recommended the project and entered the group. Epoxy is normally avoided for working health reasons.[10] The Labour Inspectorate estimated that the project offers the advantage that the period in which employees are exposed to substances hazardous to health is shorter, but they demanded that the working processes were carried out according to the rules (Letter of 5.4.1990, Labour Inspectorate, Sønderjylland). The Industrial Health Service of the painting profession has assessed the hybrid system and found that it is to be preferred to the traditional systems because of the lower evaporation of VOCs, as it is considered more difficult to protect the employees against inhalation than against skin contact (Environment project 216, 1993, p. 24). So much attention has been paid to the health hazards of epoxy that the precautions are well-known and provide a high degree of security (Mogens Kragh-Hansen from Industrial Health Service).

The project was initiated and a group was set up including representatives from The National Environmental Protection Agency, The Labour Inspectorate, managers of firms in the trade, an environmental consultant from the Trade Union, the trade organization, the suppliers of paint (Hempel's Marine Paints), The Technological Institute and ABC-Coating. The project received a grant from The National Environmental Protection Agency of 1.35 mill. kr. ABC-Coating has invested approx. 2.5 mill. kr. in climate control plant and planning, which is 10% of the total investment in halls and other plants.

The Danish Technological Institute was in charge of the testing of the coating systems. 'The real problem is to apply the water-borne paint. This is due to the fact that the water normally evaporates very quickly... The idea is to introduce some accurate spray procedures, so that you coat in the correct succession and avoid misses etc... None of us knew how to do it. We had some ideas about it, just as we only had some ideas about the level of air humidity, too.' (Niels Lund Jensen).

---

[10] Use of epoxy demands special approval: a course for the workers, special precautions in the work and use of gloves, glasses, respiratory mask and clothing. If possible, substitutions other than epoxy should be made. AT 78, AMU 80.

About 30 employees were involved, because a large order was received from The Danish State Railways. All of them had attended an epoxy course, but that did not include the necessary technique. This kind of education was not obtainable anywhere, so the company used the technicians from the paint suppliers to develop an 8-days course where everybody participated.

The employees noticed the advantages in working with the water-borne paint. 'When we entered this project, it was because the firm wanted to try something new – but it was also because of our selves, our own health. You are feeling better today, because you have put away the solvents.' (Kurt S. Christensen & Ove Nielsen). The Hybrid System calls for skill, care and responsibility from the painters. Heavy demands are also made to the organization of the work. At the ABC they succeeded in carrying through the necessary changes in accomplishing the work and organization and to find a satisfying way of controlling the climate. The experimental project showed positive results to the economy as well as to quality and environmental improvements.

One of the main reasons for the success of the development project, was the cooperation of the different parties involved, who each saw some advantages in the successful implementation of the system. '. . . they were enormously impressed for one special reason, and that was that we were so many who would agree that it was a good plan. We were able to unite trade unions, employers, people concerned with the working environment, our trade organizations and environmental authorities. They were very impressed that it was possible to do that; they couldn't imagine it being done in Germany.' (Oluf Lauridsen about German supervisors' visit). To all the parties involved, the Hybrid System seemed to imply a bettering of the environmental conditions of paint work, and therefore they supported the project.

4   What were Motives Behind the Choice of Cleaner Technology?

If the Hybrid System is assessed in a total life-cycle perspective, it offers several environmental advantages, even compared with the water-borne systems.[11] In some cases, the sand-blasting can be avoided and substituted with wire brushing or high-pressure washing. The constructions can be handled and mounted with less damage needing repair, and hence with less risk of accidents etc. Repair can be carried out with water-borne paint. The construction has a long life-time and needs less maintenance than others because of the strength and flexibility of the surface (the flexibility partly due to the water-borne topcoat). Finally, the system does not contain heavy metals. But the low-molecular epoxy and the remaining content of VOCs as well as the many additives still constitute environmental problems.

The use of purely water-borne paints seemed possible on theoretical grounds. Hempel Marine Paints had been experimenting since 1982. At the Technological Institute of Denmark studies were performed in the late '80s supported by the Environmental Protection Agency, on cleaner technology. Here it was concluded that water-borne paint systems could protect against corrosion in all classes.[12] The

---

[11] Interviews at ABC-Coating, The Danish Technological Institute, Hempel.

[12] Water-borne paints were acceptable, except class 4 (DS/R 454). For heavy protection a water-based zinc-silicate can be used, according to interview at Hempel.

disadvantages were, firstly, the necessity of painting indoors and having precise control over temperature, humidity and ventilation. Secondly, the softness of the film, which made it difficult to handle the construction. The industry was rather doubtful as to this method, and as we have seen, ABC-Coating didn't choose to adopt the water-borne system, because it would demand a longer process time than the required 24 hours, and because it would be difficult/impossible to find customers for it.

The choice of system was strongly determined by economic constraints. The crucial element of the system seems to be the epoxy primer, because only one layer is required to achieve sufficient protection. If the water-borne topcoat did not succeed, the production might continue with organic solvent-based topcoats – or even without topcoat as it is not necessary for functional reasons (C.B. Jøhrgensen & S. Nysteen, Hempel). The temperature and ventilation control is useful even if the water-borne paints were not used, as the epoxy paint will dry faster when heated. Only the humidifier would be a superfluous investment, if the experiment didn't succeed. Once installed, however, the humidifier might open the door for further experiments with water-borne paints, if better paints are developed in the future.

Oluf Lauridsen did not choose a paint with water-borne high molecular epoxy, which is less hazardous to the health, and which has a shorter drying period. Today it is twice as expensive as the low molecular paint. Besides it involves the disadvantage that it cannot be applied in a similarly thick layer and must, therefore, be applied twice in the higher corrosion classes. It implies a rise in price and a working environment problem in having to paint twice.

So, all in all, ABC-Coating chose a system, that is cleaner in some respects, though not the cleanest available, but it was a system that fitted well in the economic considerations of the firm.

5 Market Conditions for the Hybrid System

In the economic estimation concerning the hybrid system it is stated that the running costs would be reduced compared to the traditional system, if the plant is utilized to full capacity, as this would mean reductions of time flow, working hours and energy.[13] Thus, the extra investment of 2–3 mill. kr. for climate control could be considered a sensible investment, which might secure the return of investment for halls as well as plants. However, the main problem for Oluf Lauridsen has been to get the plant working near to full capacity.

The firm has not till now reached full capacity and has to work hard to get the orders and to live up to the conditions of delivery dates and price. The fact that the company did invest in two new production halls plus the necessary ventilation and climate control plant placed them in a special position compared to normal conditions in the trade. The other companies have far less depreciation and, therefore, they do not see the advantages that may lie in a quicker flow, or less energy expenditure by a smaller air shift. 'If they operate a plant that is cheap to run, then they don't see that kind of thing. Then the wages are decisive. They don't

---

[13] Environment project 216, 1990, p. 41.

think of the future. Nothing is put away for anything, development or succeeding investments.' Therefore they are able to offer cheaper solutions than Oluf Lauridsen, and they are not dependent on full capacity, but can close down or expand production according to the market conditions.

Now that the investments have been made it will, in many cases, be cheaper for Oluf Lauridsen to use the Hybrid System than the traditional system. The epoxy primer is comparable in price to a traditional primer. Therefore, it is cheaper to use the Hybrid System in the higher corrosion classes. When only a little protection is required the traditional paint is often chosen by the company. However, there is a strong incentive to use the water-borne paints wherever possible in the very improved working environment.

The real challenge to ABC-Coating is to expand the market for the Hybrid System. One possibility is to find customers who directly ask for the economic and environmental advantages which the Hybrid Systems, as described, may give... In some connections it is important that the system comprises environmental and working advantages. An example may be the large company that produces concrete mixing plants for the German market. Here they wanted an environmentally-acceptable paint, which can be maintained with water-borne paints, because the German customers require documentation of environmental circumstances. 'I think it is part of their purchase policy that they want to know how the things have been made.' (Oluf Lauridsen). This order will cover 50% of the expected future production at ABC-Coating. Oluf Lauridsen is of the opinion that it is of growing importance that the Hybrid System is more environmentally acceptable and requires less repairs and maintenance: 'Within the power stations they use the Hybrid System a lot more to day... more and more consultants want to use it because it is a well tested and documented system.'

It is a problem when aiming at a quality and environmental image that the direct contact between the supplier and the customer is rare. The paint contractor seldom has any contact with the final user. Often consulting engineers have this contact. Therefore, it is difficult to point out the advantages implied in the future working environment and the fine durability of the system, as well as the reduction of the VOC-outlet. The final user is represented by a purchaser, whose task it is to find the cheapest system, and who is not interested in environmental circumstances and durability in 30–40 years time. The Danish consulting engineer will normally just consult his technical standard[14] or do what he is used to doing. A consultant engineer will be sceptical towards coating systems like Low Organic Solvent systems that are not included in the standard, as a liability for damages may be connected to the use of a system, which does not meet the specifications.

ABC-Coating has encountered another problem, namely the fact that public invitations for tenders must not favour a single contractor for instance by prescribing an environmentally acceptable coating. Therefore, as long as ABC is the only one using the hybrid system the invitation must also contain a description of traditional coatings. Even one more coating plant using the Hybrid System would better this situation.

---

[14] DS/R 454, see DIF 1982.

This exposition of the economic conditions shows that the economic interest of the company is to have: 1. The knowledge of the hybrid system increased, as the use of the hybrid system will give the best return of the investments of halls and plants. 2. Enforced requirements made for the environment, the working conditions, quality and duration of the surface treatment, as that would increase the spreading of the Hybrid System. Thus the vital interests of the firm seem to support an offensive environmental regulation!

6   Initiatives for Diffusion of the Hybrid System

If ABC's contribution to the development of a hybrid system is going to be a way to cleaner technology for the trade it is necessary that similar process technology and water-borne paints are spread over a large part of the coating trade. In what follows we shall see how the company and parties of the development project eagerly and fairly successfully have endeavored to acquire such a spreading by means of the market, the technical standards and the authorities.

The market

Information   The company aimed at making great well-known customers take part in the experiments carried out in the project as they would in this way be able to spread the information of the system. Similarly, The National Environmental Protection Agency followed up the efforts by contributing to the installation of a corresponding plant at the Danish factory of railway carriages, ABB-Scandia, which is well-known internationally as well as in Denmark. Seminars have been held in the company, at The Technological Institute and at the paint supplier Hempel's. Here consultants and contractors have examined the plant and discussed its advantages.

Network   The market of each individual company is geographically bounded because the transportation costs are vital in connection with the large steel constructions. So, the company has entered a network cooperation of cleaner technology in which they cooperate with the more respectable coating companies from all Denmark.

Technical standards

In the wake of the development project ABC-Coating, the trade organization SME and The Technological Institute, were backed up by The National Environmental Protection Agency for Danish participation in an international work group under ISO (International Standards Organization) concerning coating systems for steel in a corrosive environment. Having presented the environmental views to the ISO work group the representative from The Danish Group was given the opportunity to propose various environmentally-acceptable coating systems. In general the proposal was well received. The draft they now work on include 25–30% of environmentally-acceptable coating systems, one of them being the Hybrid System. 'They are the ones I carried under my arm.' (Niels Lund Jensen, Techn. Institute).

The suppliers   The paint-suppliers and the big chemical factories now direct their research and development towards water-borne, high-solids or powder paints, and towards integrating other expected environmental demands on heavy metals, chlorinated chemicals etc.[15] Hempel expects a break-through for the water-borne paints anytime, and as they have well-documented and high-quality systems, they look forward to this break-through. They have a 'green' product-line with environmentally-improved products waiting for the demand. But still, only 1% of their sale is water-borne paints (C. Bjerge Jørgensen & Søren Nielsen, Hempel). The project at ABC-Coating means that they have a well-tested documentation of their products, and that many sites now have been coated with their paints, which is the best advertisement to show paints (C. Bjerge Jørgensen & Søren Nielsen, Hempel). Epoxy paints are regarded as the paints of the future at Hempel, and they market the Hybrid System as an environmentally-acceptable solution beside the water-borne paints.[16]

The authorities and the environmental regulations

Purchase Oluf Lauridsen has suggested that the special institution of a green national purchase policy in Denmark should specify environmental profiles for coating of construction works. Savings for environmental investments should be presented in tenders, and the coating supplier should work with environmentally-controlled processes. Such standards for the national purchase policy of coating supplies do not yet exist.

Working environment In the course of the development project, ABC-Coating especially had problems with the latest regulations of the working environment of paint work. They state special technical requirements of the quantity of ventilated air calculated as vertical air-stream speed. These requirements are based on products containing organic solvents and not on water-borne paint systems which need a low air-stream speed in order not to dry too quickly. Oluf Lauridsen and the trade organizations have obtained support and assistance from The National Environmental Protection Agency to try to influence the central working environment authorities to revise the requirements of the working environment due to cleaner technology options.

The air pollution control guidelines Oluf Lauridsen has also tried to influence The National Environmental Protection Agency to alter the requirements of air pollution so that the terms were the absolute quantitative limits for each individual company emitting solvents. But he did not succeed in influencing them this time. The National Environmental Protection Agency is of the opinion that the companies would bypass that kind of requirement by just dividing the companies into smaller units.

Statutory instrument for a sector However, one of The National Environmental Protection Agency's departments has been so pleased with the hybrid system that they have begun to draw up a statutory instrument for the sector (fixed functional standards of technology and emissions), for surface treatment with linked guidelines for the supervising authorities. Oluf Lauridsen and the trade organization entered this work in a group under The National Environmental Protection Agency.

---

[15] Hempel, undated a & b, and interviews at Hempel and DTI.
[16] Hempel, 1994.

According to the draft of this statutory instrument, the Hybrid System is described as an example of a water-borne solution corresponding to the requirements of all new plants, and older plants in the course of time. In this connection the present emission standards would be reduced to one-third for companies emitting solvents.

The VOC agreement   Meanwhile another department of The National Environmental Protection Agency resumed the discussions concerning the drawing up of an environmental agreement on reduction of solvents from the industry (the VOC agreement). The consequence was that the statutory instrument was abandoned in consideration of the wider range of industrial interests included in the VOC agreement. It is still unknown whether the VOC agreement will include the emission requirements for coating companies that correspond to the standards of the Hybrid System, but this is a possibility.

Classification   A very considerable part of the ready-made mixed thinners on the market duly fall under a more severe toxicity class (code numbers) than under the ones to which they are calculated on the basis of the rules. Therefore, according to Oluf Lauridsen requirements should be made of emission limits and immission contributions down to one-third of today's limits. Such requirements would render it necessary to make such large investments in cleaning plants in order to observe the standards that the Hybrid System would seem an economic attractive alternative. The result of Oluf Lauridsen and partners' efforts is not yet known.

Approvals   The developed environmental project is to day included in a list of references for the local councils as an example of cleaner technology within surface treatment. It is regarded as Best Available Technology in the area. That means that all companies who are building anything new must relate their projects to environmentally-acceptable circumstances. As a guidance they should not establish a plant that pollutes the environment above the level of what is possible when using the Hybrid System.

Enforcement   One of the most disappointing discoveries for Oluf Lauridsen has been to see how easy it is for the competing firms to go on polluting and avoid restrictions from the local authorities. Oluf Lauridsen and colleagues from the trade organization have complained to local and central authorities of the missing interventions against the sinners of the environment. Some cases have been won, but 'the decent environment company' in the trade should not count on any large effort in enforcing the environmental requirements against the 'sinners'. The company and the trade organization are thus pressing to secure development of an official regulation that will strengthen the control and requirements of the environmental aspects of surface treatment. That would strengthen their position amongst the competition.

Thus they seem to have become partners in a teamwork with the authorities in trying to obtain a strengthened preventive environmental regulation! On the other hand they may be regarded as lobbyists, trying in all ways to influence legislation, standards and public opinion to show the Hybrid System as the environmentally friendly product in this area, even as 'Best Available Technology', although cleaner products, namely the purely water-borne, are available.

A remaining question is, will the development process be turned away from the cleanest possible solution, namely the water-borne paints, by the acceptance of the High-Solids? The Hybrid System has achieved the status of cleaner technology by the state support to the development project, whereas it is obviously not the cleanest possible. The need of independent technology assessments and environmental research is highlighted by this dilemma between ecological and economic considerations.

3   FACTORS DETERMINING THE SUCCESS

Schematically our analysis of the progress of the case could be divided into three phases based upon ABC-Coating's situation in relation to the changes in the surroundings, see Table 3.

*1   Which Factors were Decisive in Choosing
    an Environmentally Improved Strategy?*

We think that the reason for the company to start thinking innovatively about environmental development to be *a combination of*: an economic structural pressure; a new strategic positioning of the company; a learning process from the relations to the authorities; a learning process of possibilities of environmental strategies; a question of personal and accidental circumstances.

The structural pressure   During the growing crisis at the construction market from the late '70s an over-capacity arises in the trade and the competition strategy on turnover up till now is intensifyed by dumping-prices. Capital is not accumulated for investments on technology development, new market aims or reserve fund. These structurally-poor conditions help to smoothen the requirements from the local council authorities as there is a widespread tendency to let the companies emit very large quantities VOC; the local politicians do not want to risk a reduction of the tax base.

The new competitive strategy of the company   A response to the market pressure aiming at quality by investments in high standard techniques, a strategy of differentiation and increasing focus on delivery of total-drafts, in which the customer's expenses and efforts of maintenance are included.
   The company learns from the customers that the working environment by sandblasting and repairs must be considered very problematic, and that if the maintenance can be reduced a new competition parameter has been found. The investments in the closed halls imply an increasing demand for reducing the pay-back period. This leads to efforts to attract more customers and multiply the number of objects at work. But here the company come up against environmental problems presented to the company by the authorities and by an awakening of a self-observation of the company's pollution load.

A learning process in relations to the authorities   In the '70s and most of the '80s the company learnt that the environmental authorities, in charge of enforcement are

Table 3   Changes in the environmental state.

ABC-Coating	Market	Rules	Authorities
1970–86: Low costs, economy of scale, strong use of VOC, no eco-care	Competition on wages, No stratification, great fluctuations	New laws, few statutory orders fulfilling the acts, 'high chimneys'	Poor knowledge, sporadic Inspectorate, vague demands
1987–89: Investments in new equipment, quality-profile, order production, reactive to eco-demands	More interest in long term perspectives, quality, quick delivery, growth	Many statutory orders, focus on minimizing VOC in workplace, 'high chimneys'	Strong efforts to ventilate VOC, focus on immission and noise
1990–: Environmental strategic positioning, LCA in product-chain, networking	Eco-minded customers, eco-dumping, decline international competition below costs	Towards cleaner technology/substitution, new demands of ventilation, focus on emissions	Central environmental authorities support R&D, innovative back-up by local authorities

not very strict. But during the expansion the company learns, however, that the lack of strictness is detrimental in the long-run when the growing noise nuisances and emissions caused by the, technically plain, expansion, suddenly produces reactions from the neighbours and the authorities. In 1989 the local council lay down more strict terms of noise and air pollution for the production at ABC-Coating, reducing the capacity utilization. The local council has thus changed its role and begun to focus on the total emissions from all the production, primarily in accordance with national enforcement campaigns and because the two approvals covering all production were applied within two years.

The company *chooses* to take the intention of these terms seriously instead of counting on the indifference of the local council. This happens when the company seeking environmental approval seriously examines its own pollution contribution. But the company also overestimate the future demands to the emissions and the enforcement against the competitors in other municipalities!

In opposition to the local authorities, the company has since 1980 had several encounters with The Labour Inspectorate and has experienced strengthened demands regarding the working environment problems. From the early '80s the local Labour Inspectorate has laid down still more demands regarding working hours, resting hours, ventilation and organization of production rooms. Through this the company slowly learns to view the VOCs as an environmental problem.

This happens concurrently with country-wide campaigns on the damaging effects of organic solvents and the possibilities to intervene. A close cooperation between active trade unions and critical researchers of the working environment led to a long series of scientific documentation from the late '70s and later, that prove organic solvents' contribution to brain damage. The result was vigorous efforts by The Labour Inspectorate in organic solvent-inspection at the companies: ventilation, observance of the limit values and pressure towards substitution.

Learning about new possibilities in choosing innovative environmental strategies
Through contact with The Technological Institute Oluf Lauridsen learns that
processes may be altered and that already a long-termed process has started to find
technological solutions to the environmental problems. Environmental requirements
are not just questions of demands from the authorities and purification, but also
concern future demands from the customers to the producers, and making the
process less expensive by thinking about the demands of regulations beforehand. It
is reasoned that developing cleaner methods of production is an on-going process,
also new types of paint are on their way from the suppliers. To consider environmental requirements in the choice of new technology may also become a parameter,
which gives a technological lead.

The question of individual characters   This eventually concerns the choices made
from the structural possibilities. We do not doubt that Oluf Lauridsen played a very
important part in ABC-Coating change of course in 1989/90. Over the years he has
acquired a very thorough knowledge of production technology, work organization,
process and product chemistry, and as a security leader he has for a long time seen
the production from a working environment viewpoint. Besides his involvement in
relation to the customers he has been the basis of ideas about a total draft, which
he generated in person.

Accidental events and circumstances   These have been of importance in this case.
The company location close to a residential area made noise a big problem in the
expansion of the firm. The burning down of the old plant gave capital to invest in
a new hall, but at the same time necessitated approval of both emissions and
ventilation according to the much more stringent rules and practice of environmental and working health authorities.

These factors contributed to the choice of a new environmentally-improved
strategy. A *front runner* was created, who aimed at gaining competitive advantage
by foreseeing and complying to future regulation of the environmental problems,
and who aimed at responding to future demands from the market of environmentally-
improved coatings and more total concepts of coating service. The circumstances,
that contributed to the development in this case, will presumably also be present in
other industries, in other localities, among other small firms. Therefore the institutional network around the firms should be aware of them and be ready to offer
support at the right time and place: is there a motivated leader? Is the firm making
new investments? Are cost-reductions by savings of waste or resources possible? Are
new market positions possible? This knowledge is valuable to strategies for introducing cleaner technology and environmental management.

2   What were the Most Decisive Factors Leading to Success of the Actual Development Project and the Diffusion of It?

Firstly, it was of the greatest importance that it was possible to unite the many
different actors in a consensus of the environmental advantages of the Hybrid
System in preference to all the other coating systems. As we have tried to point out,
it was possible for all parties as a starting point to interact with each other from
their individual positions and ways of development.

The National Environmental Protection Agency was eager to make a successful cleaner technology project, especially as it was well-known that rules were not followed in most cases of VOC-emissions. The workers and the institutions serving the working environment are keenly engaged in reducing the expositions for VOCs, and accept the compromise with the well-known risks and precautions from the epoxy paints. Also the paint suppliers, the Technological Service Institute and the customers were engaged in making the experiment a success.

ABC-Coating became an engaged partner because the environment project fitted in with the strategy of the company (cf. the above). Probably, it was a stimulus to the project in the first place that the company was granted environmental support for the installment of a ventilation plant and for working hours. It was decisive for the success of the project that it both solved ABC's problems concerning the demands from the authorities, and secured the return of the large investments of buildings and plants. Oluf Lauridsen, personally, was a contributing factor to the spreading. This is among other things due to the strong motivation in that diffusion of knowledge and more strict environmental requirements will give economic and competitive advantages to ABC. This cleaner technology project thus created a motor to initiate improved environmental conditions within the whole area instead of, as is often the case, stowing them in a dark corner when the project period expired.

The trade organization represents some of the largest companies in the trade and considers themselves to be 'the Good Society within the sand-blasting and paint contractors'. They want the future to belong to companies who observe standards and environmental regulations, and 'the pirate companies' (the many small, illegal firms) must die. Thus, they are ready to follow the example of ABC, and demand stricter and more rigorous environmental requirements which will weed out the illegal firms!

Diffusion of the Low-solvent paints is encouraged by the development of requirements of environmental documentation, caused by the wish from still more companies to obtain an environmental label, certificate or in any other way an environmentally-good reputation. The use of VOC is an approved environmental problem, which will be visible in a possible environmental estimate. Furthermore, the working environment requirements are important. The building painters no longer want to use the solvent-based paints for repairs.[17] Therefore, an increasing demand must be expected due to the effect that constructions should be maintained with water-borne paints.

REFERENCES

Acts and Statutes

Begrænsning af luftforurening fra virksomheder. (1974) Guidelines from National Agency for the Protection of the Environment, no. 7.
Begrænsning af luftforurening fra virksomheder. (1990) Guidelines from National Agency for the Protection of the Environment, no. 6.

---

[17] Building paint work is almost totally based on water-borne paints, due to the demands on the working environment.

Begrænsning af luftforurening fra virksomheder, der emitterer cellulosefortyndere og andre blandingsfortyndere til luften. (1978) Guidelines from National Agency for the Protection of the Environment, no. 2.

Bekendtgørelse om polyurethan- og epoxyprodukter og tilsvarende stoffer og materialer med lignende sundhedsfarlige egenskaber. (1978) Arbejdstilsynet.

Bekendtgørelse om arbejde med kodenummererede produkter. (1993) Arbejdstilsynets bekendtgørelse no. 302.

Bekendtgørelse om erhvervsmæssigt malerarbejde. (1982) no. 463, 3.8.

Foranstaltninger mod sundhedsfare ved malearbejde på store konstruktioner. (1982) Arbejdstilsynets vejledning nr. 369/3.

Grænseværdier for stoffer og materialer. (1985, 1988 og 1992) Arbejdstilsynets anvisninger.

Organiske opløsningsmidler og andre sundhedsskadelige luftarter. (1988) Arbejdstilsynets cirkulaere 2.

Vejledning om arbejde med stoffer og materialer. (April 1989) Arbejdstilsynets meddelelse nr. 3.02.5.

Other

Chronic effects of organic solvents on the central nervous system and diagnostic criteria. (1985) Joint WHO/Nordic Council of Ministers Working Group. Copenhagen. (WHO 1985).

Corrosion Protection of Structural Steel. (1993) Danish Technological Institute, Taastrup. (DTI 1993).

Ecology and Economy in the Development and Use of Heavy-Duty Protective Coatings for Steel. (1994) Peter Kronberg Nielsen & Jens Holm Hansen, Hempels Marine Paints A/S. (Hempel 1994).

HYBRIDSYSTEMET. (30.1.1992) Written conference material, Danish Technological Institute. (ABC 1992).

Industriel overfladebehandling. (1993) (Industrial coatings), nr. 4.

Korrosionsbeskyttelse af stålkonstruktioner (Corrosion Protection of steel constructions). (1982) Norm NP-154-R. Dansk Ingeniør forening. (DIF 1982).

Korrosionsbeskyttelse af stålkonstruktioner (Corrosion Protection of Structural Steel). (1993) Miljøprojekt (Environmental Project) nr. 216, Miljøstyrelsen. (EP 216, 1993).

Miljøgodkendelser af anlæg på ABC Coating (Environmental approval of plants on ABC-Coating) (29.6.1988 og 25.1.1989) Røde Kro kommune and other documents on the firm. (ABC 88/89).

Miljøregler skal baseres på indsigt. (1993) pressemeddelse fra Sandblæse- og entreprenøforeningen 27.11.93.

Miljøvenlige malematerialer i jernindustrien (Environmentally improved paint-products in the metal manufacturing industry). (1990) Miljøprojekt (Environmental project) nr. 126, Miljøstyrelsen. (EP 126, 1990).

Mødereferater fra møder mellem Arbejdstilsynet og ABC-Coating samt Arbejdstilsynets tilsynsreferater fra arkiverne hos Arbejdstilsynet i Aabenraa (Documents and reports at the Labour Inspectorate, Aabenraa).

Organic Solvents. (1986) Miljøprojekt (Environmental Project) nr. 72, Miljøstyrelsen. (EP 72, 1986).

Paint and Pollution—A Question of Solids. (Undated) P. Kronborg Nielsen og J. Holm Hansen, Hempel. (Hempel undated a).

Photochemical smog. Contribution of VOC's. (1982) OECD. Paris. (OECD 1982).

Polyurethan og epoxy. (1980) Direktoratet for arbejdsmarkedsuddannelserne. (AMU 1980).

Produktudvikling af mere miljøvenlige malinger (Product development of environmentally friendly paintings). (undated) Søren Nysteen, Hempel's Marine Paints A/S. (Hempel undated b).

Projektansøgning. Renere teknologi i jern- og metalindustrien. (Application for a project). (1990) Ansøgning til Miljøstyrelsen af 12.2.1990 fra Overfladeteknik, Dansk Teknologisk Institut. (DTI 1990).

Substitution of organic solvents. (1992) F. Sørensen and H.J. Styhr Petersen, in: *Staub—Reinhaltung der Luft 52*, 1992. (Sørensen et al., 1992).

Vandige malematerialer til korrosionsbeskyttelse (Water-borne paints for corrosion protection). (1990) Miljøprojekt (Environmental Project) no. 140. Miljøstyrelsen. (EP 140, 1990).

Yearly accounts for ABC-Coating a/s 1987–1993.

Interviews *(All Interviews Performed in the Period June 1993–March 1994)*

Oluf Lauridsen (twice), director of ABC-Coating.

Ole Møller Nielsen (twice), representative from the local environmental authorities in Røde Kro.

Ralph Nielsen, the head of the trade organization for sand-blaster and paint contractors.
Kurt Schau Christensen og Ove Nielsen, two painters at ABC-Coating.
Niels Lund Jensen, The Danish Technological Institute, Surface Coating Technologies.
Jørn L. Hansen, The National Environmental Protection Agency.
Sønnich Andreasen and Arne Broberg, The Labour Inspectorate in Aabenraa.
C. Bjerge Jørgensen and Søren Nysteen, Hempel Marine Paints a/s.
Mogens Kragh-Hansen, Industrial Health Service, Paint Work.

7. GLASULD: RESOURCE MANAGEMENT AND ECO-AUDITING IN GLASS-WOOL PRODUCTION

Jesper Holm and Børge Klemmensen

SUMMARY

This is a story about successful environmental innovation at Glasuld A/S, the only glass-wool insulation plant in Denmark. The story starts at a point where the company already had above-average European environmental standards in glass-wool production, far above Danish emission standards.

The environmental and technological innovations concerned the substitution of natural raw materials by recycled glass, substantial cuts in energy consumption, cuts in plastics packaging waste and chemical waste and, finally, the lowering of the emissions of natrium-oxide (NO_x). A full scale recycling of all chemical waste and solid glass-wool waste was decided, and is currently in progress. The innovations were organized by an eco-accounting and auditing scheme together with the local environmental authorities and a consulting firm. A systematic environmental account was made, providing the foundation for various environmental protection efforts. The employees have been deeply involved at all levels.

It is a story of a company's learning because of pollution problems and arguments with the authorities in the past; of it predicting future environmental demands; of it organizing the employees in environmental auditing, and establishing close cooperation with the local authorities. As a result of ecological and economic success, the eco-auditing procedure and priority-setting methodology has been distributed throughout the local networks, involving a total of 60 companies. The story provides perspectives on a new type of co-production among companies and authorities where negotiations, agreements and procedure inspection of the company's eco-management is substituted for the former norm permits, waste and emission control.

1 THE COMPANY

Glasuld was established in Copenhagen in 1936 producing glass-wool insulation slabs for the home market. After several expansions the production was relocated in 1982 to Vamdrup in the southern part of Jutland, not far from the Danish–German boarder. 'Superfoss', a Danish corporate company, bought Glasuld in 1973, and in 1988 it became a company within the insulation division of the Saint Gobain Group – a multinational corporation, which leads insulation production on the world market. The corporation employs 105,000 people in 32 countries, the turnover in 1990 was 10 billion ECU, and the insulation production accounts for 13% of the turnover. In 1990 the production of glass-wool employed 260 people in Denmark, produced 30,000 tons of glass-wool annually, and had a turnover of 340 million DKK.

The majority of the production is of bulk-products for the construction sector. The competition is strong; primarily from Rockwool, another multinational, mineral-wool insulation producer, and also polyurethan insulation producers. There is no differentiation in terms of quality for the bulk-goods. The cost of glass-wool production in the company is composed of: machinery and buildings 60%, salaries 28% and raw materials 12%. But the voluminous character of glass-wool means that the subsequent transport cost is also an important factor for the price of the product, and, indeed, for the environmental impact of the glass-wool. Transport costs were 52 million DKK in 1992, roughly equal to the costs of raw materials.

Until 1992 the quality and production manager was in charge of environmental matters. In 1992 an environmental manager was established. She was in charge of an eco-audit project which prepared an environmental plan for the company, an annual environmental account and an environmental audit, exclusively for use with the environmental authorities. An environmental management system is being developed and there are plans to apply for an ISO-9001 approval.

2   THE SUCCESS STORY IN TERMS OF TANGIBLE BENEFITS FOR THE ENVIRONMENT – IMPROVEMENTS IN THE ENVIRONMENTAL IMPACT OF THE GLASS-WOOL PRODUCTION

In the initial environmental review of the eco-audit project, which surveyed the environmental contributions from the company and thus provided the necessary background for establishing the priorities to meet the most pressing problems, a quite simple score-system was applied and three areas stood out. They were: electricity consumption at plant level, gases from the furnace during glass molting, and emissions from felt-making. Two risk-management areas, the binder production with phenolia and formaldehyde as main elements and Propane gas-storage, got high scores as well, but were considered so well under control and so manageable that they could be left to normal management procedures. On the other hand it was decided, because of pressure from the authorities and the economic and strategic/image, to include a general waste audit at plant level. Waste was thereby becoming the fourth key area in the audit program.

A number of the recommendations coming out of the audit in these four areas were valid and were implemented in the period up to 1993/1994, when this case study was made. Until mid-1994 the following areas for initiatives (with the corresponding reduction of environmental impact) are shown in Table 1. In addition to these results, experiments were undertaken in two areas to save further on electricity consumption. One was the use of recycled (windows) glass as raw-material to be mixed with the original glass raw-materials. Much less energy is used to molt glass than the 'fresh' raw-materials. The aim was to reach a level where some 60% of the raw-material was recycled glass. The problem was that the glass was not available and had to be imported from Sweden. Still, it has already meant reduced energy consumption of some 3 MWh/year, or some 5% a year.

The other experiment in energy conservation is the optimization of the compressors, where there was a substantial surplus capacity according to the way the

Table 1 Areas for initiative up to mid-1994.

Process areas	Emission			
	Consumption	Savings	%	PB/year[1]
Electricity use:				
water pumps, ventilators etc. mil kWh/year	2,0	0, 780	39	0,9
Gases from furnace:				
CO_2, tons/year	6,200	0	0	
NO_x, tons/year	191	135	70	0,5
Dust, tons/year	6	0	0	
Waste:				
Chemical-Waste, recycling in factory	199 t/ye	175 t/ye	90	2,8
Glass-wool	975 t/ye	325 t/ye	30	3,8
PE-packaging	116 t/ye	60 t/ye	50	0,5

[1] PB = pay back period.

production was organized. Here also a big electricity-saving was made—in the range of 1,5 MWh/year. The energy-saving initiatives at the Vamdrup plant have put it in the lead in terms of the lowest energy consumption per ton glass-wool produced.

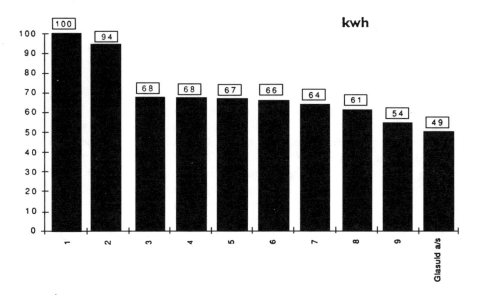

Figure 1 Comparative energy consumption per tonn of glass-wool produced within the 10 European Saint Gobain Companies, 1993.

*The Glass-Wool Production Process – Main Environmental Impact
and Improvements Achieved*

The production of glass-wool[1] is a continuous flow-process that can be divided into
three steps: 1. Production of two main categories of raw-materials: bakkelite binder
and molten glass; 2. Spinning of glass and adding the binder; 3. Drying, forming
and packaging of glass-wool in slabs, rolls etc.

Production of the binder   Bakkelite-condensates serves as a binder among the glass
fibres in the glass-wool. It is produced separately by two main substances: phenolia
and formaldehyde mixed with water (formalin). Phenol is a suspected cause of
cancer (skin, liver and kidney) and also thought to cause allergies. Formaldehyde is
mutagenic and is a suspected cause of lung cancer and severe allergy. Formalin is
also cancerogenic and acutely toxic. There is every reason to reduce the use of, and
secure the handling of, these substances.

The two main liquid substances are pumped into mixer tanks from separate
tanks, mixed with urea (ammonia sulfide) and finally mixed with bakkelite. The
principle of the production is to ensure chemical reactions among phenolia and
formaldehyde, using sulphur acid as a catalysator and ammonia to keep the process
going as long as possible. Ammonia thus ensures that chemical reactions continue
after the normal point of crystallization. Sodium lye and water (disinfected by
adding chloride acid and salt) and a high temperature keeps the bakkelite liquid.
After the process is finished bakkelite is pumped to the processing of spawned glass.

Ecologically the most important effect on the environment from this sealed part of
the production consists primarily of risks from accidents and interruption of the
work. If the cooling system breaks down it might cause damage to the tank and
1–10 tons of formaldehyde and phenolia may spill over. However this should run
into a spill-over basin.

At the stage of mixing chemicals for the binder there were possibilities of
adjusting the emissions of formaldehyde and phenolia into the air. This could be
done by changing the recipes, even though the emissions stem from the following
fibering process where binder is sprayed on the fibres. Experimenting with the
resin-recipe was one of the three main parts of the audit. It tested less 'free phenol'
in the binder solution – up to some 50%. But it was not possible to document any
significant difference in the amount of phenol actually emitted into the air at the
end. The trial period during the audit was only one day. A two-week test was
therefore suggested to get a firm picture. That test has till now not been done, first
of all because the plant is nowhere near the license limit of $5 \, mg/m^3$ for phenol
and $20 \, mg/m^3$ for formaldehyde. There are no savings in it at this point, and
therefore 'no' pay back period. The impetus has gone. The other reason was that
Saint Gobain wanted to control all tests and research on the binder in France.

Production of glass   The molten glass is produced by mixing mainly chalk, sand,
soda and dolomite adding sodium nitrate and Borax for the oxidation of organic

---

[1] The following is based upon our plant visit. The figures from Glasuld relate to 1990, that is *before* the
environmental innovations were started.

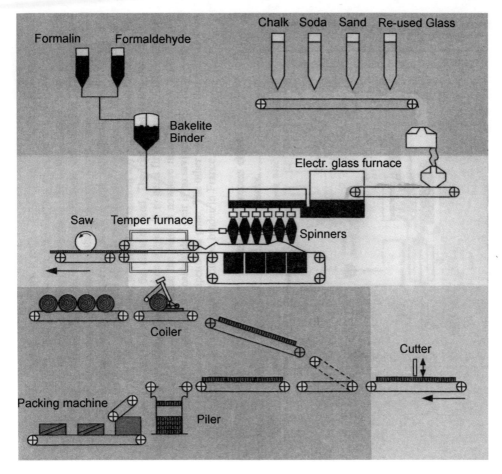

Figure 2   Glass-wool process chart.

substances. After mixing, the dry substances are pumped into an electric furnace operating at 1350°C where it is molted to a glass-fluid.

More than 50% of the total electricity consumption of the entire plant is in the molting process. The emission from the electricity production takes place at the power plant, but a reduction at this point is still important from an environmental perspective and, indeed, from the economic point of view for the company as well.

Optimizing the energy-effective running of the furnace had already taken place in the beginning of the '80s. The recent efforts to cut energy use from the molting process has been the introduction to use recycled glass. Till now some 40% of the raw material has been shifted to recycled glass. There was, however, no collection system for plate glass windows. Food and drink-glass bottles are collected, but have too high a content of organic matter if not pre-washed. The uncleaned glass contains too many organic materials giving higher emissions of CO_2 when incinerated in the glass furnace. This is both a problem in terms of the firm's environmental profile and presumably also because of the tax costs from CO_2. Machinery for cleaning the glass waste was too costly and technically problematic for Glasuld to deal with. Accordingly they have found suppliers in Sweden where state subsidies for collection, cleaning and crushing both consumer and plate glass makes it profitable! But Glasuld have an interest in using Danish glass as they want to reduce the transport costs and want more supply alternatives to press prices downwards. Accordingly, there have been negotiations with Århus County Recycling-Central, trade organizations within the automobile sector and transport companies, for organizing a collection and distribution system of plate glass. If a collection system is established Glasuld will get a grant from NEPA to build their own 'crushery' where they will be able to crush the glass to powder for the furnace. However, the saving at the present level has already been substantial, some 3 Mwh (or 10%) of the total for the molting process.

The furnace also evaporates gases, of which the NO_x is the most problematic. As was shown above, a major reduction, some 135 ton/year or 70%, was achieved here, simply by changing the oxidation additive from $NaNO_3$ to MnO_2. On top of the savings of the nitrate-oxide, there seemed to be a further energy saving of some 1–2%, which is something like 500,000 kWh. The only problem was the negative influence of MnO_2 on the working environment. After tests it was found that use of granulated matter could solve this problem apart from repairs etc., when protective clothing would then have to be used.

The spinning of glass-fibres and impregnation with binders   The stream of molten glass is kept liquid at constant high temperatures in long furnaces and finally poured down in spinning atomizers for making fibres. The jets of glass-fibres are sprayed with liquid bakkelite and silicone glass-felt is made. The pressure of jet currents from gas-burners, and impregnation by the now foamy bakkelite makes the glass-wool. Most of what is not consumed of free phenolia, formaldehyde and bakkelite is separated and treated in the scrubbers. The glass-wool is carried to a conveyer and warmed in a gas-fired furnace to temper the glass-wool and the binder, and evaporate all the water from the liquid bakkelite. Cooling water brings the temperature of the glass-wool down again.

Ecologically the most serious pollution consists of an annual emission to the air of phenol (some 10 ton/year) and formaldehyde (some 5.5 ton/year). As was seen above, the project for reducing the emission of these two substances was not carried out for economic and other company reasons. The gas-burners used to dry out the water in the binders emit CO_2 at the level of 10,000 ton/year – but no projects were run so far.

This part of the process also produces solid waste, even chemical waste, which comes from the filters and the scrubbers (where the air, taken out by vacuum from the spinning, spraying and hardening process, is 'washed' before it leaves via the chimney). This 'wet' waste contains some small amounts of phenol and formaldehyde, which makes it chemical waste, which can no longer be deposited at the waste dump. It has been sent to the neighboring competitor, Rockwool, and finally to a Saint Gobian glass-wool factory in Sweden where it is recycled. But now tests are being run to add it to the furnace again or use it for a special, hard coverplate, wherein recycled glass-wool batts are used.

Forming and packaging the glass-wool   From the temper, the glass-wool is formed by pressing, cutting and coiling. Packaging is made by therm and vacuum shrinking of PE-foils.

Ecologically the most important pollution consists of waste deposits: glass-wool: 892,000 kg, PE-foil: 113,350 kg. Approximately 300 tons of the glass-wool from failures and cuttings have been tested for recycling. The spills have been granulated and added afterwards in the fibreing phase at the spinners. The experiments have proved to be so promising that full-scale recycling of approximately all the dry glass-wool waste will be started. This will reduce the total waste disposal of glass-wool from 975 tons annually to 150 tons annually, some 80 tons annually coming from other parts of the process. This will also result in substantial energy savings. The PE-foil spill has been reduced by 60 tons by better housekeeping and more optimal use of new machinery through training of the operators.

Air and water purification processes   The most ecologically important waste that is treated by purification is produced during the processes of impregnation and spinning. From these processes most of the spill of fibres, dust and chemicals is sucked into the scrubbers by vacuum pressure. The first station is the cyclone where large particles are taken away. Then two separate scrubbers hold back 80% of the remaining particles and chemicals, while the rest is lost as airborne emissions through the 107-metre chimney. After filtration, the water from the scrubber's waste water is re-used as process water for the bakkelite impregnation. No waste water leaves the plant, but some water leaves the factory through the chimney. Water has therefore to be added to the process.

The solid part of the waste from the scrubbers constitutes the 'wet' waste, which, as mentioned above, has to be considered as chemical waste because of the phenol and formalin content. After experiments within the eco-audit this is now being sent back to the furnace for re-melting, as the main part is glass-fibres. Some 190 tons of chemical waste was thereby reduced to 50 tons or even less.

Other Environmental Aspects

Glass-wool fibres are, as other mineral-wool, suspected of causing lung-cancer and irritate mucous membranes and the lungs, all of which are features of the working environment and subject to health and safety regulations. None of the environmental efforts have made any attempt to substitute the fibres with natural non-cancerogeneous materials, which indeed leaves a long road for the firm to travel towards a sound environment.

3   GLASULD'S ROAD TO ENVIRONMENTAL CONCERN – FROM FORCED RELOCATION OF THE FACTORY AND A NEED FOR IMAGE IMPROVEMENT IN THE LATE '70s TO PARTNERSHIP IN THE REGIONAL COUNCIL'S GREENING OF INDUSTRY PROJECT IN THE '90s

1   Forced Relocation from Copenhagen

Glasuld's first factory was situated in the Copenhagen suburb of Kastrup, close to the coast, facing Sweden. Housing areas were nearby and had been allowed to creep in on the factory over the years. During the '60s and the '70s a number of police reports were made on smoke problems from the chimneys and more than 500 local residents complained about the smell, the nuisance, spoilage of plants and the noise. The local environmental authorities, were, step-by-step urged to demand filtration and higher dilution (chimneys) and the company slowly responded by installing purification equipment – just to realize that it never had any substantial and lasting effects to the emissions. Public debate arose in the newspapers about Glasuld's pollution and here there were demands for the plant's closure.

Smoke, smell and noise were not the only environmental problems linked to the Kastrup plant, although they were the most visible. From the bakkelite mixing unit waste water was poured directly into the ground until 1974. The other waste water only underwent mechanical treatment at the municipal sewage treatment plant and was then discharged into the Øresund. The installation of scrubbers and filters removed a good deal of the air-pollution, only to turn it into heavily polluted waste water, containing large amounts of phenol (from 350 mg/l in 1976 to 83 mg/l in 1982) and formalin (from 850 mg/l to 685 mg/l). Large amounts of particle emissions came from the oil-fired oven, as high as $2 \, g/m^3$, (but by filtration this was brought down to some $30 \, mg/m^3$ before Kastrup closed). Finally, the solid waste, wet and dry, with phenol and formalin and other chemical substances, was initially incinerated by the company near the factory, but later all of that ended up at the waste dump, although classified (from 1978) as hazardous waste.

When Glasuld suffered capacity shortage during the '70s, the municipal authorities categorically refused to accept any plan for expansion of the plant. A complaint to the national Environmental Protection Agency brought support for the viewpoint of the factory that increased glass-wool production would benefit the region and the economy and that the environmental burdens were not that grave. But the local authorities, supported by green grassroots and neighbours, appealed against judgment to the national Board of Appeal for the Environment, the final appeal authority. Here the verdict was in favour of the neighbours which finally extinguished any hope for expansion on the Kastrup site or of the company staying in Kastrup at all. Consequently, in 1977, the plans for expansion in Kastrup were dropped for good, and in 1979 it was decided to build the new plant in Vamdrup.

2   The Lessons Learned. Environmental Impact Became an Important Parameter in the Construction of the New Plant

The experiences from Kastrup, which the production manager, Haals, especially took to heart, played an important role in the construction of the new plant in Vamdrup. So did a revised attitude to the environmental question on the part of the former owner of the plant, Superfoss, an internationally-operating Danish corporation, producing fertilizers and some chemical goods.

Superfoss's attitude has to be understood in relation to experiences, gained primarily in other branches, and primarily in fertilizer production. Thus, many of their firms had violated the environmental license and had many image problems. In 1982 central management clearly stated that it wanted a better image on the environmental side. An important part of this was to make sure that all the necessary requirements were met, and that every single operation stayed within the limits of the license in question. And further, it was now up to the management of each and every operation to handle the environmentally-relevant issues in an accountable and fully-defendable manner.

For the new Vamdrup plant that concept was strictly applied as management expected emission/immission standards and limits to be tightened over the years to come. They wanted to be prepared for a such development in drawing the plans for the machinery etc. for the new production site. *The first* point in the environmental considerations in the construction was the choice of a glass furnace, and here the company opted for an electrically-heated furnace, while that in Kastrup was oil-fired. This reduced the emission of gases and particles dramatically at the factory, leaving it to the regional power station to protect the environment! The tempering furnace in Vamdrup is heated by natural gas, which also gave less problematic emissions than the former oil-based unit. *The second* significant improvement was the recirculation of process water, so that no process water was discharged from the plant. *The third* improvement concerned waste; the amount of 'wet' solid waste was reduced because of better air vacuum systems and the installation of efficient scrubbers. Both initiatives filtered more waste water from the waste for re-using as process water in the bakelite impregnation of the fibres. Initially the 'wet' solid waste was calculated at 990–1000 ton/year, but ended up as low as some 200 tons, reached during the '80s. The 'dry' waste was reduced primarily by new methods of shaping and cutting the wool mat on the conveyer and more optimal use of the glass-wool when cut into ready slabs or roles.

The air-emission system and the scrubbers can illustrate the application of the new attitude in the Vamdrup plant, including the extra-environmental considerations that went in. The calculations concerning the chimney suggested a height of 70 metres according to present regulations for dilution of emissions. It was, as a precautionary measure, decided to build it to 107 metres. After 12 years of increasing environmental awareness the chimney is still well within the limit.

The legally-settled emission limits of formaldehyde *was* in the final years at Kastrup 20 mg/m^3 and efforts were made to keep in line via a huge and expensive system of filters – a factory on its own. On the building of the Vamdrup factory it was made a condition in the tender that the installations proposed should be able to halve that limit – to 10 mg/m^3. Only two could match that demand, and one of them was selected as a supplier. Today it is built to operate on 10 mg/m^3 without the additional equipment, for which it is prepared. The actual emissions are about 4 mg/m^3 and the emission limit is still 20 mg/m^3.

In 1984 Glasuld was in the public focus again. It became known that the old production plant in Kastrup had it's wet glass-wool waste deposited at ordinary waste sites even though it contained large quantities of phenolia. This and other public scandals urged Superfoss to go through an environmental screening of the companies in the corporation together with the National Environmental Protection Agency (NEPA). NEPA recommended to Glasuld in Vamdrup that Superfoss stop depositing the 'wet' waste at ordinary, but controlled, waste dumps. Accordingly, the waste was sent to 'Kommunekemi', the national chemical waste treatment plant, but the charges raised the costs connected with the 'wet' wastes considerably. Glasuld therefore became interested in reducing this waste as well as finding alternative handling possibilities. The sister company in Sweden did not experience the same demands from their authorities and deposited all their waste. But actually they had the possibility of re-using waste in their type of furnace, and thus Glasuld started to export the Danish waste to Sweden!

These three examples show the precautionary economic approach to possible future requirements in the form of regulation. Investments were made, where it would be impossible (the chimney) or very expensive (the scrubbers) at a later stage. But as the investments in the scrubbers could be divided into two parts, the last to be made, only when required by the authorities in the future. And for the 'wet' waste, still remaining, nothing was done until required, and then alternative solutions were found – in this case export. Evidence for the success of that strategy is the fact that in this time period, eight years, no significant changes to, and no significant additional investment in the 1982-factory were made for the environment. They soundly estimated the situation when planning the factory (from 1979–1982), and they hit the balance quite nicely between economic interest of the company and care for the environment.

4   THE JOINT COUNTY–GLASULD ECO-AUDITING PROJECT IN 1991

The idea of starting an eco-audit project at Glasuld came from the Vejle county, responsible for environmental control and approvals, inspired by the preparation of a new environmental protection act, where cleaner technology was going to become a keyword. A cleaner technology group was established within the county and given resources in 1989 for developing ideas in this direction. It held a seminar on new challenges to environmental regulation from the cleaner technology concept. Participants were the three neighbouring counties, a consultant from COWI Consult Ltd. and a deputy director from the National Environmental Protection Agency. The conclusion was that it was necessary to develop methods for systematic self-regulation to make inspection cheaper and more effective and related to production control parameters. Base-line data needed to be established at the start through an eco-audit procedure. The director of the Vejle county environmental department therefore suggested that a pilot eco-/waste-audit be initiated.

At the same time, in 1990, Glasuld's environmental license had to be reviewed as eight years had passed since the opening of the factory in 1982. The county of Vejle – in charge of environmental inspection and licensing of Glasuld – did not foresee serious problems with that and regarded Glasuld as a responsible and open company, sympathetic towards the environmental authorities. Therefore Glasuld was asked to join the pilot waste-auditing project – and the company agreed.

Glasuld's reason for participating was, first of all, a wish to profit from a 'greenish' profile in relation to important customers. Incidently, Saint Gobain held a meeting in October 1990 with all glass-wool plants, where a general environmental policy was formed. The ICC business charter for the environment was to be signed and the national directors were held responsible for achieving good environmental profiles. The following goals were set out by the Saint Gobain corporate leadership: the depositing of glass-wool spills at waste sites should be stopped by 1994; by 1996, 50% of the raw-materials for glass production should be re-used glass; there should be no waste-water after 1993; and, finally, the German values for air-emission limits should be the standard that all plants should comply with by 1994.

This simultaneous move at corporate level was therefore another impetus for Glasuld to join the pilot eco-audit project. The managing director for Glasuld Denmark when announcing the goals of his company told the local press that, in general, the industry simply had to document their environmental data to be reliable. This commitment to public openness became an important part of environmental management at Glasuld. It took some time, however, to convince the Saint Gobain leadership that the sensitive data on the real use of energy and raw-materials could be kept closed to all except management, despite the quite detailed analysis entailed in the pilot project.

Organizing the Eco-Audit Process

At the start Vejle county organized the project. NEPA was asked for financial support and gave 1 million DKK, which financed the consultancy company COWI responsible for conducting the project, and Vejle county added funds for making the project design. The company paid for the time used by the employees involved.

A project-group was set up with three experienced environmental inspectors from the county, Mr. Halls (director of quality and environment) and the environmental director Jørgen Mølgaard from the mother-company Saint Gobain. An element of personal prestige was added in 1992, when the post of environmental manager was filled. As a manager responsible also for health and safety issues, she forged the strong employee involvement in the process.

The actual auditing was organized into four working groups at the plant, each of which was staffed with a director from Glasuld, an operator from the production, a consultant from COWI and an inspector from the county administration. After a while, the organization was simplified and the groups were joined together into one bigger group, but sub-groups were organized within each and every department of the plant to let workers discuss problems and possibilities as well as producing ideas.

The management mandate for the groups in the process was open with some limits: working conditions could not be a part of the project, although could be taken into consideration when particular experiments were made and solutions chosen. The environmental assessments were not supposed to focus on the emissions but on the potentials and sources coming from raw-materials and processes. Finally, the environmental aspects should be analyzed from a *process* point of view and *not* from a life-cycle analysis of the *product* point of view.

The audit-process had four phases: 1. Mapping of the environmental relations; 2. Waste analysis, ideas and experiments within the prioritized areas; 3. Environmental plan and operation control system; 4. An annual environmental account. Finally an environmental management system was to be implemented, but that has not taken place yet (1996).

1   Mapping of the Environmental Relations

This process consisted of total scanning of environmental relations, ranking the problems by scores and making priorities for further innovation.

Scanning The first structuring of the analysis was to divide the plant into relevant sections whence identification of problems and solutions should begin. The choice fell on a division by processes instead of by plant location. Subsequently, the working groups started to monitor systematically all the environmental relations for which Glasuld is responsible. They did this at their own plant starting from the source of the problems. In all production sections workers from the floor, engineers and management leaders started to register use of energy and raw-materials, and estimate the amount and content of spills and waste, emissions to the air and to the process water, and finally emissions from electricity use. All this was calculated for a normal process at full capacity, interruption (batch-shifts) and for risks (accidents). Finally, nuisance from smell and noise were evaluated. This period was important for the rather comprehensive information needed for the employees of the project which the environmental manager had recommended to attract participation from all levels at the plant. Later on this turned out to be very important for the number of inventions and ideas.

Scoring The different environmental relations were set up in matrixes and estimated according to three aspects: amount, diffusion and ecological effects. The relation was accordingly scored as the multiplication of the three aspects, composite materials being evaluated from the highest ranking substance. Evaluations of the experience turned out positive from all participants point of view but it was highly stressed that pragmatic reasoning and subjective evaluations had to follow this procedure; they could not assign the highest priority by themselves.

Priorities The environmental relations were subsequently prioritized according to the scoring figures.The priority process produced 13 high priority areas. Among these energy-use, emissions to the air from glass furnaces and emissions to the air from filtering processes were chosen for deeper analysis. Finally, all waste production was prioritized because of commercial and economic reasons and because the local authorities found it important to cut down deposits at their waste disposal site. The priority system seemed reasonable and practical to those involved and no deep critique has surfaced. It was seen as important for the county to participate in priority-making because it guaranteed an insurance against future interference. The county influenced the choice to go for cuts in waste production/depositing but did not have any local environmental priority scheme, enabling them to identify the region's most severe problems.

2 Waste Analysis, Ideas and Experiments on the Prioritized Fields

This phase consisted of: the identification of sources of the problems and environmental effects within the four fields prioritized; the assessment of alternative responses; the selection of solutions and starting experiments or investigations. Identifying the problems meant measuring emissions, gathering detailed information from subcontractors on content of materials, and making deep environmental assessments. As a result mass-balances were made and a better understanding of the precise causes of the environmental relations was given. The assessment of alternative responses was given as much room as possible without limiting ideas to what was actually realistic or cost-effective. Some guidelines for construction ideas were handed to the groups: Is the process necessary at all? Is the process optimized concerning loss of material? Are there any alternative processes or raw-materials causing less pollution?

The employees' attitude towards the project was, in the beginning, somewhat indifferent, but they became quite actively involved after some time at this stage of the project. They were very surprised by the amounts of waste and energy consumption that figures showed from a summing-up of the annual situation. For example, in the investigation of electricity consumption compressors, the pneumatic system, the lamps etc were surveyed. The staff was very surprised to find that the compressors were major energy consumers. Thus they changed their attitude and were eager to find solutions.

Futhermore, the authorities were quiet astonished at the amount of profound observations and good ideas: 'Normally when you do your inspections at the plant you feel quite superb. You had a long education and know all about details on pumps, chemistry etc. And of course this leads to something, but not compared to the result when engaging the employees. I'm not able to go in as an authority inspector and give fifty good ideas. And where should I get them from? Besides I'm not able to direct the company to follow these ideas. It is the ones that are working daily with the machinery who have all the ideas.' (Inspector interviewed in *Danmarks amtsråd* no. 5 March, 1993).

According to the environmental manager: '...it's the shock of getting the figures in black and white and related to each other, the same [as for the compressors] goes for electricity...there are eleven employees walking around in the night and you have light on all $60,000\,m_2$ – this can't be true?... Sometimes a lot of things happened immediately after learning the figures. Take the packaging department where they registered 88 tons of plastic waste. Here they went ahead even before we were ready to go ahead in the implantation phase, and they were down to 28 tons before we got into it...it has become an honor for the employees to reduce the environmental impacts from your own company.' (Jeanette Korsgaard interviewed September, 1993).

The chosen solutions for deeper project analysis passed through a technical and economic assessing process. To this end detailed check-lists were distributed concerning costs and pay-back for investments in equipment and process. As for the technical assessment, the guide asked for detailed information about purchase, suppliers, adjustments in production, the coping with plans and demands from authorities, training and effects to the environment.

3   Environmental Plan and Operation Control System

This phase consisted of development of environmental policy and strategy, environmental goals, an environmental program and a controlling system.

Concerning the *policy and goal setting,* Glasuld signed the ISS business charter for Sustainable Development, Principles for Environmental Management. To the general figures in the charter, the before-mentioned Saint Gobain goals from 1990 were listed. Due to the charter, Glasuld will have to take back all plastics packaging from end of 1995. Glasuld have added goals for emissions taking 1991 as a baseline: No_x down by

50% in 1995; free phenols down by 50% before 1995; free formaldehyde down by 50% before 1998 and CO_2 down by 3.5% annually until 1995.

The county's evaluation of this phase was that the project slipped out of their hands as the company took over all the strategic considerations. But on the other hand they appreciated the deep integration within Glasuld's organization of the plan.

As for the *strategy*, we will only list new principles that are not evident from what was formerly described: environmental impact assessments were to be made of all investments; the company was to open communication on environmental problems; to support the authorities in their environmental protection efforts, to register and measure the environmental impacts to both process and products; to substitute processes and materials by less environmental harming and, finally, to urge subcontractors to deliver environmentally-acceptable goods etc. As could be seen, the company became aware of environmental relations that go back to the subcontractors, to the problems from the raw-materials and, finally, to the environmental problems the product may cause. The latter is an especially relevant matter in these days where the possible carcinogenic effects from mineral fibres is discussed intensively (more on this later).

The environmental *program* concerns a survey of all activities in the following year with estimations of investments and pay-back period. The control system has been developed so far as the management parameters are found able to control the prioritized four goal-areas from the second phase: energy, waste, NO_x-emissions and phenol/formaldehyde emissions. For the last two categories the task has been to identify the correlations between normal management parameters and environmental contribution; waste is simple to count by random sampling, weight slips invoices and for energy-measuring equipment has been and will be installed.

4   An Annual Environmental Account

This has not been published but was handed over to the county's environmental department. During the process, the county changed focus from direct innovation on cleaner technology to carrying out waste analyses and ended with stressing environmental management systems. Correspondingly, they changed their focus on how to use control parameters: from emissions measures, to operation control to checking administrative routines. The project did not end with detailed analysis for using the operational parameters in control as this task was more difficult than expected; the linkages between production activities and resulting environmental data are extremely complicated and have to pass a lot of tests in order to be reproduced.

5   Background for the Choice of Environmental Strategy and Focal Points

If we start with the ideas that were expelled, why was it then that Saint Gobain decided to concentrate all experiments in large scale from fibreing, production of binders and inductions of binders to the glass, at Saint Gobain's own laboratory in France? As mentioned, this meant that many of the ideas that were thought up to reduce the emissions of formaldehyde, as the county requested were cancelled. As far as we have been able to analyse it, the strong interest in establishing glass-wool as an international recognized brand left Saint Gobain no room for local experiments.

Other ideas were expelled due to a hidden priority from the scoring system which might have passed behind the backs of the participators. Thus it is questionable whether the decision-support system has objectivized itself, as hard figures from scoring related to quantity, diffusion and effects might seem very rationalistic. But, on at the other hand, the county did not show up with any prepared priority plan except to include waste and formaldehyde; thus they missed the opportunity to concentrate on a regional ranking of environmental problems.

Generally, the bulk of choices from many options fell upon *housekeeping fields* such as discovering over-consumption of energy, reduction of spills from plastics packaging, reduction of spills from mixing and impregnating bakkelite etc. The reason for these fields of innovations were that they could be easily handled, were not costly or had short pay-back periods. Generally, the projects with clear economic benefits created a pathway for the environmental solutions.

The reason for choosing re-used glass was partly environmental and partly economic as it meant substantial cuts in energy use for the glass furnace while the costs, including transport, were the same as for normal raw-materials. Danish deliveries will make the company less dependent on one or two suppliers as they are at present, thus enabling Glasuld to lower prices.

Taking back plastics packaging were technically promising but not economic and was cancelled for this reason. As mentioned, taking back worn-out glass-wool slabs caused technical problems and demanded a considerable number of experiments. Further efforts in these two fields will have to come from market pressure or regulation demands.

Even if the working environment was not directly included in the eco-auditing project it is important to understand the reasons behind Glasuld's modest efforts to innovate further regarding the fibre problem, even though it concerns all private customers. In the past many efforts have been made to lower the emissions of fibres, such as making the fibres less harmful, by reducing the amount of fibres not attached to the binder, and by making a ready-shaped batch for specific types of insulation, as this reduces the necessary cutting of glass-wool. But the problems to the working environment from the fibres *are* technically to be handled by coating or packaging the glass-wool slabs, as is being done in Japan. Today, Glasuld delivers these kind of slabs for acoustic insulation and for some type of roofing. But the competition on the bulk market in construction is too cost-focussed to secure the necessary sales-figures, if the glass-wool batch is to be raised in price by approximately 10, which the coating process would add. In 1989 Glass wool actually tried to introduce such a new 'comfortable' coated product, but found out that there was not much scope for manoeuvre in differentiating such a bulk-good; the sales were very low. Accordingly, the motivation is missing for attaining lower amounts of fibres by coating when handling glass-wool due to lack of market possibilities or rules.

As for the authorities (Vejle county), they wanted local, concrete examples of cleaner technology and were interested in indirectly motivating the competitor Rockwool to do something for their environment. The final aim was a new type of inspection where the firm's systematic environmental adminstrative routines would ease the communication with the inspection authorities and be concentrated on process parameters instead of emission control. Thus they would be able to get current information about the environmental development at the plant. This would be an alternative to the normally very static license system that leaves no incentives for the firm to go beyond the standards in the license. The development of an environmental planning and management system would be far more dynamic and give the management constant incentives for environmental improvements.

5   PROPAGATING THE ECO-AUDIT CONCEPT IN THE REGION AND PROFILING GLASULD IN ECO-LABELLING

As a follow-up from the successful development at Glasuld, a new institution was built outside the county administration to establish an arena for voluntary participation of companies to learn from the Glasuld methods. 'Green City' was established with Peter Nissen, from the county administration, as director. He was also the leader of the eco-audit group at Glasuld. Green City was established by five municipalities and Vejle county and 50 of the most involved companies have joined. Green City established evening seminars where cleaner technology was discussed. The company members had to provide an environmental account and commit themselves to going through an eco-audit process. In each municipality it was decided to go through five companies per year. The Glasuld model of eco-auditing was used as a concept. One inspector from each of the five municipalities was trained in the methods, the local work health department joined and so did the local power plant for energy-accounting.

As a result of the audits the companies wanted to revise the conditions of environmental approval and change the parameters of self-control to operational parameters. In doing so they substituted expensive measurements and external certified consultancy with the normal operational parameters. '*But the profit perspective is not enough to engage the companies; it is a question of propagating a philosophy and focusing on the problems and possibilities. Before the priority had been to get as many goods through the plant as possible and lower the costs of salaries. No focus on types of material-flow but on flexibility in use of raw-materials with no regards to the environmental problems*' (Peter Nissen of Green City).

By building up the procedures, point of measurement and activity plans some of the companies also expected to be in the forefront in achieving the necessary conditions for getting approvals for ISO-9000 quality certifications.

In Vejle county eight companies were connected to a development group within the county for making eco-audits. All companies were invited to learn about the method used at Glasuld. Among the companies who responded were Rockwool, the most important competitor to Glasuld. They wanted to get as much public relations coverage as Glasuld regarding the environment. At Rockwool they were so eager to go ahead that they hired a specialist for the process and they skipped the detailed scanning procedures and priority-making system used by Glasuld. They wanted to win the environmental competition game, and took out a summons against Glasuld at the EU-Court for unfair environmental competition in their commercials! Rockwool have already lowered their emissions of phenolia considerably.

The green market regulation of eco-labelling have had some important impacts on Glasuld's efforts to diffuse their environmental profile in Europe. As a result of the EU directive on Eco-labelling for the whole of the EU-market, Denmark was chosen as the site for the expert group to discuss criteria behind eco-labelling thermic insulation materials for houses. The process has shown the great difficulties in achieving consensus on which parameters should be considered and how to rank the ecological effects internally. What is the worst, the contribution to the green-house effect or the depletion of the ozone-layer? The solution has been to eliminate none of the products from the start and to make a scoring system.

The conflicts among the many industrial interests in the criteria-setting for eco-labelling insulation, especially plastics versus the mineral-wool industry, were intensified when an EU-proposal for a new classification directive on products was received. The classification proposal seemed to rank mineral fibres as causing lung-cancer[2], even though discussions among scientists on this question were intense. If this ranking in terms of cancer comes into effect, then the Eco-labelling directive states that the product cannot be eco-labelled! Accordingly lobbyists from the European minerals industry have urged the speeding-up of the process before the results from the classification group were known. But all member states in the working group and the Commission have agreed to wait for the results before the labelling process can end. The minerals industry has in turn tried to argue for a differentiation between fibres and mineral-wool slabs, as the latter in themselves do not cause cancer; it is only when the slabs are handled that the fibres actually cause damage! This somewhat philosophical argument does not seem to be acknowledged by the other partners in the eco-labelling expert group or the advisory group.

Glasuld, Denmark, has the opportunity to play a role on the general policy level in the labeling process as they have their own man to represent the whole of the Danish trade's interests at the table. Glasuld is interested in getting the Danish production standards to be accepted as the European criteria for achieving an eco-label; these would be guidelines for good environmental policy for producing glass-wool at all the companies of Saint Gobain and they would know the way to adjust all the plants of the corporation. Secondly, of course, Glasuld Vamdrup is interested in getting a label on their products, as this will be a great commercial advantage quite apart from the price factor.

Recently Glasuld's export to the German market has been threatened by new rules on working conditions. In Germany, a working group under the Minister of Labour is preparing a proposal to a statutory order on handling hazardous substances and materials. In this proposal mineral wool is categorized as a cancerogenic substance that must be avoided and substituted. In Germany fire-security rules do not hinder the use of polyurethan foam ('flamingo') for insulation in houses, and this is more common there than in Denmark. German chemical corporations such as BASF and Bayer have strong interests in expanding the polyurethan insulation

---

[2] In accordance with WHO's and IAC's classification. As mentioned before the Danish Working Health Agency put mineral fibres on a list of cancerogeneus materials in 1992.

market and the head of the working group is a member of Bayer's executive committee. If the German rules do not favour Glasuld's products then the need for an Eco-label is even greater.

6   FACTORS DETERMINING SUCCESS

We will differentiate the major pre-conditions for the success without giving priority to any single factor as they are interrelated: *Basic environmental agenda setting*: learning from the neglect of environmental problems and authorities; *Pressure from environmental problems*; *The triggers*: green corporate discourses and the local authorities; *The choice of perspective and methodology*: eco-auditing, experiments, priorities and workers participation; *The continuous push and pulls*: market conditions, costs, learning about the possible environmental technological innovations and a qualified leadership.

Basic environmental agenda setting   Learning from neglecting environmental problems and from contacts with authorities. The experiences learned from Kastrup were first of all that the glass-wool production indeed caused serious environmental problems; this was learned step-by-step from the continuous claims of nuisance from neighbours, police reports and demands from a variety of local authorities. Even though not oppressive, the continuous demands from the local authorities did influence the company's environmental self-observation as they had to start measuring processes, contact experts, install filters and assess the harm caused to the nature.

Up till the relocation of the production in 1982, the company tried to rely on the public goodwill by small technical diffusion and dilution adjustments or by threats to the authorities of moving jobs. Even though the company succeeded in expanding the production by this strategy, they came to a crucial turning point when the authorities categorically prevented any further expansion of the production. By then the company learned, that in the long run it might be highly costly to ignore the environmental problems and base expansion on investments in old technology, or only invest in dillution and diffusion measures. Thus these re-active technical measures and the gentle terms from the authorities caused the growing noise and emission nuisance during the expansion which eventually hit the ceiling of what was legally possible. When the authorities finally enforced environmental rules strictly the company's options for full capacity utilization and expansion were hindered.

The pro-active environmental rule-compliance in Vamdrup since 1981 was well established and confirmed within the company and preconditioned the recent strategic turn: '*We took the choice because we wanted to be in front with respect to the law so that we were able to direct our development and our investments*' (Jeanette Korsgaard, environmental manager). The environmental manager describes the strategic turn, in leaving the position, from were new costly environmental demands unavoidably follows after some time. This was when the company finally adjusted to the investments necessary to meet the former environmental demands! They reasoned that environmental demands were here to stay and that environmental concern had to be integrated into all economic and technical decision-making. To consider environmental requirements in the choice of new technology can become a parameter, which gives a technological lead. By this step they avoided separate environmental purification investments. Secondly, they earned a good report from the authorities which left them in favorable position for future demands.

Pressure from environmental problems This is not an objective category as the identification of a problem is always a social phenomenon of discursive character. But in hindsight we may find that material conditions were the reasons for the problem. The locating of the company close to a housing sector is thus a permanent source of neighbor inconvenience and thereby strengthens environmental require- ments for outlet of the acutely toxic substances. Had the company been located away from housing areas, the end to expanding production capacity demanded by the authorities, would not have taken place. This would have caused the company to miss a vital incentive to comply with environment demands and thereby recently to make significant progress by environmental management.

It was a decisive factor in the pressure towards technological change that both the fibres and bakkelite binders were considered to be a problem to the working environment as well as a pollution problem and an indoor-climate problem. Glasuld directors and managers not only learned from the local authorities. During the '70s and '80s they learned from a broad range of complaints and demands from the employees, unions, experts and authorities, that they also had to consider the environmental problems like free formaldehyde and fibres handling and life-time use of their products.

The triggers Green corporate discourses and the local authorities. Selling bulk- products and being in a multinational corporation with central decision-making on strategy does not leave much scope for manoeuvre for a single company like Glasuld Vamdrup. On the other hand, this means that when Saint Gobain in 1991 gave orders to apply with a new environmental program, the options for environ- mentalists were opened up and this seemed to be an important trigger. Secondly, the environmental authorities from Vejle county played a very important role in starting a rather comprehensive and deep investigation into the environment at the plant. Having the county's backup also meant access to a consulting service within a field in which the company had no experience. This in turn depended on an administration that was able to take strategic decisions and go into undiscovered fields. In Vejle this was ensured by group discussions among diverse qualification profiles, a capacity for learning and the fact that nearly all companies already complied with the rules and therefore didn't demand vast time expenditure on environmental problems. It was also a precondition that Glasuld already was far below environmental standards and did not have anything to hide.

The company and Vejle county were given 1 million DKK by The National Environmental Protection Agency which gave extra room for manoeuvre for deep investigation and use of consulting expertise. The environmental manager stated that they would not have been able to go as far as they did if they had to pay for consult- ancy help themselves. NEPA's latest efforts have been to investigate environmental management and support front-runners for the development of new methods. This is the reason for grants to Glasuld and Vejle County.

The choice of perspective and methodology   Eco-auditing, experiments, priorities and workers participation. What seems to have influenced the success is the decision to perform a comprehensive investigation on all the environmental aspects of the company and go to source in the process and in the products. All divisions had to register environmental contacts and give necessary information on all aspects from purchaser of raw materials to waste deliveries. Hereafter it was possible to assess the diverse options for solving the problems. This was initiated by all partners but especially by COWI who had experience from literature and other companies. The willingness, resources and qualifications to do the brain-storming and later on the experiments was quite important for assessing options and barriers for solutions. The procedure of accounting all environmental implications had the effect that many workers became aware of what the daily, technically obvious spills and resource-use actually meant in environmental terms.

Secondly, it was important that there was a facility for making prioritizing deeper investigations, even though the serious environmental considerations were lacking in the scoring system, and that the authorities were able to influence this step. This in turn was ensured by the openness of both Glasuld and Saint Gobain to let officials into close confidence about technical and economic aspects of the firm.

The company stated that it was of great importance to the success of the process that so many workers were eager for environmental registration and for proposing solutions as they had the tacit knowledge the leaders did not have, and since the process demanded a lot of novel interventionism to be carried out successfully. In accordance, the local authorities stated that they would not have been able to point out all the ideas and problems since they did not know all the technical details. Even though the management didn't want to build upon economic motives there was a carrot for the workers: a grant for innovative environmental proposals that were accepted and a type of salary that gives benefits for minimizing waste and keeping in close touch with quality standards, e.g. regarding the amount of binders.

It would not have been possible to find the correct working procedures without the involvement and interest of the employees. Also responsibility and care are necessary conditions for the system to be used at all. Two important conditions have encouraged the cooperation of the employees and secured the success of the project. Firstly, that there have been marked and visible improvements to the working environment. Secondly, the employees received training and were involved in the organization of the work and the development of methods.

The continuous push and pulls   Costs and market conditions, learning about the possible environmental technological innovations and qualified leadership.

Costs and market conditions   It has benefitted motivation for innovation that fees from waste deposits and the threats from fees on CO_2 have lowered the assessed pay-back period of the investments. '*The environmental considerations have been integrated to already-existing economic calculations, whereby suddenly a lot of environmental parameters enter into the decision making and amplify some of the options for economic or strategic moves*' (Jeanette Korsgaard).

Many of the efforts have resulted in profits and have proved a pay-back period within two years. An important reason for the success is thus that there existed a lot of potential cost-reductions undreamt of from a normal management point of view. The priority for the most cost-effective steps of course speeded up the process but might also have opened the pathway for less profitable future innovations.

It was the German Dual System that was important for pushing Saint Gobain to consider re-use of worn-out glass-wool from the construction sector and to take back plastics packaging. This was also important to Glasuld, Vamdrup, as Germany is the most important export market for the company. It is for the same reason that Glasuld will apply for the forthcoming eco-label, if this is not made impossible due to classification of fibres as cancerogenic. It was a precondition for taking in re-used glass that there existed a market for this in Sweden and Germany, which in turn was made possible by subsidies from the state.

Learning about the possible environmental technological options Through the eco-auditing process, the management leaders and technicians learned from the process and from workers, that a new range of technical options were at hand when analyzing the production from an environmental point of view. What happened was that they stopped looking at the production from a productivity, cost-structure or technical-functionality perspective in each department. Instead they followed a material-flow perspective and came to look upon the production as technologically interconnected in the case of spills and raw-material consumptions. From very detailed observations of machinery and handling and from adjusting recipes and process parameters, many environmental results were achieved by this way of understanding the flow.

Qualified leadership The former management director, installed by Saint Gobain after the take-over in 1988, provided the strategic positioning of the company in 1990–92 by encouraging the staff, managers and directors to go into 'greening'; environmental management was the future for survival and ability to document all environmental relations was the goal. But it was Jørgen Haals, the former manager of quality, that played the very important part for Glasuld's positive environmental attitude and results at the change of course the company took in 1982 respectively 1990. Over the years Haals acquired a very detailed knowledge of production technology, work organization, process and product chemistry and he personalized the experiences during the '70s and '80s.

In the new position as environmental manager within the organization Jeanette Korsgaard has shown much enthusiasm and skill in implementing the ideas developed in the eco-auditing project, securing back-up from the technical director, profiling the environmental turn to the public and initiating dialogues with a variety of partners and groups. As a security leader she has gained very detailed knowledge of the relationship between technology and working environment, which also played a role in the eco-project, and of the adjustments of innovations to meet workers' health requirements. She was in a position that enabled her to relate shop-floor problems and observations of environmental aspects to the decisions among the other managers. It is important for the quality of the environment and for its relationship to the economy, that she, as a manager of the environment, was positioned in line-function with the technical director and had the same status as the managers of quality, purchase and production whom she could oppose for neglecting environmental concerns. Simularely it is of importance that she had the freedom to consider the environmental aspects and suggest innovations, without having to endure immediate economic censorship.

At the county level Peter Nissen also seemed to play a crucial role in getting the project on the track, for directing the course of events and for providing the perspectives for the county for diffusion of the system and for forthcoming control methods. Peter Nissen and Jeannette Korsgaard were able to discuss and argue very openly about problems and conflicting interests. The diffusion of the environmental auditing system is a success in the sense that several companies in the region have followed Glasuld and the environmental assistants at Vejle county have applied the method for use in dialogue with other companies. To this end it was crucial that Jeannette Korsgaard enthusiastically travelled around to all the interested companies and initiated energy-savings and other housekeeping initiatives. It was of great importance from a responsibility point of view and for putting drive into the new environmental management system that one of the well-known local companies was the front-runner and was able to pass the experience on from person to person. The same kind of reasons lay behind the success of broadcasting the environmental method among the Green City Network, where Peter Nissen held the chairman's post. All of Glasuld's experiences were immediately transferred within the network of companies and by informal discussion and by mutual learning the companies were motivated into become greener.

REFERENCES

Begrænsning af luftforurening fra virksomheder. (1974) Guidelines from National Agency for the Protection of the Environment, no. 7.

Begrænsning af luftforurening fra virksomheder. (1990) Guidelines from National Agency for the Protection of the Environment, no. 6.

Grænseværdier for stoffer og materialer. (1985, 1988 and 1992) Arbejdstilsynets anvisninger.

Vejledning om arbejde med stoffer og materialer. April 1989, Arbejdstilsynets meddelelse nr.3.02.5.

Bekendtgørelse om arbejde med montering og nedrivning af isoleringsmaterialer der indeholder syntetiske mineraluldsfibre. 9.6.1988 Arbejdstilsynets bekendtgørelse nr.344.

Saint Gobain in the Nordic Countries. (1993) information papers from S.G.

'Ingeniøren' (1993, 1994) pp. 18–19, no. 42, October 1993 and pp. 9, 16 in no. 11 March 1994.

Umgang mit Mineralwolle-Dämmstoffen. (1993) Handlungsanleitung, Fachvereinigung Mineralfaserindustrie a.o., October 1993.

Redegørelse om Superfos-koncernens miljøforhold. (1984) Miljøstyrelsen.

Virksomhedens Miljøredegørelse. (1993) Green City Network, Vejle.

Manual for miljøregnskab. (1994) Vejle Amt og COWIconsult.

Jydske Vestkysten. 1.10.91.

Miljøledelse. November 1993, pp. 8–9, COWIconsult.

Fagbladet. (1993) pp. 6–7, no. 38.

Danmarks Amtsråd. (1993) pp. 9–11, no. 5.

Metal. (1994) pp. 16–17, no. 3.

Miljøregnskab: *Miljøplan og egenkontrol.* (1993) COWIconsult.

Miljøregnskab: *Spildanalyser.* (1992) COWIconsult.

Miljøregnskab: *Miljørelationer og -prioritering.* (1992) COWIconsult.

Calabrese, Edvard J. (1991) *Air toxics and risk assessment*, Lewis Publishers USA.
Journal papers on Glasuld 1964–82 (control, letters, measurings, complaints, police-reports etc.) at
 Taarnby council, environmental department, not published.

Interviews

Peter Nissen, Head of Green City Network
Gitte Visby, Vejle County Environmental Department
Jeanette Korsgaard, Glasuld Environmental Manager
Jørgen Haals, former Head Manager at Glasuld

Telephone Interviews

Helle Petersen. (22.10.93, 12.4.94 and 17.5.94) Danish Dept. of European ECO-labeling on insulation,
 National Environmental Protection Agency.
Niels Christian Larsen. (17.5.94) Management Director at Glasuld Vamdrup.
Anette Christiansen and Per Bjarnvig. (17.5.94) Direktoratet for Arbejdstilsynet.
Michael Petersen. (17.5.94) Consulting Engineer for Glasuld Vamdrup *and* in the Working Group
 on Eco-labeling: the Danish Trade Organization for Mineralwool Industry.
Claus Werner Nielsen. (18.5.94) COWI consult.

8. ECO-BALANCE SHEET AND IMPLEMENTATION OF ECO-CONTROLLING IN THE KUNERT GROUP

Uta Kirschten

Translated by Mary Höcker

1 THE KUNERT GROUP: A SUCCESS STORY?

The KUNERT Group is a large German hosiery enterprise and is among the market leaders in Europe. The range of products consists largely of ladies' tights, ladies' stockings and men's and children's socks. Contrary to the original conception of the project, the present study looks at a success story in the sphere of environmental management. The KUNERT success story consists of two aspects. Firstly the fact that for five years now an annual company-wide eco-balance sheet[1] has been drawn up, forming the basis for implementing eco-controlling. The success lies in the transition from an uncoordinated approach to ecological requirements on a case-to-case basis to a systematic and on-going collection and evaluation of data. On the basis of this, weaknesses can be analysed and processed, resulting in the formulation of environmentally-oriented objectives and the implementing of measures ensuing from them. This resulted in a host of individual measures, each of which constitutes an improvement, but nevertheless does not justify the term ecological success. However, the sum total of the diverse activities couched in eco-controlling is indeed a considerable success and a major step in the direction of environmentally-oriented management. Secondly, the drawing up of an eco-balance sheet generated a process of greater awareness and reflection on environmentally-relevant implications of corporate action on the part of many members of staff and the management of the KUNERT Group. Confrontation with the status quo, the study of problem areas and the successful search for solutions generate attentiveness, heighten awareness and lead to learning processes. This heightened awareness and the altered thought-patterns on the part of the staff are the basic prerequisite for all further environmentally-oriented activities of the Group, which find visible expression in the drawing up of the eco-balance sheet and the implementation of eco-controlling.

In terms of methodology, the present case study is based not only on a comprehensive analysis of documentation but on a total of 30 in-depth interviews conducted in the period from June 1993 to September 1994. Of these, 15 meetings were

---

[1] Data collection in support of the results presented here was completed in mid-1994. Consequently, the recent trends in the 1994 eco-report (published in October 1994) could not be taken fully into account here. For the complete version of the case study the reader is referred to Kirschten 1995.

held with players within KUNERT with a view to reconstructing events, attention being paid to the importance of individual players and influencing factors. A further 15 meetings were held with external players to verify individual areas from the experts' point of view and to illuminate matters from without the firm.[2]

2   CHARACTERISATION OF THE KUNERT GROUP IN BRANCH TERMS

In branch terms the KUNERT Group belongs to the textile industry in the widest sense. In 1992 the latter produced textiles and clothing valued at some DM 24 billion in the FRG, equalling some 1% of the gross domestic product (Enquete-Kommission 1993: 167). The German textile industry which is traditionally made up of small to medium-sized firms in 1992 comprised 1,100 firms (with more than 20 employees) which employed some 190,000 workers and had a turnover of just on DM 40 billion (Gesamttextil 1993: 6 of the tables section). At the same time, more than 90% of textile firms had less than 500 employees (Gesamttextil 1993: 7 of the tables section).

With a Group turnover of DM 639 million in 1993, some 5,100 employees, and a total of 15 works, the KUNERT Group is one of the few large hosiery enterprises in Europe. The range of products consists largely of ladies' tights, ladies' stockings and men's and children's socks. The Group appears on the market with four brand names HUDSON, KUNERT, BURLINGTON and SILKONA, which compete against each other (KUNERT 1993a: 1ff.). The KUNERT Group's development is currently overshadowed by the recession. Thus gross turnover fell by 6.5% from DM 679 million in 1992 to DM 639 million in 1993. Against a deficit of just on DM 17 million in 1992, a consolidated net income for the year of DM 7.6 million after tax was recorded for 1993. Investment in 1993 totalled DM 28.5 million (KUNERT 1993a: 1). This was achieved by consistently restructuring the range of products to improve yields and a cost-cutting programme entailing works mergers (KUNERT 1993a: 6) launched in 1992.

3   RECONSTRUCTION OF THE SUCCESS STORY

The 'KUNERT' success story covers a period of some 20 years. The events are divided up into different stages in order to identify important stages of development.

3.1   Traditional Approach to Nature (1971–1984)

In individual sectors, measures, which would nowadays certainly be described as environmentally-oriented, were taken 20 years ago. Since the beginning of the '70s,

[2] At this point, I should like to commend and thank the KUNERT members of staff, in particular Mr Wucherer, as well as the non-KUNERT discussion partners for their exceptional willingness to cooperate and be of assistance.

for example, KUNERT has differentiated between different types of waste, e.g. paper, cardboard, textile waste, etc., has taken various measures to save energy (installation of heat recovery plants, the fitting of dyeing machines with economy settings with a view to multiple use of the waste water from the dyeing process) has gradually converted heating equipment from heavy oil (sulphur content of some 3%) to the more expensive light oil or oil containing less sulphur. Until the mid-'80s the enterprise's central motive in conducting all these individual measures was to achieve savings and ensure supplies.

3.2   Stage I: 1984–1989. Perception of Potential Risks

Prompted by the spectacular accidents in the chemical industry since the end of the '70s,[3] KUNERT developed a heightened awareness of risks and became more safety conscious in the field of logistics, storage and production in the course of the '80s. Suppliers, for example, were increasingly asked for information about the content and possible hazards of the raw and auxiliary materials and fuels stored. At the same time, some 'pollution problems' of their own caused by heating oil leaking into the subsoil resulted not only in a first 'environmental problem' but also in a heightened awareness of the risks.

   This increasing awareness of the risks involved and the resultant need for information about one's own potential risks led to the management of the firm of KUNERT appointing its first Environmental Commissioner in 1984. The focus of the Environmental Commissioner's work was on general tasks in the sphere of in-house technology and the identification of savings potential (energy-saving measures) rather than in the field of environmental protection. He also constituted a kind of interface and communication point between members of staff and management. The voluntary installation of an Environmental Commissioner in 1984 was something new and unusual, not only for KUNERT but also no doubt for the textile industry at that time. Within KUNERT as well there was doubt as to whether the appointment was to be taken seriously.

   In 1985 KUNERT and HUDSON were the first enterprise in the hosiery branch to declare environmental protection a corporate objective. In addition, KUNERT implemented various individual measures[4] in the course of time which were devoted to environmental protection partly or in full, but they were *ad hoc* and uncoordinated and merely processed along with other matters.

3.3   Stage II: 1989–1991. The Right Idea at the Right Time

The *idea to draw up an eco-balance sheet at KUNERT* came about at a convention in Nuremberg in 1989. This was the occasion of a chance meeting between the then

---

[3] For example the accident at the Icmesa works of the Geneva cosmetics manufacturer Givaudan (a subsidiary of the Swiss chemical firm Hoffmann La Roche) at Seveso (Italy) on 10 July 1976 in which the toxic carcinogen TCDD was released into the atmosphere (Gladwin/Winter 1980: 461–466).

[4] The conversion of dyeing machinery, for example, led to savings of energy, water, dye and chemicals (1986); and waste separation throughout the Group (1988).

assistant to the Board Chairman of the KUNERT AG, Mr B, and Mr X, who later became its external advisor. Both men knew each other from the University of Augsburg. During an informal conversation one evening, the idea of drawing up a trial eco-balance sheet at KUNERT with Mr X as scientific advisor arose. The Board assistant put this idea to the KUNERT management, which was welcomed by the Board Chairman in particular and was accepted as the 'Pilot Project Eco-Controlling' ('für uns', No. 60, 1990: 20).

Putting the idea into practice: The drawing up of the first eco-balance sheet To put this pilot project into practice a so-called 'ecology working group' (hereafter abbreviated to EWG) was set up in January 1990. Its objective was to devise a method for drawing up an eco-balance sheet at KUNERT and for taking ecological stock throughout the Group. It should be stressed here that, from the very beginning, it was decided to take ecological stock *throughout the Group*. It was primarily the external advisor and his staff who gave scientific advice and support (e.g. theory and practice of drawing up a balance sheet, collation and structuring of data, etc.). The EWG was composed of some 17 'junior' members of staff ('für uns', No. 60, 1990: 21) from all important functional sectors, such as purchasing, sales, in-house technology, environmental protection commissioner, manufacturing, etc. from the various firms in the Group. The EWG was headed jointly by the assistant to the Board Chairman and the external advisor as scientific expert. From the very beginning the members of the EWG were infused with a kind of pioneering spirit which was further enhanced by the support of the Board Chairman.

The work of the EWG was a painstaking and largely 'manual' collation and ordering of data. By January 1991, the eco-balance sheet had been presented as an eco-report. The data collated in it was structured and discussed under account heads. Problem areas were uncovered, and potential solutions and future measures were formulated as consequences and goals. A trusted member of the external advisor's staff played a major part in compiling and drawing up this first eco-report. After completion of the eco-report and a controversial discussion within KUNERT about the sensitivity of the data, the management decided to publish the eco-report in May 1991 at a press conference in Munich. The much greater interest in, and strong response of press and media to, publication of the first eco-report compared with the Annual Report for the year was a great surprise to all concerned. Publication of the eco-balance sheet made the KUNERT AG the first German enterprise to have achieved such a comprehensive operational balance sheet in cooperation with the external advisor.

3.4 Stage III: 1991–1993. Institutionalisation of the Eco-balance Sheet Operation

Not only publication of an eco-balance sheet, but also the content of the eco-report in particular which signalled a certain continuity in the environmentally-oriented activities and preparation of the eco-balance sheet, aroused public interest in the KUNERT AG. However, this also implied environmentally-oriented expectations of KUNERT. Ultimately, therefore, *publication* of the eco-balance sheet resulted in the environmental activities of the firm developing an unexpected dynamism of their own which was not intended. In the light of this situation, the various in-house expectations (continuation of the EWG yes or no?) and external ones (future environmental activities?) resulted in the management deciding to continue the eco-balance sheet and begin setting up an eco-controlling system.

Further developments at KUNERT up to 1993 were as follows. After completion of the first eco-balance sheet *the large EWG grew from the initial 'environmental pioneering group' into an institutionalised forum for processing topics of environmental relevance.* Given its growing number of participants, productive work became increasingly difficult. In addition to the large EWG, therefore, smaller result-oriented working groups for special questions and problem-related project teams were formed. An important starting point for the subsequent work at KUNERT *was the organisational extention of responsibilities for environmental tasks and their dissemination internally.* Having recognised that it was essential to communicate the environmental goals and successes internally and involve members of staff more closely, regular in-house information about the goals, activities and results of the EWG was distributed, seminars were held on the topic of environmental protection and the suggestion system was expanded. In order to be able to compare and evaluate data regularly, it was essential *to refine the eco-balance sheet concept* in terms of methodology. Taking consistent advantage of existing organisational structures, it was decided to formalise data gathering in the organogram and lay down responsibilities for certain tasks. The *development of an EDP programme (CAEC Computer-Aided Environmental Controlling) to collate and process data* in cooperation with the Institute for Computer Science at the University of Augsburg served to simplify and support the drawing up of the eco-balance sheet and was deployed for the first time for the 1992 eco-balance sheet.

By analogy with the Annual Report, an eco-balance sheet is drawn up annually. The eco-balance sheet is a Group balance sheet made up of 15 works balance sheets (including foreign works), which are compressed into three sub-balance sheets which together constitute the Group balance sheet. The special feature of the eco-balance sheet is the on-going updating of data and developments over five years so far. Although it has the makings of an eco-balance sheet throughout the Group, the environmental data are only comparable up to a point as further affiliates and works can only be integrated into the data-gathering in the course of implementation.

4   DEPICTION OF SUCCESSFUL MEASURES BY WAY OF EXAMPLE

Compared with other branches, the textile industry has hardly reduced the flow of emissions in the last 20 years at all. This is attributable to its poor economic situation (stiff competition, low turnover yields) and to the branch's lack of environmental awareness (Schönberger/Kaps 1994: 2). In terms of the spectrum of environmental pollution, waste water predominates at KUNERT. Waste-water pollution results from the rinsing oils (fibre production), the use of dyes and textile additives (substances with mainly organic materials) and basic textile chemicals (neutral salts, acids, solutions, oxidising and reducing agents) which in some cases pollute the waste water with substances that are difficult to break down biologically. In addition to waste-water pollution, the high percentage of product packaging (up to 50%) is noticeable.

Measures by KUNERT presented here by way of example show an incipient environmental orientation to date in dealing with environmental problems.

4.1   Impetus to Develop a Chromium-free Black Colouring Agent

When collecting data for the first eco-balance sheet, the substances in waste water were measured voluntarily throughout the Group in 1989 and in some cases concentrations of chromium in excess of the statutory limits were found in the waste water. After the cause of the high concentration of chromium in the waste water had been identified, namely a certain black chromous colouring agent, various solutions were open to KUNERT. A possible solution might have been to discontinue using this black colouring agent. However, as some 60% of all dyeing was done with this colouring agent in 1990, this possibility was not considered further. In continuing to use this black colouring agent, KUNERT would not in the long term have been able to avoid taking appropriate waste-water purification measures. In some works, this might conceivably have entailed the construction of waste-water purification plants and thus big investments. The third possibility was to ask the manufacturers of the dye for a black colouring agent that was of the same quality but not chromous. This was the course chosen by KUNERT.

The Central Purchasing Department therefore wrote to various well-known suppliers of dyes and inquired about a chromium-free black colouring agent of equal quality. A prominent manufacturer of dyes offered KUNERT a colouring agent which, while not excellent, nevertheless produced much better dyeing results (handling, colour-fastness, etc.) than all other alternatives offered. The KUNERT dye-works joined with the manufacturer in improving this colouring agent until the dye-works were finally satisfied with its quality where colour-fastness, handling and colour intensity were concerned. Handling difficulties apart, the new colouring agent was considerably more expensive than the old chromous colouring agent. The price negotiations conducted by KUNERT's Central Purchasing Department resulted in success for KUNERT, not least because of a colouring agent unexpectedly offered by the competition allegedly having colouring results of equal quality. This largely solved the problem of the high concentrations of chromium in the waste water and it was possible to purchase the newly developed colouring agent at a similar price to that of the old colouring agent.

Irrespective of this successful new development, however, two problem areas remained at that time. The new black colouring agent developed jointly admittedly no longer contained chromium but it still contained a heavy metal, namely cobalt. Thus, the problem of high chromium concentrations in waste water had shifted to that of considerable concentrations of cobalt in waste water, as long as black remains a fashion colour (in 1992 some 80% of all dyes). However, this was not and is still not a problem for KUNERT because German water regulations did not (1994) list cobalt as a parameter with ceilings accordingly.

The fashion for black and the trend in fashion toward mixed yarns and the use of micro-fibres in the hosiery industry presented a further problem area. For reasons of colour quality and technology, for example, certain mixtures of yarn and new fibres (e.g. micro-fibres) could not be dyed with the newly-developed chromium-free black colouring agents. Instead they had to be treated with chromous colouring agents in order to be accepted by the customer. Appropriate tests with chromium-free black colouring agents, which overall do not produce such deep black dyes, soon had to be withdrawn again because of insufficient acceptance by the customer. This in turn increased the proportion of products dyed with chromous colouring agents and thus also the concentrations of chromium in the waste water. The aforementioned problem area in particular was also unsatisfactory for KUNERT. Since 1992 therefore, work was done on developing black dyes for micro-fibres and mixtures of yarns that were free of chromium and heavy metals in general. In 1993 the corresponding test series on dyes were successfully completed, and as a result it was planned to introduce the new dyes into the production process step by step in 1994. (Ökobericht 1994: 7, 25).

4.2 Development of Nature Collections

The Group subsidiary, HUDSON, developed a collection named 'HAUTE NATURE' for the autumn/winter collection 1993 comprising socks, tights and leg warmers made of undyed fibres and those dyed with plant pigments. The fibres used were cotton, pure wool, alpaca and silk. They were dyed with natural pigments such as, for example, indigo, betel, cochineal or gallnut (see HUDSON-Prospekt 1993). The KUNERT AG developed a sock collection by the name of '100% Natural'. In 1993 this collection comprised products made of 100% pure wool with no pigment of any kind being applied. The only sock colours were the natural shades of wool from natural white to black. In the second 'generation of socks' patterned products were also put on sale. (Ökobericht 1993: 7).

Both collections were well received, but the sad thing is that in 1994 they accounted for less than 1% of KUNERT's overall range of products.

4.3 Incentive to Develop a Wood-fibre Material

Due to the initiative of a member of staff of KUNERT's Product Development Department, cooperation came about between KUNERT, the Technical University of Dresden and a firm of Saxon consulting engineers. They were successful in jointly developing a wood-fibre material made of 50%–60% hosiery waste, which is readily malleable, light-weight and break-resistant. The particular feature of this material is that it is manufactured without the use of binding agents. Since 1994, appropriate ways of deploying this new material (furniture industry, caravan interiors) are being sought. The striking thing here is the innovative attempt to recycle hosiery waste. Since 1994 it has not been decided how to organise and set up appropriate collection points and procedures for old hosiery products.

4.4 Selected Quantitative and Financial Successes

Here some quantitative successes and the resulting cost savings are presented by way of example.

Throughout the Group, *total energy consumed* in 1992 fell by more than 15%, whilst production overall sank by 26%. These unequal percentage declines are attributable to so-called fixed amounts of energy consumption and increased fuel consumption due to production increasingly being shifted abroad. In 1992 it was possible to reduce total oil consumption by 7% to 45.8% of total energy consumption. Overall the *conversion of heating systems from heating oil to natural gas led* to a reduction of the pollutants emitted. The reduction was in the region of just on 100% for sulphur dioxide, 61% for nitrogen oxide and 27% for carbon dioxide. For the Immenstadt works alone this has meant savings of DM 210,000 since 1991. Admittedly the environmental pollution caused by methane gas must be borne in mind.

In terms of quantity (in kg), *the use of cover folios, cardboard inserts and semi-tubular foils* has fallen by some 30% to 40% disproportionately to the fall in production and is the result of various measures in 1990 and 1991 (Ökobericht 1993: 25). *Thinner cover folios and foils in packaging* since 1990 resulted in a fall in paper consumption of 176 tons per annum and a drop in foil consumption of 140 tons per annum. Economically, these reductions in consumption result in savings of DM 400,000 year for cover folios and savings of DM 500,000 per year due to the use of less foil. The *replacement of price labels by direct printing onto the packaging* resulted in savings of 100 million items per year. At the same time, by reducing the variety of the material by abandoning paper and adhesives it became possible to recycle the polypropylene packing material because of its purity. This produces annual savings of some DM 450,000. During the period from 1990 to 1992, the *average proportion of packing materials in the products sold* was reduced by 5% from 37% to 32%. In *transport packaging* the conversion to folding cartons brought a saving of almost 20% in terms of materials and weight. The price per folding carton is also some 50% lower than that of the kind used hitherto. Overall this produced savings of some DM 317,000.

In 1992 *water consumption* throughout the Group fell by 21% compared with 1991, almost in proportion to the drop in production. Drinking water consumption dropped by 50,000 m per year or 25% in 1992. In the Immenstadt works, leaks in the drinking-water pipes were discovered by a comparison of input with output and repaired. In addition, two separate water networks were installed for drinking and industrial water. This produced savings of some DM 200,000 per annum for the Immenstadt works. In 1992 the conversion to *silicone-free softening agents*, which has largely been completed, reduced the pollution of waste water with silicones which are difficult to break down, and led to savings of DM 40,800 per annum. By successfully *developing chromium-free colouring agents* in cooperation with a large manufacturer of dyes in 1992, KUNERT was able to avoid investing in a waste-water purification plant. In addition to reducing the pollution of waste water with chromium, the cost of a waste-water purification plant running at DM 800,000 was saved.

5   ANALYSIS OF THE RELEVANT PLAYERS, THE INFLUENCING FACTORS AND THE INTERPLAY OF MAJOR DETERMINANTS

Based on an analysis of the players reflecting their problems, interests, power constellations and perceptions, the influencing factors are highlighted and the interplay of the major determinants and conflicting positions commented upon.

5.1   Players

Mr X: External advisor   During an informal discussion, the external advisor offered the assistant to the Board Chairman of KUNERT a specific idea for processing environmental questions systematically. Given that the enterprise was basically willing to engage more fully in environmental protection activities, the drawing-up and implementation of an eco-balance sheet for KUNERT was a welcome idea for giving more weight to environmental protection in the enterprise and a more systematic approach to it. Consequently, the right idea (eco-balance sheet) was suggested to the right man (assistant to the Board receptive to environmental protection) at the right time (there was already a willingness to do so) at the right place (a social occasion). Mr X deserves credit for having suggested this specific idea. In suggesting the drawing-up of an eco-balance sheet, the external advisor also put forward the scientific/theoretical approach to eco-controlling at KUNERT. In addition to specifying the theoretical idea of an eco-balance sheet at KUNERT, the cooperation between the advisor, his staff and KUNERT also entailed implementing organisational structure changes and assistance in putting the idea into practice at KUNERT. The external advisor received a great deal of support from his two members of staff with its practical implementation.

Mr B: Initiator[5]   Mr B initiated the drawing up of the first eco-balance sheet as part of the 'Pilot Project Eco-Controlling' at KUNERT. As personal assistant to the Board Chairman, he introduced the external advisor's idea to the firm by suggesting it to the Board Chairman. The personal assistant's willingness to take up the idea of an eco-balance sheet can be explained by his open-minded and environmentally-oriented approach at the time.[6] While Mr B in initiating the 'Pilot Project Eco-Controlling' performed a pioneering function, as his career progressed and he became Head of Sales, he increasingly 'applied the brakes to ecology'. Mr B does not appear to have succeeded in making the Sales Department more receptive to ecology, if that was ever even attempted. Economic facts also seemed to have been a stumbling block for him, for example the respective sales figures for environmentally-oriented products and the worsening economic situation.

Mr M: Power promoter   The Board Chairman of KUNERT is the most important power promoter. After his attention had been drawn to the idea of an eco-balance sheet by his assistant and it had later become his conviction, he advocated implementation of this pilot project and gave implementation of the idea full support from the top. By advocating ecological measures within the enterprise he at the same time became an environmental model with which members of staff could identify. This is emphasised by the fact that he declared ecology to be a corporate objective on the motto of 'ecology is a policy matter'. The ecological objectives were made binding upon the managers of the enterprises in the Group and the appropriate

---

[5] These observations are based on reports of other players and an analysis of the documentation as, despite intensive efforts, Mr B was not willing to give an interview. This is all the more regrettable since he plays a key role in this 'success story' and as a primary source on important events (in which he himself played a major part), he could have given information but was unwilling to do so.

[6] This is borne out by the fact that the subject of his thesis from the University of Augsburg published in 1988 was 'Ecologically-oriented marketing'.

sector heads and monitored. At the same time, he called upon his staff to take appropriate measures and signaled that he would give them support and a kind of backing from above. This made him the driving force and the pillar of the entire Group in an ecological sense. Despite that, profit and the survival of the enterprise remain his first priority.

Mr Z: Technical promoter, 'controller'   The Head of the Controlling Department is one of the central technical promoters. The controller had not been involved in drawing up the first eco-balance sheet. He probably had simply been forgotten. Later the controller made data collection a lot simpler because in data processing there were a great deal of data and evaluation opportunities which other departments (Purchasing, Sales) were not aware of. The controller is an important technical promoter because he has the controlling know-how which is needed to draw up an eco-balance sheet and institute eco-controlling. In addition, he is in receipt of all the main corporate data so he has a very good fund of information and performs a key ecological role. In addition, he has a position of authority in the organisational structure and hierarchy. This makes it easier for him to implement the eco-controlling system that he personally supports even in the face of other interests.

Central purchasing and Mr V: Ecological gatekeeper   The Purchasing Department at KUNERT AG has two special features: firstly, there has been a Central Purchasing Department throughout the Group since 1974. Secondly, Purchasing has also been responsible for refuse and waste disposal (e.g., the sale of materials and old machinery), as a result of which recycling aspects were already taken into account at the purchasing stage. In addition, the head of Central Purchasing is an important power promoter. As part of his field of duties he seeks cost savings, he takes account of the susceptibility of materials to recycling and, at the same time, the cost of recycling when purchasing. Within the corporation, his power position means that ultimately he decides which materials are to be purchased. Outside the firm, Central Purchasing is a strong negotiating department in as much as suppliers and their products are judged according to the environmental compatibility of their products; in the meantime, a so-called Black List of products and substances not to be procured has been developed at KUNERT. (Ökobericht 1994: 8, 58f.). The strength of Central Purchasing proves a disadvantage in so far as, hitherto, receptiveness for environmentally-oriented ideas had mainly focused on purchasing and waste recycling. At the same time, for example, the company has neglected to optimalise its actual manufacturing process, although there is a great deal of room for improvement from an environmental point of view.

Ecology Working Group (EWG): Motor and monitoring body   The EWG[7] is the body in the company that draws up the eco-balance sheet and implements eco-controlling. At the beginning, it served as a kind of 'testing ground' in which ecological questions were processed as a new field of activity with, in part, new working methods. The EWG is a forum for identifying environmental problem areas

---

[7] The Working Group meanwhile comprises 21 member of staff (Ökobericht 1994: 71).

and providing the appropriate solutions. By formulating concrete objectives, it ensures that the changes can later be monitored. It thus becomes a body for monitoring the successes achieved. This made it necessary to come up with a success that was not mere lip service but instead required active cooperation and the implementation of measures. In the course of time, a division of labour had come about between the large EWG and the small working groups: whereas the large EWG devotes itself to the more general reporting and an exchange of information as well as providing new ideas, concrete measures for achieving the goals that it has set itself are meanwhile carried out extremely effectively in the small working groups.

Mr W: Coordinator As part of a *de facto* environmental protection department, Mr W, in addition to his other tasks, is responsible for coordinating environmental activities throughout the Group and developing eco-controlling, and this entails, among other things, heading the EWG and compiling the annual eco-report. The drawing-up of an eco-balance sheet annually was implicitly result-oriented and required the KUNERT Group to present successes. Both factors were manifest in the goals and consequences explicitly formulated in the eco-reports by which the enterprise imposed obligations upon itself. Mr W's function is to take appropriate action to implement the self-imposed goals. His main forum of activity is the EWG in which weaknesses are identified and appropriate solutions proposed.

Sales: The brake In general Sales acts as a brake as its response to the environmental activities of the Group and the work of the EWG varies from very reserved to negative. The Sales Department is necessarily market- and fashion-oriented. The reason for its reserved attitude toward the enterprise's environmental activities is that both environmental protection as well as the environmental compatibility of hosiery products are (not yet) viable sales arguments for the Sales Department because they are obviously not yet viable sales arguments in retail circles and, moreover, they need explaining to the consumer. Sales thus represent the two faces of Janus. On the one hand, its hands seem tied because environmental compatibility is not yet a sales argument for the retailers. Consequently, the environmental work of the Group is not at all helpful to the Sales Department. On the other hand, it is precisely Sales which would have the opportunity to introduce environmental compatibility to the trade as a new sales argument and to implement it. This presupposes, of course, that the members of the Sales Department are convinced by the environmental activities of their Group and can not only present arguments for the advantages of certain products in sales negotiations but can also prove them in practice. Whether lethargy in adopting new arguments or the marginal nature of the improvement in products are the reasons for the reticence in the past is something that cannot be evaluated here.

5.2   Influencing Factors

A major influencing factor is first of all the *location of the group headquarters* of the KUNERT AG in Immenstadt in Oberallgäu. Thanks to its attractive landscape the Oberallgäu has been a tourist area for many decades. The fact that KUNERT is manufacturing in an idyllic holiday landscape calls for very much higher standards in one's treatment of this rural and ecologically sensitive environment. In addition, the City of Immenstadt is more interested in sustaining and expanding tourism than in extending industrial activities. The growing awareness of environmental damage (e.g., water pollution) and spectacular accidents in the chemical industry have attracted public attention and, secondly, triggered *public discussion of environmental pollution and potential risks caused by industry* which in turn resulted in a heightened sensitivity toward environmental pollution and the potential risks in its own industrial production at KUNERT. Thirdly, publication of the first eco-balance sheet served *to enhance KUNERT's image*. For the first time, the public perceived KUNERT as an environmentally-oriented enterprise. In the course of time, this image was boosted by the award of various environmental prizes as well as the broadcasting of diverse television and press reports about the firm's activities. As a result, KUNERT was under a certain obligation to uphold this image, i.e. continue presenting the eco-balance sheet and developing eco-controlling. This does not preclude reducing environmental protection activities in bad times, such as, for example, the present recession. A fourth influencing factor is *the new generation of managers* heading the KUNERT Group in the '80s. The present Board Chairman has brought about a clear shift of emphasis in two sectors: firstly, jobs have increasingly been going abroad since 1978 because of the considerable difference in costs, in particular lower wage costs abroad. Secondly, environmental protection is increasingly being taken into account. Whereas Mr Julius Kunert, the firm's founder, grew up in an industrial city in Bohemia (today the Czech Republic) and for him a flourishing industry was symbolised by smoking chimneys, *the present Board Chairman was influenced by his early years and life in Immenstadt and the landscape of Allgäu*. This is the reason for the present Board Chairman's receptiveness to environmental thinking and his greater willingness to take new demands for environmental protection into account.

5.3   Interplay of Major Determinants and Clashes of Interest

The interplay of various corporate sectors and departments is also of significance for development at KUNERT. *The interrelationship between the Product Development, Manufacturing and Sales Departments, for example, is decisive for actual innovations.* Taking new ideas generated by the Product Development Department as a starting point, the Manufacturing Department must determine whether production changes are necessary to manufacture the new products. Ultimately, however, it is for Sales to judge the marketability of potential innovations. This interplay can be traced in numerous ideas of which only a portion could be put into practice. The subsidiary HUDSON, for example, was able to assert itself in developing the autumn/winter collection 'HAUTE NATURE'. Conversely, developments in the packaging sector did not prove marketable on image grounds; for example, wrapping made of 100% recycled paper is considered cheap. In this interplay, the Product Development Department acts as the initiator of new ideas, whereas some proposals encounter resistance on the part of the Sales Department which argues that they are unmarketable.

Mention should also be made of *the interplay between dye works, the Purchasing Department and the manufacturers of dyes over the substitution of chromous black colouring agents by chromium-free ones.* After identifying the problem, namely that use of the chromous black colouring agent was causing considerable water pollution with chromium, the Purchasing Department wrote to various manufacturers of dyes to inquire into the possibility of chromium-free black colouring agents. Concrete dye tests were conducted in the dye works using different black colouring agents. After a substitute of the same quality had been developed in close cooperation with a dye manufacturer, it was the Central Purchasing Department that conducted the price negotiations for the new dye. Here, the division of work between the dye works and the Purchasing Department in looking for a dye substitute of equal quality becomes apparent. Whilst the dye works discussed the technical requirements and quality desired with suppliers, it was the Central Purchasing Department that conducted the price negotiations with them. Due to the company's purchasing power and negotiating skills, it was possible to keep the price differential between the old chromous black and the newly developed chromium-free colouring agent to a minimum.

Central clashes of interest exist in the following sectors   A central guideline exists in the statement that *activities in the sphere of environmental protection must not be ruinous for KUNERT.* There are limits to its environmental commitment where it does not result in cost-saving or additional sales potential or where it actually costs money. Thus, whilst there is a *de facto* environmental protection policy section, it is not a separate department but one coming within public relations and board matters. Moreover, this *de facto* environmental protection policy section has no budget of its own, i.e. it can use the resources that are available anyway (staff, infrastructure, office equipment, etc.), but it cannot incur any expenditure of its own, for example to employ an additional member of staff temporarily as part of an environmental project or purchase an extra computer. Where funds for particular environmentally-oriented measures appear imperative, recourse has to be had to other department budgets. This naturally imposes severe limits on the leeway for environmentally-oriented measures and successes. This explains amongst other things why, throughout the Group, there is concentration on implementing cost savings by environmentally-oriented measures. One example of the struggle between the pronounced financial restrictions and the striving after ecological improvements is the investment in a waste-water purification plant in the Portuguese works of Vila do Conde. This met with fierce criticism on the part of the Finance Director and was only approved by virtue of the authority of the Board Chairman and the fact that the EU paid for half of the investment (on application). The more difficult the times become, the narrower the leeway for environmentally-oriented investment, particularly if it cannot be fully recouped. On the other hand, when assessing the severe financial restrictions imposed on environmental protection measures, one must take into account *the extremely poor economic situation of the textile industry and in particular of the hosiery industry in the Federal Republic of Germany.* In this context, the conflict between economy and ecology becomes apparent, at least where the provision of funds is concerned. Nevertheless it should be seen as positive that, despite these poor economic parameters, an enterprise should take such an active interest in environmentally-oriented measures and attach such great importance to the subject of environmental protection.

The severest restrictions on environmentally-oriented modifications to products (tights, stockings, socks, etc.) are imposed by the *dictates of fashion*. In the hosiery industry, new fashion trends determine not only colours, shapes and patterns, but also the materials used. A pertinent example is the new trend in micro-fibre. Originally, it was developed for the sports clothing sector but meanwhile it has come to be used for many other types of clothing because of its various advantages (e.g. it wears well and is elastic, etc.). It does have grave disadvantages where dye is concerned, as micro-fibre, by virtue of its structure, can only be dyed by using more colouring agent and other additives. This in turn results, amongst other things, in increased waste-water pollution. As fashion declares this new micro-fibre to be a trend-setter, KUNERT cannot go against this trend without suffering considerable sales losses. A further obstacle to relating environmental activities to products is with a few exceptions *insufficient sales/marketing successes*. There are several reasons for this. The central problem lies in the products themselves. Most stockings, tights and socks, etc. are largely made from synthetic yarns and frequently of mixtures of yarns as well. This is a dictate of fashion and the standards of quality required by the consumers. In addition, there are narrow limits to the durability of the products (at least where stockings and tights are concerned). For these two reasons, these products *per se* are not very environmentally-friendly. This does not apply to the new near-natural products developed by the KUNERT AG and HUDSON. Admittedly their importance is marginal as the proportion of near-natural products accounts for less than 1% of the total range of products.

6   SUCCESS FACTORS IN KUNERT'S ENVIRONMENTAL DEVELOPMENT

The following aspects are central success factors in this case study: the *Board Chairman's function as power promoter* is one of the main success factors in promoting environmentally-related initiatives and implementing them within management (the Board) as well. The Board Chairman's positive attitude toward environmental protection and his fundamental willingness to take environmentally-oriented action constitute both the fundamental prerequisite and, at the same time, the fertile ground. Concrete proposals and ideas are submitted to the Board Chairman by *some extremely committed members of staff and departments*. The decisive idea for systematising environmental protection efforts at KUNERT was the proposal to draw up an eco-balance sheet which led to on-going work on environmental problem areas as well as the introduction of eco-controlling. In addition, *cooperation with an external advisor* was decisive for successful development in the direction of an environmentally-oriented enterprise management. The external advisor contributed the necessary specialised and academic competence (know-how in form of devising a balance sheet) to the cooperation. In addition, he and his staff succeeded, as outsiders and therefore not blind to in-house operations, in identifying problem areas and possible improvements. In addition, the translation of ecological successes to cost savings in particular played a great part in ensuring the support of the management. The central motive for most of KUNERT's ecological measures is *cost savings or the expectation of increased profits*. 'We are not eco-freaks; ecology has to pay off!' is a remark that is often heard. Many measures lead to cost savings and so in many cases it was possible to combine ecology and economy at KUNERT. In the meantime, many activities are standardised and better organised and so members of staff no longer have to sacrifice so many working hours, and cost savings also result. The amount of time spent on environmentally-oriented measures is reduced as a result. Initially, eco-controlling was initiated as a pilot project. *By publishing the eco-report as well as by virtue of its contents* the entire project acquired a considerable dynamism of its own in that the eco-report implied a certain continuity of activity which KUNERT had not envisaged. There was a tremendous and unexpectedly positive response on the part of the public and the media to KUNERT's first eco-balance sheet. One reason was undoubtedly the public's interpretation of the eco-report to the effect that KUNERT was proposing to continue drawing up an eco-balance sheet and conduct on-going work on environmental problems. This produced pressure for action on the part of KUNERT which was not intended, and was a major contributory factor in ensuring that the pilot project 'Eco-Controlling', which originally was certainly not envisaged as long term (over several years), was not simply ended. On the contrary KUNERT suddenly found itself faced with public expectation that it would continue work in the field of environmental protection. Publication of the first eco-balance sheet thus constituted the 'breakthrough in terms of an on-going eco-balance sheet' for KUNERT. The eco-balance sheet acquired, as it were, an unintended automatism all of its own. In setting itself goals and drawing conclusions in the eco-report published annually, KUNERT *de facto* imposes an *obligation on itself* to meet its self-imposed environmental goals by taking suitable product and process-related actions. In so doing, KUNERT places itself under pressure to present environmental improvements every year. This self-imposed pressure to present successes applies to the entire Group and on the one hand motivates individual departments as a challenge, whilst on the other imposes constant pressure on the members of staff to produce results. Overall this obligation on the part of the KUNERT Group must be seen as an important success factor in implementing individual measures as well as in advancing environmental management. A further success factor in substituting certain additives and the deployment of other materials, which should not be underestimated, is *the Central Purchasing Department in Immenstadt as a power and technical promoter*. As a result of its negotiating strength and the volume of purchases throughout the Group, it assumes the role of a power promoter. It acts as a technical promoter in as much as it ultimately decides on all the preliminary products, materials and substances which enter the enterprise. In a metaphorical sense, the Purchasing Department is thus an ecological gatekeeper for the entire KUNERT Group. In addition, the Purchasing Department has an interest of its own in seeing that the materials purchased are readily recyclable. In addition, the Purchasing Department in some cases has the power to bring pressure to bear on suppliers and call for certain additives to be substituted as well as for information about products. At the same time, the Purchasing Department seeks closer cooperation with the initial suppliers with a view to buying less environmentally-hazardous raw materials. The fact that all *environmental measures were organised and made the responsibility of one central department* was a further major success factor. At KUNERT such a central department has *de facto* been set up as an environmental protection policy section within the Department for Board Matters and Public Relations. (KUNERT does not yet have a separate department for environmental protection.) This central department receives all the information about environmental activities throughout the Group. It develops and implements ideas and measures, and draws up the eco-balance sheet as well as the annual eco-report.

7 ENVIRONMENTAL IMPLICATIONS

Government environmental policy had no bearing on the decision to implement an eco-balance sheet and devise an eco-controlling system at KUNERT. Admittedly, government environmental policy can be seen to have a direct and indirect influence in generating and implementing many measures: the *water regulations* in textile processing and in particular in dyeing, for example, play a very important role for the Group. Thus, the fact that limits were exceeded under the Indirect Emissions Ordinance was a decisive factor in successfully substituting chromous colouring agents by chromium-free ones. The *proposed amendment to the 38th Appendix to the Framework Waste Water Administrative Regulation pursuant to §7a WHG* would have meant considerable waste-water purification investment for KUNERT (see Schönberger/ Kaps 1994: 520f.). One first waits therefore until the new ceilings are decided upon before planning or indeed implementing appropriate water-purification measures. In addition, *the EC Ordinance on the Award or an EC Environmental Symbol* in 1992 was decisive for devising a product-line analysis at KUNERT. The discussion surrounding the introduction of certain labels in the sphere of textiles (MUT, MST) is being followed by KUNERT with interest. The increase of refuse fees by a total of 240% by the Bavarian *Waste Association* encouraged KUNERT too to make a greater effort to reduce and separate waste (e.g., development of a so-called recycling policy for fluorescent lamps, cost savings of 50%). The *provisions of the Packaging Ordinance* with regard to the return of various kinds of packagings have had an indirect influence on one's approach to one's own packaging and that supplied. In 1993, for example, progress was made on a transport system deploying reusable packaging between KUNERT and its suppliers. Within the KUNERT Group, however, reusable packaging had been in existence since the beginning of the '70s.

The concentration process in the German textile industry which is proceeding at vast speed due to the high wages and ancillary costs is an incentive to shifting production abroad. Tighter environmental regulations exacerbate the situation, especially if uniformity throughout the EU fails to come about.

8 RESUMÉ

The KUNERT success story consists in the systematic and annual company-wide eco-balance sheet forming the basis of implementing eco-controlling.

This led to a remarkable development. Firstly, it was possible to analyse and process many weaknesses. Secondly, the drawing up of an eco-balance sheet generated a process of greater awareness and reflection upon environmentally-relevant implications on the part of many employees and the management of the KUNERT Group. Thirdly, environmentally-oriented activities were anchored in the corporate structure.

Although various problems in the KUNERT Group still await resolution, the progress made toward an environmentally-oriented corporate management must be viewed as a considerable success.

LITERATURE

Burghold, J.A. (1988) Ökologisch orientiertes Marketing. Diss. Schwerpunkt Marketing Bd. 25, hrsg. von Prof. Dr. Paul W. Meyer. FGM-Verlag, Augsgurg.

Enquete-Kommission (Hrsg.) (1994) Die Industriegesellschaft gestalten. Perspektiven für einen nachhaltigen Umgang mit Stoff- und Materialströmen. Hrsg. von der Enquete-Kommission 'Schutz des Menschen und der Umwelt' des Deutschen Bundestages, Economica Verlag, Bonn.

Gesamttextil (Hrs.) (1993) *Jahrbuch der Textilindustrie* 1993. Textil-Service- und Verlags-GmbH, Eschborn.

Gladwin, Th.N. and Walter, I. (1980) *Multinationals under Fire. Lessons in the Management of Conflict.* New York, Chichester, Brisbane, Toronto.

Kirschten, U. (1995) Ökobilauzierung und Aufbau eines Öko-Controlling in Kunert-Konzern. Report, Free University, Berlin.

KUNERT (1990) 'für uns' Werkszeitschrift, Nr. 60, Juni 1990.

KUNERT (1993) Ökobericht 1993, Immenstadt.

KUNERT (1993a) Geschäftsbericht der KUNERT AG 1993, Immenstadt.

KUNERT (1994) Ökobericht 1994, Immenstadt.

Schönberger, H. and Kaps, U. (1994) Reduktion der Abwasserbelastung in der Textilindustrie. Im Auftrag und herausgegeben vom Umweltbundesamt, UBA Texte 3/94, Berlin.

9. HERTIE AND THE BUND: THE HISTORY OF AN UNCOMMON COOPERATION

Simone Will

Translated by Mary Höcker

INTRODUCTION

The present case study describes and analyses the process of ecological reorientation introduced by one of the largest department stores in Germany.

Contrary to what was originally planned for the project, the following pages do not focus on one single substantial ecological improvement such as might emerge for example as a result of a technological innovation in the production process. Instead, the study is based on a number of smaller individual measures that Hertie had initiated in the course of recent years and which constitute the success of the case under study only when seen in their entirety. This approach seems justified for at least two reasons.

As a department store, Hertie belongs to the retail branch and this enables the firm to give its activities an ecological thrust in two respects. On the one hand the retail trade, just like the manufacturing industry, is a direct cause of negative environmental effects. This includes, for example, the energy and water consumption for operating the department stores as well as the whole complex of negative consequences for the environment arising out of logistics as a whole, etc. On the other hand the retail trade must be seen as an indirect cause of negative environmental effects in that, by virtue of its position as 'gate-keeper' between manufacturer and consumer, it can exert a certain influence and pressure and can thus have a lasting impact on the prerequisites for ecologically-responsible action both on the part of the manufacturer and also on the part of the consumer. However, the exercise of this gate-keeper function in an ecologically-responsible manner only provides very limited opportunities for incorporating the resulting ecological improvements in operations. The change in role perception can better be described by qualitative aspects.

Secondly, what makes the Hertie case study stand out is above all the ecological reorientation of the firm which is an unusual *process* for commercial enterprises. Against the backdrop of the company's first faltering steps to make its activities ecologically more compatible, Hertie in 1989 contacted the Federation for Environment and Nature Protection Germany (BUND),[1] the largest environmental

---

[1] The Federation for Environment and Nature Protection in Germany (BUND) was founded in 1975. In the spring of 1993 the environmental organisation had more than 216,000 members and membership was slightly on the increase. The BUND engages largely in lobby work and other political work, campaigns

organisation in Germany, in order to seek advice on ecological questions. The close conceptual cooperation between the two organisations that arose out of this initiative has plainly been pioneering in nature for both the company and the environmental organisation.[2] The cooperation which has been placed on a contractual basis since 1991 entails giving Hertie ecological advice – mainly on goods stocked – and this has already resulted in numerous ecological corrections being made to the items stocked.

Against the background of this specific given situation, this study is primarily interested in the process of Hertie's ecological reorientation taking special account of the influence of the BUND. Secondly, it must be recognised that, in view of the large number of smaller ecological successes, one cannot fully reconstruct all the measures and can at best highlight major individual activities by way of example.[3]

Despite the operational difficulties described above, an attempt will nevertheless be made, given the ecological activities of almost all retail firms meanwhile, to identify the special quality of Hertie's ecological commitment, justifying the selection of Hertie as a case study in preference to other retail firms engaged in environmental protection.[4]

With a consistency hitherto without parallel in German department stores, Hertie, under the influence of the BUND, has meanwhile corrected some 3,500 of its department store items on ecological grounds since 1989. Hertie's pioneering approach for some ranges of its store items in particular emphasises the exemplary nature of its purchasing policy compared with other department stores. Its ecological adjustments include both the deletion of certain items altogether as well as the substitution of particular articles by alternative products which are more environmentally-friendly. Hertie's express demand that its manufacturers make ecological product modifications as well as the requirement that all suppliers provide information on their packaging come under this heading. In this context special mention should be made of the comprehensive catalogue of communication measures to inform the client serving to educate the consumer in ecological questions. Where the role of the retail trade as a direct polluter of the environment is concerned, Hertie's efforts in the sphere of logistics, which have been awarded the '1993 Environmental Prize of the German Retail Trade', deserve particular mention. More than 70% of total transport traffic between manufacturers and Hertie has meanwhile been transferred to rail.

---

on specific themes, such as waste, energy and climate, traffic, genetic engineering and in concrete nature protection projects. The BUND employs some thirty-eight full-time staff in its headquarters in Bonn. The Federation has a federal structure with local and district groups and autonomous *Länder* associations. The BUND is financed mainly by contributions from members as well as general and purpose-linked donations, fines and bequests. With an annual income of some DM 15.2 million and overall expenditure of DM 14.4 million, only a small annual surplus remains. See BUND (1994).

[2] The cooperation practised by various retail concerns (e.g., Karstadt, Otto-Versand, etc.) with the World Wildlife Fund (WWF) can more correctly be described as eco-sponsoring. As a rule the exchange there is limited to 'image for money'. The WWF is not proposing to cooperate substantively with the enterprise concerned. On eco-sponsoring see Bruhn/Dahlhoff (1990).

[3] A detailed presentation is found in the long version of the case study which is available as a working paper. See Will (1994).

[4] On the ecological activities of various retail companies see Hopfenbeck (1991: 425f.).

The present case study covers the period from 1987 to the autumn of 1993 in the main. Developments occurring after this time are only taken into account if they are of major importance for the process of ecological reorientation. The merger between Hertie and Karstadt, the largest German department store concern to date was begun after conclusion of the interview period in September 1993 and doubtless constitutes an important marker for such a process. The final consequences of the merger are only taking shape now.[5] To this extent the case study is essentially a retrospective one.[6] After a recent meeting between the new Hertie Board and representatives of the BUND, there is no longer any doubt that both parties to the cooperation wish in principle to continue it. Whether the cooperation will endure in the medium- to long-term, however, depends primarily on the successful conclusion of a current project and implementation of organisational structures and responsibilities so far existing on paper only.[7]

The case study is based on a total of 22 qualitative interviews with relevant players in the Hertie concern mainly situated in middle management at the second and third managerial levels, as well as with experts outside the concern whom the author interviewed in the period from June 1993 to September 1993.[8] The group of experts comprises members of the BUND as well as representatives of the Branch Association. The material collected was supplemented by a comprehensive study of literature.

The case study begins by describing the historical background of Hertie. Section 2 focuses on the economic importance of the retail trade in which the entire branch is concerned and gives a survey of its general opportunities for ecological action as well as the relevant parameters. Section 3 briefly summarises the ecological improvements achieved by Hertie. Based on the ecological measures introduced by Hertie in the course of time, Section 4 seeks to identify the stages of the process. Section 5 provides an analysis of the players, explaining the interaction of the relevant players at Hertie from the point of view of the firm and finally, with an eye to the BUND, highlighting the latter's influence. Section 6 deals solely with the relevant factors influencing the course of events both within and without the company and the parameters governing action. The final section summarises some important aspects of the case study once more.

---

[5] The former Board of the Hertie GmbH has meanwhile been reshuffled several times. The comprehensive organisational and personnel measures are far from being complete. In the environmental sector in particular responsibilities have been and are still being redefined. For example, there will no longer be the current Environmental Department in its present form. The proposed structures still have to prove themselves.

[6] As of January 1994 Hertie, hitherto a family concern, merged with the Karstadt AG, hitherto Germany's largest department store concern. In future Hertie holds some 30% of the total shares of the expanded concern, making it the largest single shareholder. The new expanded retail enterprise employs some 100,000 staff and has an annual turnover of some DM 28 billion (position at 1993). It has been agreed that in future Hertie will be an autonomous entity within the Karstadt concern.

[7] The BUND wants the 'Eco-Cleaning Materials Project' successfully completed by March/April 1995 by the latest, a date that Hertie finds difficult in view of the stubborn resistance of powerful suppliers. The BUND considers, however, adherence to the deadline a precondition for having this year's Delegates' Assembly approve continuation of the cooperation.

[8] The interviews lasted for forty-three hours. For details of the method adopted see Will (1994).

1   HISTORICAL CONTEXT OF THE COMPANY

The history of the Hertie concern can be traced back to the year 1882.[9] After Hertie rose to become the largest department store in Europe during the twenties, the world recession and above all the advent of the National Socialists brought about the company's decline. The Jewish owners, the Tietz family, were expropriated and had to emigrate. The Second World War destroyed more than 80% of Hertie's real estate; a large proportion of the remaining department stores and properties lay out of reach in East Germany at the end of the War. Hertie's new owner, Georg Karg, did not succeed in catching up with the other department store concerns such as Karstadt and Kaufhof until the '60s. When Karg died in 1972, having ruled the concern in patriarchal style for forty years, the 'leaderless' concern found itself in an extremely critical situation, not least because of the crisis that department stores began to find themselves in about the same time and from which the company did not recover until the mid-'80s after numerous closures of various department store branches.[10] After a comprehensive strategy of diversification during the eighties – in which Hertie acquired numerous specialised shops and holdings, above all in the consumer electronics branch[11] – the concern sought in the nineties to again strengthen its core 'department store' business.[12]

Hertie GmbH employed some 32,000 members of staff on average in 1992/93 (this equates to some 25,000 full-timers).[13] With a turnover of DM 6.9 billion (DM 34 million net income for the year), Hertie ranked third amongst the German department stores with the greatest turnover. At 43.2% the fashion sector (clothing, materials, shoes, perfumes, clocks, jewelry, leatherware, etc.) earned the lion's share of turnover. The technology sector (consumer electronics, audio equipment, cameras, electrical equipment, etc.) accounted for 21.7% of turnover; the remaining turnover is divided up amongst foodstuffs/gastronomy (13.9%), leisure (10.6%) and other items (10.6%).[14]

From 1972 until the merger with the Karstadt concern, the department store firm was conceived as a foundation for reasons of estate duty. The main shareholder in the Hertie Waren- und Kaufhaus GmbH (capital DM 300 million) was the Gemeinnützige Hertie-Stiftung (Hertie Non-Profit Foundation) with a capital share of 97.5%.[15] Despite that, all economic decisions affecting the Hertie concern were in the hands of the Hertie Family Foundation, although the latter only had 0.5% of the capital at its disposal. The Supervisory Board and the five-man Board

---

[9] For details of the firm's history see Rehmann (1986).

[10] By the mid '80s Hertie had accumulated overall losses of DM 600 million. See Pellinghausen (1993: 21). 1987 was the first year in which, following the losses of the '80s, Hertie made a profit of DM 4.2 million. See FAZ of 11/1/1989, p. 13.

[11] For associated enterprises and major holdings see Hertie (1993: 46f.).

[12] The merger with the Karstadt concern must also be seen in this context. Hertie and Karstadt look to the merger for synergies in logistics, purchasing and administration in particular. In addition, two concerns' branch policies complement each other in many regions of the Federal Republic.

[13] Unless otherwise stated, this and the data below relates to the 1992/93 business year and thus to Hertie in its original boundaries.

[14] See Hertie (1993: 24f.).

[15] The remaining 2% were held by the community of joint owners Karg/v. Norman. A detailed survey of the firm's ownership structure as well as its decision-making bodies is found in Rehmann (1986).

of Directors responsible for the management were supplementary decision-makers of the GmbH.[16] The influence of the Hertie Non-Profit Foundation was confined to use of the surplus income in keeping with the non-profit purpose of the Foundation.[17]

2   THE RETAIL TRADE: A BRANCH-RELATED APPROACH

2.1   The Economic Importance of Retailing in Germany

The retail landscape in Germany is above all characterised by very large and, as a rule, broadly diversified retail concerns.[18] The extremely small number of department store concerns is the result of a process of concentration that has been occurring since the '50s.[19] The planned merger between Hertie and Karstadt as well as the announcement of the takeover of the fourth largest German department store concern, Horten, by Kaufhof, the number two in the branch, are the latest climaxes in this process which is still continuing. In 1992 the German retail trade had a turnover of some DM 600 billion. That equates to some 22% of Germany's gross domestic product. At 6% (in nominal terms) for 1992 the growth in the retail trade branch is on the moderate side. After the record turnovers occasioned by German reunification, with annual growth rates of 8.3% (in nominal terms; 1990), department stores in particular must meanwhile be content with a growth rate of approximately 0.6–1.2% (in nominal terms; 1992). This is an annual decline in turnover in real terms.[20] Compared with other forms of retail trade, the department stores have lost appreciable market shares since the beginning of the '70s. In 1980, for example, department stores still had a share of 7.2% of the total retail trade turnover in Germany. In 1993 it has fallen to some 4.3%.[21] Accordingly, competition amongst the department stores is very intense.

2.2   Makings and Parameters of an Ecologically Compatible Orientation of Retail Activities

As already outlined above, the retail trade is particularly important as a player in the environmental discussion by virtue of its position between manufacturer and consumer in which it performs various functions as a mediator and organiser. In

---

[16] It was only after Hertie's crisis years in the '80s that the Executive Board succeeded in 'emancipating' itself from the determining influence of the Chairman of the Supervisory Board and running Hertie's day-to-day business independently. See Glöckner (1993).

[17] The main purpose of the Hertie Non-Profit Foundation was to conduct research into multiple sclerosis (MS) and support MS sufferers. See Rehmann (1986: 299).

[18] On this see the survey in Kursawa-Stucke/Lübke (1991: 165f.).

[19] The department store landscape in 1993 was as follows: in terms of turnover and net income for the year (position 1993) the four large German department store concerns are led by Karstadt with an annual turnover of DM 20.6 billion/224 million net, followed closely by Kaufhof (DM 20.5 billion/222 million). Hertie comes a distant third (DM 6.1 billion/34 million). Horten takes the bottom position with DM 3.4 billion turnover and DM 24 million net income for the year. On this see the Annual Reports of Horten and Hertie. For details of Karstadt and Kaufhof see Glöckner (1993).

[20] See BAG (1992: 36).

[21] Ibid.

their function as 'gate-keeper', the retailers with a high turnover in particular can have a major influence on the flow of goods and information also in respect of their ecological quality.[22] By setting concrete conditions, for example, retailers can exercise a direct influence on the manufacturers and suppliers. As part of this so-called *ecology-pull-strategy* toward manufacturers and suppliers, the following forms of action are conceivable for retailers: selection of suppliers from an ecological viewpoint, monitoring, assistance programmes, purchasing guarantees, ecologically oriented product selection, etc.[23]

With regard to the consumer, the retail trade can exert an influence as part of the so-called *ecology-push-strategy*. This can be done by deliberately including or excluding products from an ecological viewpoint or as part of an ecologically-oriented communication policy: for example, detailed product information can be passed on with the additional ecological benefit highlighted. Other fields of activity might be the environmentally-friendly design of prospectuses and catalogues, advice to the client by the sales staff, ecologically-oriented after-sales activities and services, ecologically-oriented brands of one's own.[24]

The following observations reflect the environmental policy and socio-economic parameters where the ecological leeway of retailers is concerned.

For a long time, the importance of the retail trade in the discussion of environmental policy was underestimated. In the meantime, the Packaging Ordinance has placed obligations on retailers at least in Germany where the problem of waste is concerned. In view of further ordinances supplementing the Waste Law of 1986 (Electronic Waste Ordinance, Old Paper Ordinance, Battery Ordinance, etc.) as well as the Law on Recycling and Waste that has meanwhile been passed, the retail trade will in future become a central addressee of governmental environmental policy on the problem of waste disposal. The cooperation of the retail trade in building up the Dual Waste Disposal System Germany (DSD) already binds the resources of the branch to a great extent.

The problem of waste apart, various empirical studies show that, compared to other branches, the retail trade is still affected below average on the whole by government environmental policy measures.[25]

The growing environmental awareness of consumers noted in Germany since about the mid-'80s has remained at a consistently high level[26] and offers the retail trade increasing opportunities of finding favour with clients because of the ecological compatibility of its stocks even if manufacturers and retailers in particular often bemoan the fact that their clients do not always translate environmental awareness into environmentally-compatible buying and consumer habits accordingly.[27]

---

[22] On the 'gate-keeper' role of the retail trade and the resultant opportunities of retailers to take action see Hansen (1988, 1992); Mattmüller/Trautmann (1992).

[23] See Hansen (1992: 745f.).

[24] Ibid.

[25] See for example FUUF (1991).

[26] For empirical studies on environmentally-aware consumer behaviour see for example Balderjahn (1986); EMNID Institut (1987); Wimmer (1988); Adelt/Müller/Zitzmann (1990); Diekman/Preisendörfer (1992) *et al.*

[27] See for example FUUF (1991).

Even if the public has so far not been the prime mover on ecological questions, since discussion of the meaningfulness of the dual system of Germany, the public increasingly sees retailers as bearing responsibility for the packaging problem. Several comparative studies on the environmental activities of the retail trade intended to make its importance as a player in environmental protection an issue have at most stimulated the discussion temporarily amongst the experts.[28] In addition, the various studies provide partially contradictory results and so a reliable evaluation is scarcely possible. There is however consensus that the majority of retail companies are relatively unsystematic in their approach to environmental protection, that the measures are largely isolated in nature and that often there is no consistent overall policy.

The attitude of the Association of Department Store Concerns (BAG) to ecology can be described as varying from wait-and-see to resistant. The Association tries, where possible, to prevent or delay laws and ordinances in the making. The Association categorically rejects the approach of Parliament in environmental policy which is to use the retail trade as a demand-lever *vis-à-vis* manufacturers.[29]

Where the environmental protection challenges in retailing are concerned, it can be said in summary that, overall, the retail trade is affected relatively little by ecology in market, political and social terms. Where its ecological commitment is concerned, the branch has hitherto displayed a largely voluntary, opportunity-oriented approach in which as a rule only those ecological measures which promise additional acquisitive potential are implemented. At present environmental policy-induced cost pressure has had an impact on the retail trade in the sphere of packaging at most.

3   DEMONSTRABLE IMPROVEMENTS IN ENVIRONMENTAL PROTECTION

The difficulty about reconstructing a detailed course of events where the Hertie case study is concerned is the absence of a particular salient ecological improvement, as already stated above. It is not possible here to comment in depth on the many small ecological steps taken by Hertie which only assume relevance in their entirety. The following survey merely highlights a few ecological improvements by way of example.

As part of its work on stocks, Hertie has corrected more than 3,500 products across the board in recent years on ecological grounds. The ecological corrections to stocks were concentrated mainly in the cosmetics and perfumes, stationery and household goods branches.

---

[28] For comparative studies see for example Kolvenbach (1990); Kursawa-Stucke/Lübke (1991). See also the feasibility study supported by the Federal Ministry for the Environment, Nature and Reactor Safety entitled 'Environmental logos in the retail trade' which takes a new approach to the topic of evaluating ecological achievements in retailing. See imug (1993). See also remarks in Hopfenbeck (1991); FUUF (1991); Meffert/Kirchgeorg (1993) etc.

[29] See BAG (1993: 74f.).

Salient examples of work on stocks that attracts good publicity are the deletion of liquid laundry-softening agents from the order books, the complete substitution of all pesticides by biological agricultural products and the deletion of all stationery and office equipment containing PVC and solvents.

In addition, in response to the problem of packaging, Hertie called upon its major suppliers to observe the packaging regulations developed with the help of the BUND, threatening a monetary charge for failure to do so. In close cooperation with a Berlin manufacturer of stationery who supplies the major part of Hertie's department store stocks, blister-free packaging was arranged for the entire programme under the influence of the BUND. In imported textiles Hertie reduced the volume of packaging drastically; several hundred thousand individual foil wrappings as well as disposable hangers made of plastic were dispensed with.

To supplement the on-going work on the range of goods stocked, Hertie arranged special exhibitions in an attempt to draw public attention to particularly environmentally-compatible products with a view to acqainting the consumer with environmentally-compatible alternatives already in existence. In its communication policy Hertie committed itself to educating the consumer. To complement the above-mentioned special exhibitions, information brochures printed in large numbers were distributed to customers in which Hertie in some cases explicitly disassociated itself from part of its range of goods and encouraged the customer to purchase certain environmentally-compatible alternative products. In addition, a large proportion of the prospectus material as well as the newspaper inserts were printed on recycled paper.

As part of a comprehensive reorganisation of the entire logistics system, Hertie moved some 70% of its total volume of goods traffic between manufacturers and Hertie's own distribution centres to rail traffic. The logistics activities, which were awarded the Environmental Prize of the German Retail Trade in 1993, are attributable in the main to pilot projects with German Rail.

In the vocational training sector the Personnel Department of Hertie developed a comprehensive training programme for members of staff. In addition a video, some thirty minutes long, on the subject of 'Environmental Protection at Hertie' was shown to all members of the staff from the various branch stores. Hertie's ecological responsibility was integrated into the corporate image in 1991 and subsequently formal responsibility for environmental protection was laid down at Board of Management level as well. The organisation of environmental protection, which includes the contractual cooperation with the BUND, will be discussed once more below when reconstructing the interaction of the relevant players, with special attention being paid to the BUND.

4   HERTIE'S ENVIRONMENTAL ORIENTATION –
   A PROCESS-RELATED ANALYSIS

Three main stages can be discerned in Hertie's environmental orientation (1987–1995):

- Growth in general awareness (1987–1988);

- Orientation and information stage (1989–1990);
- Structuring and institutionalisation (1991–1995).

The *stage in which Hertie acquired a general awareness of ecological questions (1987–1988)* was the one in which the enterprise first addressed the topic of environmental protection without already introducing concrete activities at this stage. Interest in the topic was triggered by:

- public discussion about the connection between products containing fluoro-carbons and depletion of the ozone layer;
- public opinion polls showing increasing environmental awareness on the part of the consumer; and
- observations of its own at the 'point of sale' showing plainly that the environmental compatibility of products is increasingly becoming a relevant purchasing criterion; but mainly also
- marketing successes of competing retail firms which already use environmental protection to good public effect to enhance their competitiveness.

When at the end of 1988 Hertie decides to set up an interdepartmental working group entitled 'Active Environmental Protection' which is to consider ways of making constructive and effective ecological improvements within the enterprise, responsibilities for environmental protection are formally laid down – albeit in addition to the actual functions of the approximately 10 members – but without the working group having been given powers to issue directives.

The subsequent *orientation and information stage (1989–1990)* is marked by Hertie's intensive efforts to gain a competitive edge by seeking out and taking advantage of ways of gaining prominence in environmental protection. Starting from a comprehensive analysis of the status quo which once more highlights the need for an ecological reorientation on Hertie's part, above all in the interests of its own competitive position, the working group draws up a first rough guide as to how the enterprise's activities can be given an environmentally-compatible thrust. In addition to increasing dialogue with customers, improvements in ecological product counselling and the introduction of a classification system for environmentally-friendly products, the working group suggests, for example, dropping environmentally-damaging products from the order books and demonstratively promoting environmentally-friendly suppliers. Here the inadequate ecological knowledge and judgement of the members of the working group proves an obstacle to making a reliable evaluation of allegedly ecologically-more-compatible alternative products. In the face of grave doubts as to their ecological judgement on the one hand and the duty felt by some working group members toward consumers and public to act in a *credible* ecological manner on the other, initial contact is made with the BUND via a member of staff in marketing – at the same time an active BUND member. Hertie looks to the environmental association in particular for competent assistance in preparing, organising and coordinating various environmental protection activities.

By providing criteria for evaluating the ecological quality of products etc., the BUND largely compensates Hertie's deficits in ecological knowledge. In addition the BUND suggests numerous projects to Hertie, recommending for example the introduction of meat from animals reared in conditions approved by veterinarians, the sale of carrier bags made of cotton instead of plastic and the dropping of one-way beverage containers from the order books. Hertie accepts most of the projects suggested and gives them a trial run. However, as not all the measures prove compatible with Hertie's economic objectives, some of them are dropped after the test stage.

Overall it can be said that the environmental measures initiated by Hertie during this stage tend rather to be unplanned and uncoordinated. There does not yet appear to be a strategic thrust to the ecological activities during this stage. Hertie is well-disposed toward the measures suggested by the BUND and shows that it is willing in principle to implement the various proposals on a trial basis. Hertie's activities overall during this stage can be interpreted as the enterprise's willingness to take risks for the moment in seeking ecologically-meaningful fields of action that are at the same time economically-acceptable.

During the stage described as *structuring and institutionalisation (1991–1995)* Hertie lays down largely formal in-house responsibilities for environmental protection under the dominant influence of the BUND. In addition to making the Board member responsible for purchasing the 'Head of Environmental Protection', a purchasing director at the second managerial level is appointed Environmental Commissioner. In appointing a full-time Environmental Commissioner Hertie recognises moreover that environmental protection can no longer be dealt with on the side.[30] With the appointment of the full-time Environmental Commissioner, the existing institutionalisation of environmental protection in the form of the working group, which meanwhile has some thirty members, is reviewed. Ultimately the working group is dissolved in favour of several small project groups working on specific problems.

Cooperation with the BUND becomes an established part of Hertie's environmental work by being given contractual form. The influence of the BUND results in the ecological analysis of stocks becoming more systematic. At Hertie's request the BUND draws up well-founded material on which to base decisions, identifies problematic items of stock, develops solutions, and plans, and coordinates and monitors their implementation. Since 1991 at the latest a clear focus of cooperation is apparent on the subject of packaging. On the basis of a comprehensive review of stocks which the BUND conducts at a Hertie head office, numerous ideas for presenting goods with a minimum of packaging are mooted.

As a matter of principle the BUND presses for joint action to be planned in the long term and, above all, thoroughly. This is also in Hertie's interest, particularly after a store leaflet highlighting allegedly environmentally-compatible products had legal consequences for Hertie. Important examples of projects planned in the long term are the dropping of pesticides and, later, of stationery items containing solvents from the order books at the suggestion of the BUND, which were then implemented and publicised after an appropriate period. Where Hertie's discontinuation

---

[30] Unfortunately the critical event that preceded this insight cannot be commented on within the framework of this summary. Please see the long version of the case study for this (Will, 1994).

of items of stationery containing solvents was concerned, the fundamentally-different goals of both organisations resulted in a crisis which put their continued cooperation at serious risk. The reason for this was Hertie's undertaking, obviously given prematurely, to discontinue all items of stationery containing solvents by the end of 1991. Hertie gave the undertaking on the assumption that manufacturers would offer environmentally-friendly alternative products by the agreed date. When it became clear that the manufacturers would not be able to adhere to the date agreed, the purchasers responsible at Hertie feared significant sales losses. At the same time a fall-off in turnover indirectly threatened the purchasing staff with negative effects on their own income. To prevent this, the purchasing manager tried to prevent or at least to delay the agreed deletion from the order books. The BUND in turn threatened to make use of its contractual right to withdraw from the cooperation in the event of the agreement not being adhered to. It was only through the mediation of the Board Purchasing Director, who was also Head Environmental Commissioner, that the dispute was settled and a compromise regarding the date found.

The current 'Eco-Cleaning Materials project' with the BUND is also one in which Hertie is encountering stubborn resistance on the part of suppliers who are unwilling to accept Hertie's demands. As already-intimated above, continuation of the cooperation with the BUND depends on the project's successful conclusion even against resistance from the suppliers. At present Hertie does after all seem inclined to impose sanctions on uncooperative suppliers by excluding them from the order books.

5   INTERACTION OF THE RELEVANT PLAYERS

The following remarks focus on the analysis of the players relevant to the process of ecological reorientation at Hertie. First of all, brief reference will be made to the in-house players supporting or obstructing the process up to 1993. After that the importance of the BUND in the course of events will be described.

5.1   The In-house Perspective

Apart from the formal competences laid down for environmental protection – mention has already been made of an environmental commissioner at the second managerial level in Purchasing and later also in Sales, the responsibility of one member of the Board for ecological questions, the appointment of a full-time environmental commissioner, etc. – one member of staff in Marketing, in particular, proved in retrospect to have been the central figure for the entire process of environmental orientation at Hertie.

This *technical promoter* at the third managerial level was the first to even see the need to look critically at the claims of manufacturers and suppliers that their products were environmentally-compatible. The concrete occasion for doing so was a special exhibition of particularly-environmentally-compatible washing machines planned by Hertie. It was important to the technical promoter to be able to talk to consumers and the interested public with an easy conscience justified by having a credible commitment to environmental protection. In this connection he contacted the BUND which he felt had the necessary ecological competence and also fundamental willingness to offset Hertie's inadequate ecological knowledge. It was he who ultimately instituted the contractual cooperation by continually making clear the potential opportunities that the competence of the BUND made available to Hertie. He also had the idea of the 'shop check' – a comprehensive and thorough inspection of *all* the department store items from an ecological viewpoint had been carried out by the BUND – which the Environmental Commissioner responsible for Purchasing ultimately issued on his insistence.

Throughout the entire duration of the cooperation he was the BUND's discussion partner; he planned, coordinated and supported the BUND measures at Hertie. After the unpleasantness that he incurred when stationery containing solvent was dropped from the order books – he was accused by the Purchasing Department of having given Hertie's contractual assurance without their knowledge – he has distanced himself noticeably from the ecological work on stocks despite the amicable settlement reached in the end. Given the personnel changes at Hertie, however, he will in future be officially responsible for the cooperation with the BUND.

Although the Board purchasing director responsible for environmental protection performed mainly representative functions – for example he represented Hertie at the joint press conferences with the BUND – his formal responsibility was nevertheless an important indicator, especially for the members of staff. His overall favourable attitude toward the cooperation with the BUND was of decisive importance for the continuation of the cooperation in the above crisis, for he assumed the role of *power promoter* at Hertie. Moreover, during the decisive working group meetings, his presence alone ensured that the decision to abandon pesticides was definitely passed after the purchaser responsible for the gardening sector, fearing a drop in sales, had called the decision into question again in the working group. Despite several proposals by the above-mentioned technical promoter to reorganise the system of material incentives – the buyers tend to look on environmental protection as being in conflict with their income objectives – the Board Purchasing Director had not yet defused this basic source of conflict and will not do so in future either because he is no longer with the company.

Apart from this promoter duo, the buyers already mentioned above must be seen as players applying the brakes to the process. In addition to the lack of material incentives to behave in an ecologically compatible manner, the proximity to manufacturers and the lack of willingness on the part of the buyers to question the manufacturers' claims about the ecological quality of their product made it difficult to give stocks a consistent ecological thrust and still does.

The first Purchasing Director to be appointed Environmental Commissioner for Purchasing was very important on occasion for implementing the measures decided by the working group 'Active Environmental Protection'. In respect of his own responsibility for purchasing in particular, he was able to bring influence to bear to see that the appropriate measures were implemented quickly. Admittedly his relatively authoritarian approach was not always conducive to ecological success. On the contrary, given his motive of wanting to gain stature *vis-à-vis* the Board by

achieving ecological successes, his authoritarian manner provoked a negative response in some members of the working group. At no time did the Environmental
Commissioner for Purchasing recognise the opportunities resulting for Hertie from
the cooperation with the BUND. On the contrary he repeatedly 'forgot' that he had
requested the BUND to draw up decision-making material, such as the shop check,
for example. His thinking in formal competences had a markedly demotivating
effect, above all for motivated, but not formally authorised, members of staff.

By contrast, the Environmental Commissioner for Purchasing who succeeded him
in the spring of 1993 tended rather to take an opportunity-oriented attitude to
environmental protection. He, for example, was more inclined to make down-
payments for environmental protection. Moreover, unlike his predecessor, he was
the first to bring ideas of his own and concrete proposals for projects into the
cooperation with the BUND. In addition he succeeded in motivating the members
of staff to cooperate. The interesting thing in this connection is that this new
Environmental Commissioner had formerly been critical of Hertie's environmental
activities and, as the Purchasing Director responsible for the Stationery Department,
had even almost brought the cooperation with the BUND to an end. He too is no
longer with Hertie either, having moved to the Karstadt head office early in 1995.

With the full-time Environmental Commissioner, the so-called Head of the
Environment Department, environmental work at Hertie acquired a somewhat more
professional touch. Occasioned by the Packaging Ordinance, the Head of the
Environment Department concentrates principally on the problem of waste. When
he dissolved the working group in favour of smaller project groups, he abolished an
important communication platform for environmental protection throughout the
firm. The Head of the Environment Department does little to counter the resulting
lack of information on the part of most Hertie members of staff – at least some of
the working group members who were previously well informed consider this a
deficit. He tends to be scientific-technical in orientation.

Unlike the Works Council representative at the Hertie headquarters – who had
previously been released from his normal duties and who, in the decisive situation,
had been able to mediate between Hertie and the BUND because of his active
membership therein – the employee representation at Hertie has been of absolutely
no importance in the process of ecological reorientation.

The environmental working groups set up in individual branch stores concentrated principally on operational opportunities in their own particular stores. Only
in exceptional cases have individual department stores in the past devised activities
of importance for the headquarters too.

5.2   The Role of the BUND

Only in exceptional circumstances has there also been an economic incentive for
Hertie from the outset to implement ecological measures in the form, say, of lower
costs, additional turnover or other competitive advantages. Only in a few cases did
Hertie have the appropriate ecological know-how to make meaningful ecological
adjustments to its range of products independently. Instead these were based as a
rule on ideas and projects proposed by the BUND. The decisions taken by Hertie

in the past to delete items from order books are the result in some cases of long and bitter disputes within the enterprise or tough negotiations between Hertie and the BUND.[31]

For a thorough understanding of the structural conflicts built into the cooperation between Hertie and the BUND, it would first seem worthwhile to explain how the BUND perceives itself and what its prime goals are.

The BUND looks upon the major objective of its work as helping to achieve the 'eco-social market economy' and it pursues this objective by the following means:

- consumer education;
- influencing the state;
- pressure on enterprises by consumer campaigns and public-relations work.

It is only since an official resolution of the BUND in 1992 that these areas can be supplemented by direct influence on firms under certain circumstances:[32]

> 'Enterprises that meet the challenge need support and critical assistance. The BUND is willing to assume this role and is open to any discussion that accelerates the social learning process. (...) Firms that refuse to submit to this learning process will continue to be subjected to pressure and publicly attacked by the BUND.' (BUND, no year: 23)

The meaningfulness of cooperation with economic enterprises is nevertheless still a controversial issue within the BUND. Internally, therefore, the cooperation with Hertie is the focus of criticism by virtue of its pilot nature. The controversy centres mainly around the different attitudes to the possibilities and limitations as well as the opportunities and risks of such a cooperation. The advocates of cooperation see it as a complementary field of activity in which ecological improvements, which could not be achieved in the BUND's traditional sphere of action, can be achieved with relatively little effort. By contrast the opponents of cooperation fear that the BUND could be misused by enterprises as a 'green fig leaf' and that it risks jeopardising its hard-won credibility as an independent and competent environmental body. In addition the opponents stress that, by entering into a cooperation, one, as a matter of principle, risks becoming financially dependent.

The ecological successes resulting from the cooperation indicate that the BUND, with its ecologically-sensitive approach, was able to influence Hertie in such a way that both Hertie's ecological and economic leeway were expanded. The incompatibility or insufficient compatibility of ecological measures with the economic objectives of the firm, as seen from Hertie's perspective, on several occasions proved unfounded when implemented – as the BUND was able to insist it did by reference to Hertie's contractual obligations. When, for example, pesticides were substituted by ecological horticultural products and publicised accordingly, there was an unexpected increase in turnover in the department in question.

---

[31] For further details see Will (1994).
[32] See BUND (1992).

Given the opposing primary objectives of the two cooperation partners however, the reasons for retaining the cooperation for both sides are of interest.

Under the given circumstances (market and social incentives for environment-oriented behaviour, lack of ecological know-how of one's own, insufficient transparency of the ecological quality of the product for the consumer, etc.) competitive considerations make it perfectly reasonable for Hertie to retain the cooperation with the BUND. On the one hand it gives Hertie access to the necessary ecological knowledge, etc., whilst outwardly, that is to say *vis-à-vis* competing enterprises as well as the consumer and the interested public, the close cooperation with the BUND indicates the credibility of Hertie's own commitment. Hertie looks to this exclusive seal of quality to give it additional sales potential with the segment of environmentally-friendly consumers. In addition the cooperation reduces the uncertainties for Hertie in that the vigilant eye of the BUND recognises potential legislation at an early stage already. Moreover, as the BUND is also one of the major social lobbies in environmental protection in Germany, Hertie can evaluate social trends in environmental protection more reliably.

As long as the BUND has an unsullied reputation as a committed and competent environmental association and, by virtue of its large membership, constitutes a popular authority with a broad social base that continues to give environmental protection a high priority, Hertie views a continuation of the cooperation with the BUND as worthwhile.

From the vantage point of the BUND the reasons for the decision differ accordingly. In pursuit of its overall ecological goals and aware that its resources on the whole are limited, the BUND has to take alternative forms of environmental commitment into consideration. As long as the ecological improvements achievable by cooperation are seen as higher than the ecological successes achieved with comparable resources outside enterprises or on occasion on a deliberate confrontation course *vis-à-vis* individual firms or branches, it is also worth the BUND's while to continue the cooperation. Admittedly one always has to ensure that, by cooperating and by the image transfer involved, the BUND's reputation does not suffer and that at no time does the BUND become dependent, resulting in it having to remain in the cooperation even under conditions inconsistent with its actual objectives.

After this brief survey of the incentives underlying the cooperation, there now follows a summary of the influence of the BUND on Hertie's environmental work:

- As a supplier of ideas for several joint projects, the BUND served as an *initiator of action* on many occasions.
- By sharing its ecological knowledge and powers of judgement in concrete decision-making situations and drawing attention to prospective environmental developments, the BUND served to *heighten Hertie's sensitivity.*
- After initially having mixed feelings about cooperation with Hertie, the BUND urged that joint activities be prepared more thoroughly. The BUND demanded that in future cooperation be solely project-oriented and it therefore performed a *systematising* function.

- When joint projects were being planned and above all implemented, the BUND repeatedly gave its *support* by preparing background material and offering its ecological expertise, etc.
- Where publicity was concerned the BUND, against the backdrop of its own demands and experience, had a *disciplinary* influence and ensured that the campaign material focused more on educating the consumer than on advertising.
- Given the ongoing cooperation and the realisation of the Hertie staff involved that environmental protection was not only a cost factor but also provided opportunities for increased revenue, the BUND opened up individual players to learning processes which opened their eyes to ecological-economic opportunities. In this sense, where environmental orientation was concerned, the BUND heightened its degree of *acceptance* to Hertie.
- The drawing-up of contracts for individual projects (stating the service to be performed and a specific date) has the effect of placing Hertie under an *obligation*. Their obligating nature results in particular from the fact that, in the event of agreed dates not being observed, the BUND is entitled to terminate the cooperation.

The above considerations will hopefully have made clear that on both sides the projects resulting from the cooperation with the BUND, such as for example the dropping of liquid laundry-softening agents from the order books together with that of pesticides and stationery containing solvents, etc., were the result of fundamental deliberations. The advantages of cooperation were carefully weighed against the economic disadvantages or risks of a specific situation containing conflict. From Hertie's point of view, every concrete adjustment to its range of items stocked gave rise to fears of unacceptable financial losses resulting from lack of acceptance on the part of the client. Internally the fact that the salaries of Hertie's purchasing staff were linked to turnover proved an obstacle to implementing these measures. Although in retrospect almost every major deletion from the order books resulted in increases in turnover, fears of a fall in turnover repeatedly led to obstruction on the part of the purchasers. In an extreme case the discontinuation of items of stationery containing solvents, that had been agreed jointly with the BUND, put the cooperation to a severe test.

6   INFLUENCING FACTORS AND FRAMEWORK CONDITIONS

Based on a comprehensive analysis of the relevant parameters, the main factors influencing the course of events are set out below.

In terms of *competition* the following influencing factors have been identified. As the *environmental awareness of members of the public and the consumer* grew, the staff in Purchasing and Marketing at Hertie became aware of the subject of ecology. In view of the *market successes of fellow competitors*, above all in food retailing, Hertie recognised that environmental protection certainly provided a means of gaining a competitive edge and that in some cases sales could be significantly increased by adopting an environmentally compatible approach. Where some segments of their range of goods were concerned, Hertie realised that an ecological approach was going to be necessary to prevent clients from seeking out suppliers in ecological niches.

The incentive to take advantage of the competitive opportunities identified by the 'environmental pioneers' of the branch can be explained not least by the *extremely stiff competition* in the retail trade. The huge shakeout within the retail sector – and in the shrinking department store sector in particular – forces department stores in terms of *competitive strategy* not only to keep an eye on costs but also to consistently take advantage of ways of *standing out* from their fellow competitors. Against the backdrop of the competition structure of the department stores, Hertie's ecological reorientation can be seen as the attempt of an organisation that has been plainly overtaken by the big two in the field to regain the initiative by, amongst other things, gaining a reputation for the environmental quality of goods and services in the minds of consumers. Given the merger in particular, Hertie looks upon its ecological commitment as a means of distinguishing itself from Karstadt.

The initial reserve in Hertie's environmental commitment can be explained by the severe crisis which it suffered in the '80s in which it incurred a loss of more than DM 600 million in all. Consequently with this concern Hertie only had limited scope for taking advantage of new uncertain trends in order to stand out from its competitors. The company at first adopted a wait-and-see approach and only took advantage of the opportunity to enhance its image as an 'early follower' after the trend had continued unabated with consumers.

In *structural* terms, the *savings opportunities and yield prospects* (savings in energy and water costs in the department stores, savings from abandoning or reducing packaging, increases in turnover by environmental campaigns and environmentally-more-compatible ranges of goods, tremendous savings potentials in the field of logistics, etc.) proved a major prerequisite and at the same time a motive for continuing Hertie's efforts to bear ecological aspects more closely in mind in its own work. In this connection mention should be made of the *virtual absence of investment* necessary for an environmental orientation, investment being an obstacle to environmental orientation in manufacturing in particular.

An unprecedented *situation-related* factor that had a positive effect favouring the process of environmental orientation was the *record turnover in the retail branch due to reunification*. This considerably alleviated the lot of the department store branch where competition was extremely stiff and provided a temporary stimulus for retailers to take risks and be innovative.

Another *branch-related* factor is the *willingness of manufacturers* to entertain Hertie's ecological demands or give their products and services an ecological thrust of their own volition. Here we see that in branches in particular where business relations for both sides, manufacturer and retailer, mean high turnover, ecological demands can be enforced with relative ease even when the manufacturer has little ecological self-motivation.[33]

---

[33] Mention has already been made of a successful cooperation of this kind, namely with Herlitz, the manufacturer of stationery and office equipment with whom Hertie has worked closely, particularly where the environmentally-friendly presentation of goods is concerned.

In addition to the market opportunities associated with an environmental orientation, *legislation*, in particular the guiding influence of *government environmental policy*, should not be underestimated. The need to act caused by the passing of the *Packaging Ordinance* presents a considerable challenge not only to manufacturers and the packaging industry but in particular to retailers as well. Now that the Packaging Ordinance has made industry responsible for the costs of recycling, the retail trade has appreciable financial incentives to exploit the possibilities of reducing shelf, and above all transport, packaging. Against the backdrop of further foreseeable ordinances supplementing the Waste Law, the retail trade will have to bear further product-related responsibilities in the future which could influence service and after-sales activities accordingly.

Another *situation-related factor* that impedes the process of ecological reorientation is that there is frequently competition in terms of economic and ecological goals due in part to the *wide range of goods* in department stores and partly also to the customer's *unwillingness to pay* for expensive environmentally-friendly products. This conflict can scarcely be solved, at least not in the short term. *Manufacturers* prove an obstacle to themselves where they use their dominant market position to refrain from altering their insufficiently ecology-oriented market strategies even on pain of sanction on the part of retailers.

At the same time it must be said from the *competition* point of view that, due to the recession, competition in the retail trade has become stiffer still and this has limited the leeway for long-term costly and risky image measures and overall has increased the cost-push in the retail trade. The willingness to accept short-term falls in turnover in favour of giving an ecological thrust to one's range of goods has markedly decreased.

The somewhat *wait-and-see attitude of the retail association* is an important element applying a brake to environmental orientation. As already stated above, the Association mainly interprets its function as largely keeping the retail trade from bearing ecological responsibility in foreseeable ordinances by mitigating, or even preventing, the need for the trade to internalise the negative external effects which the political parameters require.

In terms of *in-house* influencing factors, one is the *corporate size* of the Hertie department store concern which produces an inertia which played a major part in making it difficult to put environmental activities into practice internally, and still does. On the other hand, Hertie's purchasing power resulting from its corporate size is an effective tool *vis-à-vis* suppliers where work on its range of goods is concerned.

These have been the major parameters of Hertie's environmental orientation. To conclude, the following remarks sum-up once more the main statements made about the analysis of the players and the environment.

It is market and political factors that are identified as the principal relevant influences. A critical public plays a subordinate role, at most, for Hertie's environmental orientation. Given a high level of environmental awareness on the part of the consumer and increased environmental pressure in the packaging sector, coupled with insufficient knowledge of ecological causes and effects, Hertie in the past entered into a cooperation with the BUND which deeply committed Hertie because of its contractual nature. The situation which gave rise to the cooperation with the

BUND – a member of Hertie's staff from Marketing, who at the same time was an active member of a BUND local group, suggested the BUND as a competent advisor – shows that the course of events can be influenced by individuals to a large extent. The function performed by the *technical promoter*, who is very prominent at Hertie, was also obviously of great importance to the course of events. Without him the suggestion made by the above member of staff would probably never have been taken up, nor would the cooperation with the BUND have achieved such intensity. Without the willingness and commitment of the technical promoter to assert himself against opponents in his own enterprise, Hertie's environmental orientation would probably have developed less consistently. The role of the *power promoter*, by which we mean the Board Purchasing Director responsible for environmental protection at Hertie, also has an important influence on the process. It was only through his intervention during the above-mentioned critical stage of cooperation that the relationship could be continued with the BUND. In other ways, too, the Director showed his interest in continuing the cooperation.

The appointment of the new Environmental Commissioner for Purchasing in the spring of 1993 is a pointer to the potential stimulus that such a position can give. To link formal responsibility for environmental protection to a managerial function carrying great authority would seem extremely promising when exercised by a member of staff who perceives environmental protection as an opportunity to gain a competitive edge rather than as a cost factor. The investment necessary for environmental activities, in particular, can be achieved much more easily by a person endowed with such competences.

To conclude, reference is again made to the self-commitment resulting for Hertie from the cooperation, which in view of the arbitrary nature of the course of events, gives Hertie's environmental orientation the necessary stability and coherence.

7   CLOSING REMARKS

What is so special about the Hertie case study is undoubtedly the unusual course taken by the department store concerning ecological reorientation. Unlike many other enterprises, Hertie draws on the ecological expertise of one of Germany's largest environmental associations in endeavouring to give its corporate activities an ecological thrust, instead of adopting the course, not unusual for enterprises, of adopting a negative stance toward the ecological demands of social lobbies.

From an ecological standpoint, positive mention should be made in particular of the influence of the cooperation in expanding the leeway for action. On several occasions it emerged that the conflict between ecology and economy initially anticipated by Hertie was not so pronounced in reality; consumers, for example, were more willing than expected to pay for the additional ecological benefits. In other cases, however, the limits of compatibility between economic and ecological objectives emerged. Where the consumer did not accept the additional ecological benefit or where the ecologically-more-acceptable alternative product was unacceptably less convenient to use, e.g. disposable beverage bottles, the influence of the BUND had its limits.

It is impossible to say at present whether, and to what extent, the cooperation between Hertie and the BUND will be continued. That apart, it is clear that the above cooperation model is only applicable elsewhere up to a point, given the limited resources of the BUND as well as its explicit objectives. It would be conceivable however to set up institutions – possibly with government support – which, as in the cooperation project described, provide incentives to firms without necessarily repeating the disadvantages of the cooperation. The setting-up of a broad-based jury on which the BUND, amongst others, might sit and which could award a seal of approval in the retail trade on a branch basis would be a conceivable institutional solution which could achieve similar effects at least in terms of market incentives.[34]

LITERATURE

Adelt, P., Müller, H. and Zitzmann, A. (1990) Umweltbewußtsein und Konsumverhalten, -Befunde und Zukunftsperspektiven, In: (eds.) Szallies, R. and Wiswede, G. *Wertewandel und Konsum: Fakten, Perspektiven und Szenarien für Markt und Marketing*, Landsberg/Lech 1990, pp. 155–184.

BAG (Bundesarbeitsgemeinschaft der Mittel- und Großbetriebe des Einzelhandels e.V.) (1992) *Vademecum des Einzelhandels 1992*, Cologne.

BAG (Bundesarbeitsgemeinschaft der Mittel- und Großbetriebe des Einzelhandels e.V.) (1993) *Geschäftsbericht 1992/93*, Cologne.

Balderjahn, I. (1986) *Das umweltbewußte Konsumentenverhalten*, Berlin.

Bruhn, M. and Dahlhoff, H.D. (eds.) (1990) *Sponsoring für Umwelt und Gesellschaft. Neue Instrumente der Unternehmenskommunikation*, Bonn.

BUND (1992) Zusammenarbeit mit Wirtschaftsunternehmen. Beschluß der BUND Delegiertenversammlung 1992 in Leipzig, Bonn.

BUND (1994) *Jahresbericht 1993*, Bonn.

BUND (no year): Bundesarbeitskreis Wirtschaft und Finanzen, Kurz, R. *et al.*: Ökologische Unternehmensführung, BUND positionen 22, Bonn.

Diekmann, A. and Preisendörfer, P. (1992) Persönliches Umweltverhalten—Diskrepanz zwischen Anspruch und Wirklichkeit. In: *Kölner Zeitschrift für Soziologie und Sozialpsychologie*, **44**, 2, pp. 226–251.

Emnid-Institut GmbH (1987) Privater Umweltschutz 1987—Trendkommentar, Bielefeld.

FUUF (Forschungsgruppe Umweltorientierte Unternehmensführung) (1991) Umweltorientierte Unternehmensführung: Möglichkeiten zur Kostensenkung und Erlössteigerung.—Modellvorhaben und Kongreß—, Forschungsbericht 10901041 des Umweltbundesamts Berlin, Berlin.

Glöckner, Th. (1993) Großer Knall. In: *Wirtschaftswoche Nr. 41*, edition of 8.10.1993, pp. 184–188.

Hansen, U. (1988) Ökologisches Marketing im Handel. In: (eds.) Brandt, A., Hansen, U., Schoenheit, I. and Werner, K. (1988) *Ökologisches Marketing*, Frankfurt/M., pp. 331–362.

Hansen, U. (1992) Umweltmanagement im Handel. In: (ed.) Steger, U. *Handbuch des Umweltmanagements. Anforderungs- und Leistungsprofile von Unternehmen und Gesellschaft*, pp. 733–755.

Hertie Geschäftsberichte (1988, 1989, 1990, 1991, 1992, 1993) Frankfurt/M.

Hopfenbeck, W. (1991) *Umweltorientiertes Management und Marketing. Konzepte—Instrumente-Praxisbeispiele*, 2. ed., Landsberg-Lech.

Horten Geschäftsberichte (1988, 1989, 1990, 1991, 1992) Düsseldorf.

imug (Institut für Markt—Umwelt- Gesellschaft e.V.) (1993) Umweltlogo im Einzelhandel—Machbarkeitsstudie, im Auftrag der Verbraucherinitiative e.V., gefördert durch das Bundesministerium für Umwelt, Natur und Reaktorsicherheit und das Umweltbundesamt, Hannover, Bonn.

---

[34] On this see the proposal for an environmental logo in the retail trade developed by imug. See imug (1993).

Kolvenbach, D. (1990) *Umweltschutz im Warenhaus. Thesen und Realität*, Bonn.

Kursawa-Stucke, H.-J. and Lübke, V. (1991) *Der Supermarktführer. Umweltfreundlich einkaufen von allkauf bis Tengelmann*, Munich.

Mattmüller, R. and Trautmann, M. (1992) Zur Ökologisierung des Handelsmarketings. Der Handel zwischen Ökovision und Ökorealität. In: GfK 2/92, *Jahrbuch der Absatz und Verbrauchsforschung*, pp. 129–155.

Meffert, H. and Kirchgeorg, M. (eds.) (1993) *Marktorientiertes Umweltmanagement*. Grundlagen und Fallstudien, 2. rev. ed., Stuttgart.

FAZ vom 11.1.1989, p. 13.

Pellinghausen, W. (1993) Gemeinwohl auf Zeit, In: *Capital*, 11/1993, p. 21.

Rehmann, K. (1986) Die Hertie-Stiftungen. In: (eds.) Haner, R., Rossberg, J. and Pölnitz-Egloffstein W. Frhr. v. *Lebensbilder deutscher Stiftungen*, vol. 5., Tübingen, pp. 287–302.

Will, S. (1994) Hertie und BUND—eine Kooperation im Spannungsfeld zwischen Ökonomie und Ökologie, FFU-Report 94-6, Berlin.

Wimmer, F. (1988) Umweltbewußtsein und konsumrelevante Einstellungen und Verhaltensweisen, In: (eds.) Brandt, A., Hansen, U., Schoenheit, I. and Werner, K. *Ökologisches Marketing*, Frankfurt/ M., N.Y. pp. 44–85.

10. THE ENVIRONMENTAL FACTOR IN THE DESIGNING OF PASSENGER CARS BY NEDCAR

Geerten J.I. Schrama

1 INTRODUCTION

NedCar is a relatively small Dutch passenger car manufacturer. The firm's 'success story' concerns the achievements of NedCar's engineering and design department in environmentally-conscious designing. The case focuses in particular on specific aspects of the engineering and design process: the choice of materials plus the facilitation of future disassembly and recycling of the materials used. NedCar is not an absolute front-runner in this field; this is impossible for such a small car manufacturer with limited means. The essence of the success story lies in the fact that – thanks to the substantial internalization of environmental consciousness within the engineering and design department, and the experience and skills acquired through the years – NedCar has been able to keep up generally with the front-runners within the European automotive industry, and even to be in the forefront in some respects.[1]

Research for the case study was conducted during 1993–1994. The findings reported here are based on interviews with four respondents: the firm's Environmental Officer, who is also NedCar's specialist on polymers, the Head of the Vehicle Engineering Department, the Director of Product Design & Engineering, and a representative of Volvo Car Corporation in the Netherlands, a person responsible for the environmental aspects of NedCar's design efforts. The environmental specialist from NedCar (the contact person for this study) and his colleague from Volvo Car Corporation were interviewed a number of times. Additional information has been obtained from internal and external written sources as well as external informants.

2 ORGANIZATIONAL CONTEXT

In this section the organizational context is reviewed. Coverage focuses on NedCar, the company under study; its relationship with Volvo Car Corporation, its most important shareholder and 'principal'; and the Product Design & Engineering unit, the focal department in this study.

---

[1] An extended version of this case study is published as Schrama (1995).

NedCar

NedCar Netherlands Car BV is the sole Dutch producer of passenger cars. Its formal status is as an independent company. Ownership is equally divided among three shareholders: Volvo Car Corporation from Sweden, Mitsubishi Motor Corporation from Japan, and the Dutch state. The government intends to end its partial ownership before 1998.

NedCar is the successor to the former Dutch passenger car producer DAF. In the 1970s DAF was transformed into a company called Volvo Car Netherlands BV (short: VolvoCar),[2] owned by the Dutch state (70%) and Volvo Car Corporation (30%). Although its products were successful, the company itself has never shared in that success. It has been supported financially by the Dutch state in a number of ways to remain a (formally) independent and integrated or full-fledged car producer. At the end of 1991 VolvoCar was transformed into NedCar, when Mitsubishi entered as the third shareholder.

NedCar made an ambitious start in 1991. First of all, it was unique to produce models of two different brands within the same plant, as was done at NedCar. Furthermore, the shareholders agreed to invest hfl.700 million each in the reconstruction of the production site in Born, and in the development of two new models, a Volvo and a Mitsubishi. The annual production capacity had to be doubled to 100,000 cars of each model.

Both headquarters and production site are located in Born (province of Limburg). Currently, three models in the Volvo 400-series are in production: the Volvo 440 (a five-door family car), the Volvo 460 (a four-door, more luxurious version of the former), and the sports model Volvo 480, especially developed for the US market. A total of 490,000 Volvo 400s from this facility had been sold by 1994.

NedCar went through difficult times in 1992 and 1993. On total sales of about hfl.2 billion, the net results showed unprecedented losses of about hfl.240 million in each of these years. A drastic cost reduction programme was set up, and this effort resulted in more favourable preliminary figures for 1994, probably without any losses. VolvoCar employed more than 9200 in 1990. Through the sale of subsidiaries and other measures, this number was reduced to 4300 by the end of 1993. Further reductions, to 3900 in 1998, have been announced.

Because NedCar is eager to be an 'integrated car producer', the engineering and development section (until recently a distinct division) is equipped for the full development and testing of passenger cars. The departments concerned are located at Helmond (province of Noord Brabant, near Eindhoven).

While the case study was conducted, NedCar was preparing the introduction of two completely new models. Meanwhile, the *Mitsubishi Carisma* has been introduced in 1995. The design has been adapted for the European market by NedCar from the original Japanese model. The Volvo 40, developed by NedCar's Product Design & Engineering and successor to the Volvo 400-series, was introduced to the market in 1996 as the S40 and V40 models.

---

[2] The name 'VolvoCar' is used in this study to refer to NedCar's predecessor. In order to avoid misinterpretations, the Swedish shareholder is referred to as 'Volvo Car Corporation.'

Structure of the NedCar Organization

NedCar's administration consists of its management board plus seven directorates: Finance & Economy, three production directorates, plus three directorates that formerly constituted an Engineering and Design division. When VolvoCar was transformed into NedCar, a number of units were shifted to Volvo Car Corporation. One of the respondents characterized these units as *'the head and the tail of the car production'*. Most of the organizational units involved are still located in Helmond as part of the Snava Holding, a subsidiary of Volvo Car Corporation.

Product Design & Engineering

Product Design & Engineering is more or less the core of NedCar's development activities. Its organizational chart reflects a functional division of departments. The activities, however, are mainly organized through project teams, with participants from all the relevant departments.

Environmentally-conscious design has been incorporated into the revision projects of the Volvo 400 models, as well as in the design of the new Volvo 40. Implementation of NedCar's policy in this regard is coordinated by the 'specialist for polymers and environment', who is a member of the Vehicle Engineering Department.

Most of the engineering and development projects performed by NedCar and its predecessor during the last ten years concerned the Volvo 400-series. These models were introduced on the market between 1986–1989. In 1993 both the Volvos, 440 and 460, received 'face-lifts'. Other projects concerned the adaptation of new engine types, supplied by Renault, with specific characteristics such as turbo compressors, catalysts, injection systems or diesel gas, and the introduction of transmission systems, as well as the smaller annual model modifications. Much effort is currently being invested in the development of the new Volvo 40 and the Mitsubishi Carisma. Most of the effort focuses on the Volvo 40 project, which was initiated by VolvoCar

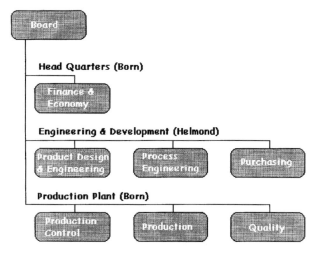

Figure 1   NedCar Organisation Chart: seven directorates at three locations.

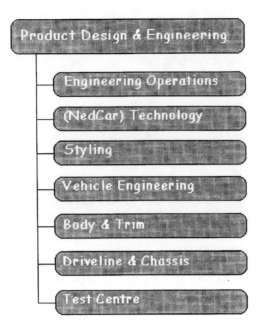

Figure 2   Product Design and Engineering: organisation chart.

and has continued. A special Volvo 40 project team has been formed, a group containing specialists on all relevant disciplines; the team includes the specialist for polymers and environment. The project leader, however, is a representative from Volvo Car Corporation. Volvo is thus able to monitor the proceedings very closely.

In the negotiations about the transformation of VolvoCar into NedCar, the prospects of Engineering & Development were discussed extensively. Dutch industrial policy specialists feared that NedCar would eventually be confined to assembly activities. From this perspective it was important that NedCar did not become just a 'screwdriver factory' (a frequently-used phrase), without any engineering and development capacity of its own. Such an outcome would mean the end of another Dutch basic industry. During the last few years, the number of employees was reduced by one half to 600. Under pressure from the Dutch state, the shareholders of the newly-founded NedCar agreed to maintain the Engineering and Development Department at its present size at least until the end of 1995.

In line with this agreement, Mitsubishi was the first to guarantee future engineering and development tasks to NedCar. These will likely be limited, nonetheless, to the adaptation of Japanese designs to the European market. Recently Volvo Car Corporation followed suit. These developments do not mean, however, that NedCar will ever develop another complete Volvo model. Volvo is expected to treat NedCar in the future like any other external supplier, and to base its 'make or buy' decisions on economic considerations (interview). The commitments expressed thus far are probably insufficient to employ 600 people after 1995 (Volkskrant, 1993). To keep its engineering and development section at the present size after 1995, NedCar is searching for ways to become less dependent on its industrial shareholders.

3 ACHIEVEMENTS IN ENVIRONMENTALLY-CONSCIOUS DESIGN

This section gives a survey of the achievements triggered by NedCar's policy of environmentally-conscious design. Due to the limited discretion granted to NedCar's engineering and development staff (discussed below), the successes are confined to a number of tasks related to the design of the car body and the interior trim. The official 'Guidelines for Environmentally-Conscious Design' is reviewed in this section. In addition, 'Project Goes' is discussed, a pioneer effort for the disassembly and recycling of passenger cars.

Choice of Materials in Vehicles

The average amount of plastics applied in European passenger cars, with an average total weight of 1000 kilogrammes, is about 13%–14%. About 90% of these plastics are 'commodity plastics', nearly 10% of them 'engineering thermoplastics', plus a small fraction of 'specialities'. Any further increase of this amount is said to be dependent on breakthroughs in plastic production technologies and in basic design shifts in the concept of the automobile (Den Hond et al., 1992; Groenewegen and Den Hond, 1992).

The main environmental impact of automobiles, of course, is fuel consumption and the related emissions of exhaust gases. These issues, however, are only partly affected by the choice of materials. New developments concerning environmental properties of motor and catalyst technology, fuel composition, and aerodynamics are of much greater importance in this respect. Meanwhile, the substitution for steel by plastic has brought about a relative weight reduction, and hence an improvement in fuel economy, which is generally considered to be an environmental improvement. An important argument in favour of plastic, in addition to weight reduction, is the fact that the production of plastic parts requires far less energy than is used in the manufacture of similar steel parts. Arguments against plastics include, for instance, the use of finite resources and recycling problems (Den Hond et al., 1992; Groenewegen and Den Hond, 1992).

Guidelines for Environmentally-Conscious Design

Within the automotive industry, environmental considerations are only of minor importance with respect to the trade-off between steel and plastic as such. From an environmental point of view, it is much more interesting to look in more detail at the way distinct polymers and other materials are applied. The environmental properties of NedCar's products have developed over time in response to the policy objective of environmentally-conscious design. By and large, these developments are in line with those of the European automotive industry in general.

Although NedCar is a minor car producer, with limited engineering and design facilities, it pays much attention to the environmental aspects of automobile design. The 'Guidelines for Environmentally-Conscious Design' is a major administrative tool for Product Design & Engineering's environmental policy and has the status of an official NedCar standard. The guidelines apply to the design of new models as well as to 'face lifts' (major revisions of existing models) and design activities by sub-contractors. With respect to the latter, some environmentally-related requirements were added to the selection criteria for NedCar's suppliers.

The first version of the 'Guidelines' was formulated by the environmental officer in 1990. Requirements were based on several sources: Volvo Car Corporation's corporate policy, earlier NedCar research, and plain common sense. The updates include new experiences, such as the results of the 'Project Goes' or Life Cycle Assessments.

The guidelines are not coercive in nature. Apart from the inclusion of a list of 'forbidden materials', which has an obligatory status, they are formulated mainly as points of attention or as suggestions for the designers. Precise instructions are provided only by way of examples. The introduction mentions that no design can comply with all of the guidelines to the same extent, and that a designer will always have to compromise.

Below, the individual items – corresponding to the chapters of the document – are reviewed. This discussion also includes actual achievements with respect to these items.

Weight reduction Designers are advised to base the material thickness on experience with previous products, and to optimize through stiffness and strength calculations. Initially, the material should be as thin as possible, since it can be adjusted later on if necessary. Designers are encouraged to find creative solutions in construction and with choice of materials.

With respect to the impact on the vehicle weight, the sharp increase of the amounts of plastics, from 7% in a 1976 Volvo 300 to 11% in a 1988 Volvo 440, constitutes a major feat. NedCar's Product Design & Engineering considers weight reduction to be the decisive factor in favour of plastic (NedDoc). The unit expects

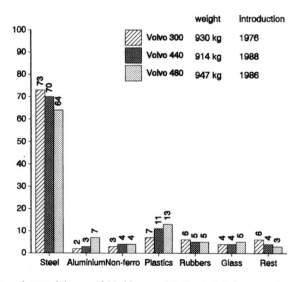

Figure 3 Weight and materials use of NedCar models (materials in percentages of weight). (Source: NedDoc)

the plastic content of their products for the European market to rise to 15% in 1995, and will stabilize at 15% in the year 2000 (Groenewegen and Den Hond, 1992).

It is difficult to make unequivocal statements on the trend in Volvo model weights in the NedCar facility. The Volvo 440 is somewhat less heavy than its predecessor, the Volvo 300, although it is larger and has some extra safety devices. The new Volvo 40, however, will probably be heavier than the Volvo 440, due to additional safety provisions and class upgrading (interview).

The Volvo 480, designed specially for the US market, shows some remarkable differences in materials composition compared to the Volvos 440 and 460, and is heavier than these models. Due to US regulations, the Volvo 480 has to meet more stringent requirements on safety and emissions. The bumper, for instance, must remain undamaged after a collision of 8 km/h, whereas the European standard is only 4 km/h. Thus NedCar has developed a special PUR energy absorption system for the bumper of the Volvo 480 (a typical example of a safety measure that leads to a weight increase). Other deviations are due to market demands for higher levels of luxury and comfort.

Production process   The design of products has an important impact on the ensuing production process. Environmental considerations include emissions, energy use, waste, and so on. In general, plastic parts are preferable to metal parts in view of the energy used in their production and during product life.

A major impact on the environment from the production of cars has always been the emissions of organic solvents from the paintshops. The application of metallic paint, which is more durable, leads to about four-times as much emission of solvents than traditional paints. In the case of metallic paint, the application of sustainable water-based alternatives involves an extra cost of about hfl.200 per vehicle. Based on marketing research, NedCar has decided to switch to water-based metallic paint. Its introduction, however, has been postponed until the start of the production of the Volvo 40, so as not to hinder production or to absorb the costs of refurnishing the paint shop in Born twice (interview).

The elimination of harmful materials[3]   NedCar uses lists of forbidden materials, and of materials to be avoided. They are part of Volvo Car Corporation's environmental policy. Both lists are added as annexes to the 'Guidelines for Environmentally-Conscious Design'. '*Forbidden materials shall not be used for new models in processes, components, assemblies and products.*' The lists contains eleven items: CFC/HCFC, asbestos, some heavy metals, a number of chlorinated chemicals, and halogenated flame retardants. '*Materials to be avoided should, preferably, be replaced or not used. Possible use must be responsible and as limited as possible.*' This list contains additional items, mainly heavy metals and toxic or hazardous chemicals.

The actual impact of these lists, however, is limited. Although an assessment was made of the forbidden materials used in the present models of the Volvo 400-series, and some adjustments have been made, the product still contains some forbidden materials. The lists were not applied during the recent 'face lifts' of the Volvos 440

---

[3] The elimination of harmful materials is not a separate chapter of the 'Guidelines for Environmentally-Conscious Design'.

and 460. The new Volvo 40 will be free from 'forbidden materials', but not completely free from 'materials to be avoided'. The latter are sometimes used for technical or economic reasons (interview).

Nevertheless, NedCar has recently carried out a number of elimination programmes to get rid of harmful substances. According to Volvo's corporate policy, all cadmium and asbestos were recently substituted. Cadmium had sometimes been used as a pigment. Asbestos had been used, for instance, in friction materials in the brakes, clutches, and gaskets. Stimulated by a voluntary agreement between the Dutch Ministry of Environment and the Dutch automobile importers, NedCar gave priority to the elimination of asbestos and accomplished this goal within the terms of the covenant – before Volvo Sweden did so (interview).

From model year 1993 onwards, all Volvo Car Corporation models have been produced completely free of CFCs and HCFCs. The elimination of CFCs is in line with the Montreal Protocol, an agreement involving a worldwide 50% reduction over the 1986–1998 period, and is a common policy in the automotive industry.

NedCar started its CFC project in 1989. The targets were the same as those prevailing in Volvo's corporate policy. But NedCar's environmental specialist, who was responsible for the implementation of this project, saw no reason to follow the same gradual approach. He tried to ban all CFCs at once. An inventory produced by him resulted in a list of about sixty items in which CFCs were involved. In cooperation with the purchasing department, the relevant suppliers were approached for information about CFC-free alternatives. The use of CFCs in the 1990 models of the Volvos 440 and 460 was only 42% of that in the 1986 models of the Volvos 340 and 360. In 1991 this figure dropped below 15%. By the end of 1993 all CFC applications had been eliminated (NedDoc; interview).

The main CFC application in the vehicles was the refrigerant in the air conditioning units, which are only available as optional devices in the NedCar models. The main CFC application in the production process was in the production of PUR foam. In the case of the rear spoiler of the Volvo 440, the PUR foam was replaced by a somewhat harder foam type that does not require CFCs. Additional advantages are that the new spoiler is made entirely from a single recyclable thermoplastic, and the shift reduces vehicle weight by two kilogrammes. Other PUR foam applications involved were the seat cushions, the dashboard, and the steering wheel (NedDoc; interview).

Another design policy is to look for substitutes for PVC applications. In the 1988 Volvo 440 PVC amounted to nearly 2% of the total vehicle weight, the largest proportion of all types of plastics. Recently some PVC applications in the Volvo 400 series have been eliminated. In 1993 PVC was removed from the retainer profiles. At the end of 1993 the PVC underbody coating was replaced by a another type of coating containing 'active filler materials', which release only 5% of chlorine at incineration. According to a Life Cycle Assessments, this option was preferable, because alternative materials free of chlorine were heavier.

The approach to PVC is somewhat different from that for dangerous substances, for PVC is not harmful in itself, provided the risk of chlorine emissions from PVC waste are controlled. The respondents from the engineering department indicate that there is no absolute need to ban all PVC, as long as PVC waste is taken care of.

PVC is still applied in a number of parts in the Volvo 400 series. In the Volvo 480 both the foam and the foil for the dashboard are made of PVC. In the 440 and 460 models PVC is used only in the foil portion, which is made of ABS and PVC. A dashboard is a complex component with many functions, which has to meet all kinds of requirements with respect to stiffness; energy absorption; noise suppression; comfort; and resistance to temperature, material degradation, and expansion. Product Design & Engineering is working on a PVC-free alternative, which will also satisfy other environmental requirements (NedDoc; Den Hond *et al.*, 1992).

Interior door panels, usually constructed of a wooden frame with PVC foil, are subjected to a Life Cycle Assessment. The PVC foil has the advantage of matching well the rest of the interior trim and allowing the attachment of textile. It has now been decided to replace the PVC, and two replacement options are suggested by an eco-based analysis. A wooden frame with PP/EPDM foil cannot be recycled and generates only harmless waste, while a PP frame with PP/EPDM foil can be recycled. Both alternatives are more sustainable than the present component. Weight might be the decisive factor in determining the final choice (NedDoc; Den Hond *et al.*, 1992).

Choice of materials   In general the recyclability of steel is much better than that of plastics, but this is no major consideration in the choice between them. However, when it comes to the choice among different types of plastics, recyclability is more important. Another supplement to the 'Guidelines' contains a survey of the environmental effects of different kinds of polymers. A general recommendation is to avoid materials harmful to the production personnel, the consumer or the environment. Designers should: replace foams with other materials when possible, consider the use of natural materials like wood or cardboard, and use eco-balance techniques.

In choosing among plastics, NedCar prefers thermoplastics above thermosets for reasons of recyclability, although there is no policy of phasing out thermosets altogether. The Volvo 480 has a sheet moulding compound bonnet (a regular technique of producing plastic car body panels from thermosets), while those of the Volvos 440 and 460 are sheet steel. The design choice for the 480 was driven by several considerations. Styling requirements were hard to meet in steel, and the shift yielded a weight reduction of three kilogrammes. NedCar also wanted to gain some practical experience with thermosets applied in relatively small series. Between SMC and RTM, the economic advantages of SMC – its lower cost and shorter production cycle – were preferred above the lower weight and higher stiffness of RTM. The design of this bonnet was contracted to a US-design bureau. However, Product Design & Engineering is not completely satisfied, because obtaining the required surface quality takes too much effort, and the strength and recyclability of SMC are not optimal (Den Hond *et al.*, 1992; Groenewegen and Den Hond, 1992).

Similar materials   With respect to recyclability, modules should consist of just a few types, or ideally of one single type of material. For duplex materials this means, for instance, using foam and foil of the same material – sun visors made from PP foam with PP foil are an example. Combinations of different types of plastics should be screened as to their mutual compatibility for recycling. Dissimilar materials should be easy to separate during disassembly or scrapping phases. Separating metals from plastics, for instance, is easier than separating different types of plastics. This consideration has to be taken into account in the design of connections, such as metal clips for attaching plastic parts. Finally, for putting a text or code on plastic parts, imprints, reliefs, or labels are preferable to adhesives.

After restyling, the bumper of the Volvos 440 and 460 consisted of a cover and filler material made of PP/EPDM, joined with a steel beam in an easily detachable way. Many options were considered during the design process. Steel or aluminium were not suitable for the cover because of their lack of elasticity. Alternative polymers were rejected because of various technical considerations, so PP/EPDM remained the only option. Originally PUR was intended as padding. When it turned out that this was not required to meet European collision standards, PP/EPDM was selected. Considerations included the fact that it was also used in the cover, and it was somewhat lighter than PUR. Prospects for recycling the nine kilogrammes of PP/EPDM used in the Volvo 440 are under study (Den Hond *et al.*, 1992; NedDoc).

Endurance   According to the NedCar design standards the service life of complete cars, as well as that of individual parts, must be as long as possible. The option of recycling parts when the car itself must be dismantled should be taken into consideration. Wear of the parts must be avoided as much as possible. If not, simple maintenance or replacement must be possible. Surfaces sensitive to corrosion must be avoided wherever possible, and the life of protective coatings should match that of the entire vehicle.

Serviceability   Service friendliness yields environmental benefits because pollution or waste during maintenance are avoided, and the service life of the vehicle is extended. At the design stage, attention must be paid to draining and refilling liquids, the ease of disassembling parts, the ease of re-assembly after repair without the need of special tools, and the use of detachable connections, preferably with re-useable click-on connections.

Detachability   The possibility of easy disassembly is also relevant to recycling, especially with respect to the costs of labour. NedCar designers are advised to: design assemblies as modules which are recyclable as a whole; make detachable components easily accessible; use standard connections in order to reduce the need to change tools; prefer screwing, clicking-on, and clamping over welding, soldering, and gluing; make bolts, nuts, and screws accessible to electric or pneumatic tools; and place cable harnesses behind removable panels instead of locating them in closed parts of the coachwork.

Material identification   Recycling materials during the dismantling stage can be facilitated by marking the parts with material codes. NedCar uses the same codification system for plastic parts as Volvo Car Corporation and Mitsubishi. All parts weighing over 100 grammes are marked with a polymer type code. This system, developed by the German Car Manufacturers Association (VDA), is based on the ISO 1043 and ISO 1629 standards on plastic symbols (Groenewegen and Den Hond, 1992 : 20 + 116).

Application of regenerates   Another NedCar design policy is to apply regenerates wherever possible. Apart from other quality standards, the main obstacle to the use of regenerates is their appearance. Therefore, regenerates are mainly used in the Volvo 400-series in out-of-sight parts, such as the wheel-arch liners and several air deflector plates (all made out of old battery casings), heat shields (made from PET soft drink bottles), rear light housings, noise insulation materials (partly made from plastic production waste), and the rear spoiler (partly made from damaged and discarded. bumpers). Plastic battery cases are among the few plastic parts from automobiles for which a logistical system of recycling is already operational, due to the fact that the batteries themselves have a high value.

In this respect NedCar can be said to be one of the front-runners in the European automotive industry (informant at the Ministry of Environment).

'Project Goes'

Several European car producers have set up pilot projects regarding the treatment of discarded cars. Common attributes are emphasis on the recycling of plastics, the assessment of disassembly costs, and design clues for recycling or dismantling.

One of these pilot projects was 'Project Goes', executed in 1992. This was initiated and funded by the Dutch Ministry of Environment, and conducted by the industries involved. NedCar participated as the representative of the RAI, the Dutch branch organization of automobile producers and importers. Other participants were GE Plastics, as the representative of the Dutch plastics industry, and the StiBa, the branch organization of the car wreckers and dismantlers.

In their review study of seven European pilot projects, Den Hond and Groenewegen (1993) presented a classification of the strategies involved. Two dimensions were distinguished: the level of innovativeness with respect to technologies and the organizational infrastructure. 'Project Goes' was classified as incremental or adaptive on both dimensions.

NedCar was one of the initiators of this project, and its environmental specialist participated on the Supervisory Board as well as in the working group. Volvo Car Corporation also participated through Supervisory Board involvement on the part of the environmental specialist in Product Planning and Design. Furthermore, NedCar made an important contribution by submitting 130 Volvo 440s for dismantling. These vehicles originated from the 'nil series', built for testing purposes and not for sale. On the basis of the actual dismantling of these vehicles, the most favourable sequence of disassembly and the times required for each component were assessed. The aim was to recover 90% of the total weight of the cars.

The economic aim of the project was to get a maximum amount of materials available for re-use at a minimum of disassembly costs. For all components a ratio of disassembly costs per kilogramme was assessed, and this review yielded a priority ranking. As a next step, an analysis was made of the market value of the resulting materials, estimates which turned out to be negative in most cases because of transportation and treatment costs. Plastics and, to a lesser extent, electric cables and components were positive exceptions. The disposal costs of the remainder might be reduced further by separating those materials that do not need to be treated as hazardous waste. The extra costs can be compensated by lower removal costs.

The project yielded only limited benefits to NedCar. Some clues were obtained for smoother deconstruction of the Volvo 440, and these served to adapt the 'guidelines for environmentally-conscious design'. Furthermore, NedCar developed disassembly manuals for the Volvo 440 and other models, an effort which is also in line with Volvo's corporate policy. The respondents stress the long-term benefits. These experiences will probably be useful at the point when automobile manufacturers are held responsible for their own products when they are discarded.

4   DISCRETION LEFT TO NEDCAR

In this study it has become quite clear that Volvo Car Corporation is the major stakeholder with respect to all strategic choices made by NedCar. This also includes NedCar's environmental policy. The automotive industry is very competitive, and manufacturers devote considerable effort to marketing their products. Because the NedCar makes are marketed under the name of Volvo, they have to fit with Volvo marketing concepts.

Volvo Car Corporation's Environmental Policy

During the 1980s, Volvo Car Corporation started to give more attention to environmental issues. Through its public relations Volvo stated that automobility was not their top priority. In 1988 the corporate environmental policy was modelled more explicitly by the firm's announcing that environmental care ('care for people') was a fourth cornerstone, in addition to safety, quality, and reliability. Hence, the principle of minimizing any environmental burden stemming from Volvo's products and production processes was formally announced as corporate policy. The implementation of this policy is considered to be successful (according to Rothenberg and Maxwell, 1993).

Of course, commercial considerations do play a major role. Environmental care was proclaimed as fourth cornerstone at a time when public awareness of environmental issues was perceived as paramount in the Netherlands as well as in Sweden. This policy was also very actively publicized.

The Volvo marketing concept is based on a so-called product profile, which is attuned to the specific target group of Volvo buyers. The four cornerstones refer to features of automobiles for which Volvo aims to deliver the best performance compared to models of other brands in the same market segment: safety, reliability, endurance, and environmental impact. A remarkable achievement has been that Volvo was the first brand to apply regulated catalysts with lambda-sondes (interview; Rothenberg and Maxwell, 1993). As a consequence Volvo is not emphasizing other features, such as power or performance (speed, acceleration rates and so on). Although these latter aspects are not neglected, Volvo has consciously refrained from competition with brands like Mercedes and BMW on them.

Volvo Car Corporation's Product Planning

In the initial phase of development projects for new models, market investigations, technical innovations, and legal demands have to be balanced. At Volvo Car Corporation, this phase is called product planning. Its execution is assigned to different organizational levels.

At the corporate level, the overall Product Planning Department is responsible for scanning and monitoring market demands. This department issues the Volvo product profile based on the four cornerstones, and the profile defines the specific characteristics of a 'Volvo'. At the next level specific 'business areas' are distinguished – for instance, for the Volvo 400-, 800-, and 900-series. Each business area has its own Product Planning Department.

After the decision is taken to develop a new model, Product Planning is responsible for formulating the vehicle specifications or 'FKB' (the Swedish abbreviation for 'functions and characteristics description'). This is an important document which serves as input to the Engineering & Development Department, where the actual design takes place. In the case of the Volvo 40, the FKB is also part of the contract between Volvo Car Corporation and NedCar.

In 1981, at the outset of the project that resulted in the Volvo 400 series, the Product Planning Department involved was part of the VolvoCar organization. When NedCar was formed in 1991, however, Volvo Car Corporation did not include the strategic functions for product planning and marketing in the new company. Some of the tasks involved were transferred to Sweden. Another portion stayed in Helmond and was assigned to Product Planning and Design. This unit, which is formally part of 'Snava Holding', has about 20 employees and reports to Goteborg. In practice it serves more or less as an intermediary between Volvo Car Corporation and NedCar.

Environmental aspects in the FKB   The FKB contains the vehicle specifications concerning a large number of characteristics. These are formulated in measurable quantities – either as demands that need to be satisfied, or as aspirations or goals to be pursued by the designers. Often the specifications are formulated in terms of a margin beyond the prevailing legal requirements. For some specifications, minimum current levels as well as more ambitious future goals are stipulated. The FKB involves all characteristics of the vehicle design, organized in individual sections. Special attention is paid to the four cornerstones, on which more ambitious specifications are formulated. The environmental specifications are dealt with in chapters on environment, noise, and fuel economy and emissions.

Because environment is one of the four cornerstones, the specifications in the FKB with respect to environmental aspects are aimed at entailing requirements and targets that secure Volvo's place among the front-runners in this respect. The vehicle specifications involved, however, are not formulated in terms of 'demands' but in terms of 'aspirations'. According to spokesmen from Product Design & Engineering, this approach provides their design team with some room to manoeuvre, thus appealing to their creativity. The absence of any severe and absolute requirements is explained by the fact that environment has only recently become a cornerstone. Future FKB's are expected to include more specific and restrictive directives on environmental items, as the experiences of the various Volvo designing teams and the solutions they have found come to be institutionalized.

Meanwhile the designers at NedCar's Product Design & Engineering claim to be ahead of their Swedish colleagues on certain environmental questions. This standing is due to the limited 'carry over' in the Volvo 40 project – there have been few barriers to applying the best available techniques.

NedCar's Role

Although Production Planning is responsible for the outcome, the design process is not completely top down. NedCar's Engineering and Development Department was closely involved. Sometimes they have been challenged: 'Do you think you are able to realize this?' And sometimes they have taken the initiative themselves, putting forward their own ambitious objectives. During this phase, there has been intensive communication between NedCar's specialists and their Swedish colleagues.

Small producers tend to contract out more design activities than larger ones. This generalization also applies to NedCar. In some cases existing products from the supplier are adapted for application in NedCar's products, as has been the case for the engines. In other cases the design of components according to FKB specifications is contracted out to external designers, who are not always from the same firm as the one producing the components. This approach was used in the case of the SMC bonnet for the Volvo 480 (Den Hond *et al.*, 1992).

For the Volvo 400 series, there is not just a single FKB. Some characteristics apply to all models, others only to specific models. These differentiations continue down to the level of each individual variant, for instance the recently introduced turbo-diesel versions of the Volvos 440 and 460. The Volvo 480 holds a special position. The differences in, for instance, legal requirements on crashworthiness for this model's market have resulted in quite different designs of the components concerned.

The vehicle specifications in the FKB are not permanently fixed. The FKB is called a 'living document'. In 1986 the Volvo 480 was introduced as the first model of the 400 series. In 1988 the Volvo 440 followed suit, and in 1990 the Volvo 460. In 1993 the Volvos 440 and 460 got extensive face-lifts that resulted in new appearances. Meanwhile, new versions are being introduced almost every year: a Volvo 480 Turbo in 1988, a Volvo 440 with automatic transmission in 1990, a Volvo 440 1.6 litre in 1992. Developments in EU-regulations have required the introduction of models equipped with unregulated and, later, regulated catalysts. Furthermore, existing models are regularly updated at the time of 'model year changes' to adapt them to technological developments and changes in market demands.

Each of these events has resulted from changes in or supplements to the FKB. Sometimes such developments are anticipated in the FKB, when ambitious targets have been formulated at the start of a project, for instance on environmental or safety issues; these are not intended to be realized at the first introduction of the model, but as a working goal for the designers, to be realized later on, at some model year change. According to one of the respondents, the FKB is usually one step ahead of technological developments.

5   ASSESSMENT OF THE SUCCESS STORY

This case study is about a relatively small, and far from independent, Dutch-based manufacturer of passenger cars, a company active in the Western European market and having some access to the US market as well. Since virtually no information is available on the interaction between units or individual members of Product Design & Engineering, nor on the interaction between Product Design & Engineering and the rest of the NedCar organization, no clear distinction can be made between Product Design & Engineering as level of analysis and lower (units or individuals) or higher (NedCar as a whole) levels. Hence, NedCar, *in casu* Product Design & Engineering, appears in this analysis as an internally-consistent organizational actor.

In this section the concept of 'scope of choice' is used, which refers to the strategic options left open for an organisation under consideration of situational constraints. Three perspectives are to be distinguished: general conditions applying to all firms in the same market, specific circumstances applying to the unique firm under study, and the orientation through which the firm perceives its own scope of choice.[4]

General Conditions

The first question concerns the scope of choice open to NedCar and refers to general constraints imposed by market situations and legal requirements. The legal require-ments considered here concern the properties of NedCar's products, i.e. the automobiles. Compliance with these requirements is realized through the design and engineering process. Market situations often have a negative impact on the sustain-able behaviour of companies due to production costs and market prices. In other cases a positive impact is manifest, because 'pollution prevention pays', or the market favours environmentally-friendly products.

The European automotive industry can be characterized as a highly competitive market. Competitors watch each other very closely. Each promising initiative of any first mover is followed quickly by almost all other players. Issues and fashions alternate rapidly: engine technology (turbo, injection systems), safety (ABS, airbags), emission levels (lean engines, regulated catalysts), fuel economy, produc-tion costs. No one, not even the largest players, can be in the lead in an area such as environmental properties in all respects or for a long period of time. Addition-ally, margins are very narrow. No one can stay behind on any important issue or fashion and still survive. As a consequence the characteristic distinctions between the different brands with respect to technology, design, and style become smaller each year.

In this market Volvo Car Corporation's position, as a medium-sized player, is rather insecure after the cancellation of the merger with Renault in 1994. A trend

---

[4] This distinction is based on the theoretical framework based on the concept of 'available scope of choice' (e.g. Schrama, 1994).

towards enlargement of scale can be seen worldwide. Even the largest companies are looking for liaisons, preferably with Japanese car producers.

In this situation, the market and the law (EU-directives and national regulation) press the automotive industry to achieve a high level of environmentally-conscious design. Although a comparative analysis is beyond the scope of this study, Volvo, with its ambitious environmental policy, including the 'fourth cornerstone', is probably doing more in this respect than most other companies. Their environmental achievements cannot be explained as being driven simply by external pressures.

For NedCar, the restrictions discussed above have their impact mainly through Volvo's corporate policy, and more specifically through its Product Planning Department and its FKBs. NedCar's achievements in the field of environmentally-conscious design are not a direct response to market or legislative pressures. Naturally, this point does not alter the fact that it takes substantial devotion, skills and creativity to keep up with the front-runners even in certain limited areas, especially for a small player such as NedCar. Table 1 provides an indication of the 'available scope of choice' for each of NedCar's achievements discussed in this case study. Many items are characterized with the phrase 'general trend', indicating that NedCar is keeping up with broad developments in this field.

Specific Circumstances

The feasible scope of choice concerns the extent to which company-specific circumstances have affected NedCar's behaviour. In this case study, the limitations of the

Table 1 Indication of the objective scope of choice.

NedCar's 'Guidelines for environmental friendly design'	
Weight reduction	General trend, advanced by the proliferation of plastics counteracted by the proliferation of safety devices
Production process	General trend, special attention for CFCs and water-based paints
Elimination of harmful materials	General trend, swift banning of CFCs
Choice of materials	General trend, special attention for the use of Life Cycle Assessment systems
Similar plastics	General trend, special attention for recyclability of plastics
Endurance	General trend
Service friendliness	General trend
Detachability	General trend, central issue in 'Project Goes'
Material identification	General trend
Application of regenerates	NedCar among the front-runners
'Project Goes'	General trend, although each pilot project in the automotive industry has its specific pioneering characteristics

number of viable options due to endogenous factors (related to Volvo Car Corporation) are more relevant than those due to exogenous factors (market situation and
legal restrictions). NedCar's 'feasible scope of choice' is determined mainly by its
relationship with Volvo Car Corporation. Although the policy of environmentally-
conscious design has been fully internalized and is fully supported by Product
Design & Engineering, the company does not have much choice but to act according
to Volvo's corporate environmental policy. Furthermore, Volvo's 'product profile',
with its specific accents based on the four cornerstones, is also imposed on NedCar.

Another kind of limitation stems from the fact that Product Design & Engineering is not large enough to develop all essential components by itself. The fact that
NedCar and its predecessors long used Renault engines has already been mentioned.
The magnitude of this order gives NedCar some market power *vis-à-vis* Renault.
This leverage was used to adapt the engines involved (interview), but this sort of
influence is limited to marginal improvements of (in this case) environmentally-
relevant specifications. The relationship with Renault should be seen in light of the
pivotal and very insecure relationship between Volvo and Renault. Although the use
of Renault engines may be reconsidered in the near future, it is very unlikely that
environmental considerations will be of decisive influence.

Environmental Orientation

The final question concerns the way NedCar perceives its own scope of choice. Its
environmental strategy can be characterized as proactive and innovative. Major
incentives result from international legal regulations concerning emission levels,
waste disposal, and harmful materials. Market influences come to NedCar through
Volvo's corporate environmental policy and the FKB's or vehicle specifications. In
line with Volvo's 'four cornerstones policy', these incentives are perceived as
challenges. The challenges are translated into targets concerning the highest performance within the relevant market segments. Targets concerning the activities of
Product Design & Engineering itself are attributed to the specialists concerned, while
in other cases the suppliers are requested to take up the issue.

The success story under study concerns specific aspects of the engineering and
design process of passenger cars. As mentioned before, NedCar is not an absolute
front-runner in this field. Such a position would be impossible for such a small car
manufacturer with such limited means. The essence of the success story is found in
the fact that – thanks to the extent of internalization of environmental consciousness within the engineering and design department, and to experience and skills
acquired through the years – NedCar is able to keep up with the front-runners
within the European automotive industry in many respects, and to be in the front
line with regard to some.

This high internalization shapes the 'perceived scope of choice'. It makes the
Product Design & Engineering Department very sensitive to environment-related
market developments, pioneering activities of other, usually large, car manufacturers, new legal regulations, and customer demands. Such a culture not only helps
to conceive external demands as challenges instead of treats, but also makes an
organization sensitive to options which are both sustainable and can also further

other corporate goals. In this specific case, Volvo's four cornerstones are primarily marketing concepts: 'sustainability sells'. Besides, in NedCar's engineering and design practice some items of environmentally-conscious design also have yielded cost savings.

The relationship with Volvo, where environmentally-conscious design is also highly valued, constitutes an extra dimension of the perceived scope of choice. Communication on this issue between NedCar and Volvo is stimulating to both parties. At the same time Product Design & Engineering has come to realize that their skills and experience in this field may be a valuable asset in their struggle for survival. The respondents mentioned this issue as a potential 'unique selling point' for the acquisition of new assignments from third parties. In the course of the field work for this case study the researcher came to realize that this asset may be a factor as well in acquiring new assignments from Volvo. This point may suggest an extra incentive for NedCar to perform even better where environmentally-conscious design is concerned.

CONCLUSION

The market situation and legal regulations tend to confine the scope of choice to a large extent, thus forcing any automobile producer to remain very sensitive to environmental issues. NedCar's specific situation involves even more constraints because of its dependence on Volvo Car Corporation, the pressure for compliance with Volvo's environmental policy, and NedCar's small scale and limited means. As a result NedCar's possibilities for pro-active behaviour are confined to those aspects of vehicle engineering and design that are performed by its own department, such as choice of materials, and design for deconstruction and recycling.

On the other hand, the perceived scope of choice does not seem to entail any further restrictions. On the contrary: because of the high level of sensitivity developed regarding environmental questions, the Product Design & Engineering Department is likely to identify all sustainable options that are open to NedCar. This is the essence of NedCar's 'success story' of environmentally-conscious design.

LITERATURE

Groenewegen, P. and Hond, F. den (1992) *Report on the Automotive Industry*. Sast Project No 7: Technological Innovation in the Plastic Industry and its Influence on the Environmental Problems of Plastic Waste, EUR-14733-EN. Luxembourg: Commission of the European Communities.

Heidemij (1992) *Project Goes. Materiaal Hergebruik door Selectieve Demontage. In het Kader van Implementatieplan Autowrakken. Hoofdrapport.* z.p.: Heidemij Adviesbureau.

Hond, F. den, Groenewegen, P. and Vergragt, P. (1992) *Milieuvriendelijker Produktontwerp door samenwerking. Een case-studie in de automobielindustrie.* Den Haag: Nederlandse Organisatie voor Technologisch Aspectonderzoek.

Hond, F. den and Groenewegen, P. (1993) 'Solving the automobile shredder waste problem. Cooperation among firms in the automotive industry.' In: (eds.) K. Fischer and J. Schot *Environmental Strategies for Industry. International Perspectives on Research Needs and Policy Implications*, 343–367, Washington DC: Island Press.

Netherlands Car B.V. *Annual Report 1992*. Helmond, 1993.

NedDoc = Internal documents from NedCar.

Rothenberg, S. and Maxwell, J. (1993) 'Volvo. A case in the implementation of proactive environmental management.' Paper presented at *the Greening of Industry Conference 'Designing the Sustainable Enterprise'*, Boston, 14–16 November, 1993. Boston: Massachusetts Institute of Technology.

Schrama, G.J.I. (1994) 'The available scope of choice for corporate environmental strategies. Preconditions, environmental orientations, and motivations for pro-active environmental management.' Paper presented at *From Greening to Sustaining: Transformational Challenges for the Firm. Third International Research Conference of the 'Greening of Industry Network'*, 13–15 November, 1994, Copenhagen. Enschede: Centre for Clean Technology and Environmental Policy, University of Twente.

Schrama, G.J.I. (1995) The environmental factor in the designing of passenger cars by NedCar. Case study report submitted to the European Commission. Enschede: Centre for Clean Technology and Environmental Policy, University of Twente.

Volkskrant (1993) = *De Volkskrant*, 7 September 1993 (Dutch newspaper).

Volvo Car B.V. Annual Report 1988. Helmond, 1989.

Volvo Car B.V. Annual Report 1989. Helmond, 1990.

Volvo Car B.V. Annual Report 1990. Helmond, 1991.

11. THE INTERNALISATION OF ENVIRONMENTAL MANAGEMENT AT GE PLASTICS EUROPE

Geerten J.I. Schrama

1 INTRODUCTION

This case study focuses on GE Plastics Europe, the European branch of an American chemical company. Although the environmental policy of this company at the European level is reviewed in this chapter, the success case is demonstrated by examining specific achievements at the Bergen op Zoom production site.[1]

It can be asserted that the success story of GE Plastics derives from a profound internalisation of environmental management on the part of this firm and its management. In this respect GE Plastics, as a medium-sized chemical company, is among the front-runners of the European chemical industry. Understanding the case requires attention to events in earlier years, because GE Plastics in Bergen op Zoom had experienced numerous mishaps, one result of which was a very problematic relationship with the local authorities and population.

Demonstrable environmental improvements achieved at the facility during the years following these incidents include the following:

- emission reductions with respect to waste water, air pollution, and noise and odour nuisance;
- investments in major risks reduction, especially in the chlorine and phosgene plants, and in energy reduction, especially in the cogeneration installations;
- involvement in special projects which are very promising with respect to the recycling of plastics.

The study is based on frequent contacts with the Manager of Environmental Programs at the HSE-Europe department; interviews with two other respondents from the company's environmental management cadre; an additional interview with a management team member of GE Plastics Europe; written documentation submitted by the company; information from external sources, such as business and general newspapers, the archives of the regional ecological movement; and an interview with two officials of the Province, the regional authority involved.[2]

---

[1] An extended version of this case study is published as Schrama (1995).
[2] The case study was conducted in 1993 and 1994. Official data were updated to 1993. In the editing phase, early 1995, some actual data were added.

2   ORGANISATIONAL CONTEXT

The Company

General Electric Company is a giant industrial conglomerate, and one of the largest companies in the world: ninth on the Fortune list by sales, and first by assets (Fortune 1994). GE Plastics Business Group is its chemical division. As a medium-sized chemical company, it has concentrated on one segment of the chemical industry, engineering thermoplastics (ETP). Worldwide it is the largest producer of polycarbonate (PC).

GE Plastics Europe is the European branch, founded in 1969. Its headquarters and main production site are located in Bergen op Zoom in the Netherlands. The number of employees in Bergen op Zoom is about 2000 (January 1995). Until 1988 this was the only production site. Through the acquisition of the Borg Warner ABS-division, three other production sites were acquired in Amsterdam, Beauvais (France), and Grangemouth (Scotland). A completely new production site is being built in Cartagena (Spain).

Compared to other chemical companies, GE Plastics places a strong emphasis on marketing and customer service. GE Plastics is following a pro-active marketing strategy, thus leading the company into far-reaching explorations of new applications of GE Plastics' products (with the R&D department), and active contacts with (potential) customers in, for instance, the automotive, the electrical, and the packaging industries. Almost all large car producers are customers of GE Plastics.

The Products

Polycarbonate is a very rigid transparent plastic, applied, for instance, in car lights, noise screens, and bullet-proof glass. It is GE Plastics' main product, sold under the name *Lexan*. It is also used in blends, such as *Xenoy* (a blend of polycarbonate and PBT) applied, for instance, in car bumpers, or *Cycoloy* (a blend of polycarbonate and ABS), GE Plastics fastest growing product.

The polycarbonate production process in Bergen op Zoom is the traditional one, which is relatively hazardous because of the large amounts of chlorine and phosgene required. In the various plants on the Bergen op Zoom site bis-phenol-acetone (BPA), chlorine, carbon monoxide, phosgene, and finally the polycarbonate resin are produced. Through compounding and extrusion the resin is transformed into the final product, 'granules'.

GE Plastics Business Group is trying to develop alternative production processes without using chlorine, phosgene, or any solvents, in a pilot plant in Japan. Application of this production process in Bergen op Zoom, however, will be postponed until the present PC-plant is at the end of its economic life, because reconstruction is too expensive.

The second main product of the Bergen op Zoom site is polyphenylene-ether (PPE). The PPE polymer is based on the 2,6-xylenol (dimethylphenol) monomer, and is a GE Plastics patent developed in the '50s (Annema, 1990). PPE is traded under the brand name *PPO*. This ETP, which has an unattractive appearance, can be transformed into an attractive and high quality plastic by blending it with the commodity plastic PS. This is one of the most important ETP blends, and GE Plastics' major patent, developed in the '60s. This blend, called *Noryl*, is GE Plastics' second major product, and is used, for instance, in car dashboards and in all kinds of electronic applications. A variation on this product is a blend of polyphenylene-ether and PA 6.6 (*Noryl GTX*), developed for application in car coach-work, which is also available in sheet form.

3   ENVIRONMENTAL ACHIEVEMENTS

This section reviews the concrete environmental improvements that have been achieved in recent years at the Bergen op Zoom site. Coverage is organised on a category-by-category basis. The accounts are based mainly on GE Plastics' annual reports on health, safety and environment of 1991, 1992, and 1993. The section concludes with a discussion of two innovative projects undertaken by GE Plastics, each of which demonstrates promising ecological approaches.

Waste Water

The waste water of the Bergen op Zoom site consists of process water, cleaning and rain water, and occasionally chemicals. Apart from a few outlets into the public sewer, all waste water coming from the plants is collected in one of the two water treatment installations on site. After treatment, the waste water is drained into the Westerschelde through a sixteen-kilometres-long waste water pipeline.

GE Plastics has been implementing a waste water quality programme aimed at reductions in the amount of pollution in its waste water. Activities have included the building of a biological waste water treatment installation and several process-integrated measures taken in the plants. Figures are available beginning in 1986. Emissions through the public sewer have been rather stable, varying between 1700 and 2900 units (population equivalent units, a general Dutch standard for the amount of pollution in waste water) per year. The amount of pollution drained through the waste water pipeline reached its highest level in 1987 (11,360 units) and was then reduced considerably. In 1991 the lowest level (3465 units) was reached, while 1992 (3820) and 1993 (3635) showed somewhat higher levels (GEP 1993).

Apart from these general figures on the site level, special attention has been paid to the solid substances in the waste water from the Noryl and the Flexible Compounding plant. These emissions are difficult to control (and sometimes exceed the prevailing emission requirements by marginal amounts). These solid substances contain, among other things, heavy metals stemming from additives. The installation in 1992 of a 'dissolved air flotation' installation in the Noryl plant has yielded considerable improvements (GEP, 1992, 1993).

Air Pollution

At Bergen op Zoom GE Plastics has achieved remarkable results in reducing air pollution over the last few years, a result that can be demonstrated from a number of different perspectives.

For four specific combinations of substances and plants the emissions into the air are monitored by means of mass balances. The results are expressed as emissions in tons per year. In all cases, substantial reduction of emissions in absolute terms were realised over the 1987–1992 period:

- toluene from the BPA-I plant reduced from 44 to 10 tons;
- toluene from the PPO plant reduced from 213 to 21 tons;
- methanol from the PPO plant reduced from 151 to 35 tons;
- methylene chloride from the polycarbonate plant reduced from 153 to 57 tons.

The emissions have been reduced both in absolute amounts (tons per year), and in relative amounts (related to the production level). These figures are also expressed as indexes (1987 = 100), which show an impressive reduction, down to 25% or even less.

A more specific way of monitoring emissions into the air is to look at specific outlets or 'emission points'. The environmental permits mention emission limits for fifteen outlets, expressed as kilograms per hour. Figures on these outlets are given in the HSE annual reports for the 1987–1993 period. Most figures remain far below these limits, seldom amounting to even a fraction of them. In two cases during the past two or three years the limits have been exceeded. Furthermore, no phenol or acetone emissions whatsoever have been allowed from the BPA-II plant; the measurements have registered some emissions.

Besides the monitoring activities mentioned above, ever since the 1970s GE Plastics has been operating a system for measuring the air quality at the boundary of the site. This system includes four stations at which about 1500 measurements are made each year. In the HSE annual reports, figures about toluene and methylene

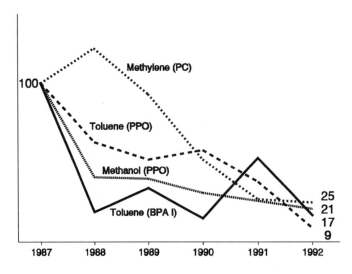

Figure 1 Indexes of emissions per kilogram product at four critical combinations (substance + plant) (1987 = 100). (Source: GEP, 1992a)

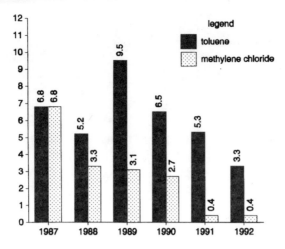

Figure 2 Average concentrations (expressed in ppb) measured at the border of the site. (Source: GEP, 1992a)

chloride are reported. The average concentrations of both substances were reduced substantially over the 1987–1993 period (GEP, 1991a, 1992, 1993).

To a large extent the reductions of emissions mentioned above have resulted from a number of specific measures taken during recent years. Some examples are:

- Several modifications in the resin plant in the past few years have resulted in a substantial reduction of methylene chloride emissions (to 43 tons in 1993).

- In 1991 a waste gas washing installation was installed in the PPO plant, thus removing 95% of the toluene from the waste gas (about 110 tons per year). Another contribution to the reduction of toluene emissions can be attributed to the 'central ventilation system'.

- In the PPO plant a special project to detect the sources of 'fugitive emissions' of toluene and methanol was executed successfully. Similar projects are planned for other plants.

- On the site there are two cogeneration units (COGEN-I and -II) for the combined generation of heat and power. In the case of the older one (the COGEN-I), the installation of 'dry low NOx burners' (developed by one of the GE divisions in the US) in 1991 resulted in a 75% reduction of the NOx emissions. For the other one (the COGEN-II) steam injection is applied, a system which was adapted in 1992. The emission levels of both installations have been below the prevailing limits of 135 gr/Gj for a long time. Since 1992 for COGEN-II and 1993 for COGEN-I, the units have also remained below the new limits of 65 gr/Gj that came into effect on 1 January 1994.

- Additional measures to reduce further the fairly low toluene emissions from the BPA-I plant were taken in 1992.

- In 1991 measures were taken in the Phosgene and Polycarbonate plants to reduce the emission of 1.5 tons of CCl_4 to virtually zero. (GEP, 1991a, 1992, 1993)

Soil

Since GE Plastics began operating on the Bergen op Zoom site, some minor contamination of the groundwater has occurred. In 1986 GE Plastics, at its own initiative, set up an inquiry into this issue. The results were communicated to the authorities, and a press bulletin was issued. With the assistance of outside experts a sanitation programme was formulated. In addition, an extensive groundwater measurement system was installed throughout the site. This system enables monitoring of the results of the sanitation programme, as well as detection of new contamination (GEP, 1991b, 1993).

Waste

In 1992, GE Plastics formulated a policy goal of waste reduction at a rate of 5% per annum. Targets at the level of the units within the plants have been established. In 1993 a reduction of only 1.1% was realised, due to increases in production level, special attention to the disposal of waste water sludge, and stagnation in the sales of specific byproducts. The amounts of both hazardous waste and landfilled industrial waste have been reduced significantly compared with 1990, while the amount of hydrocarbon fractions incinerated on site with recovery of energy has been multiplied several times.

Other aspects of the waste reduction programme concern the careful controlling and registration of (hazardous) waste removed from the site, and the handling of packaging. Wood, cardboard and PE-folio used for packing shipments from suppliers is collected and sold for reuse. Further, GE Plastics has decided to give priority to (recycled) plastics as packaging material for its own products. It aims to phase out all wood and cardboard by 1995. In 1992 the number of different materials used for packaging was reduced from 34 to 9, resulting in savings of over $2 million. (GEP, 1991b, 1992, 1993; Milieumagazine, 1993).

Noise and Odour Nuisance

Noise and odour nuisance have figured prominently among the issues which the local residents used to complain about. GE Plastics made a major effort to realise the required reduction of the noise level, a precondition for their expansion plans of 1980, and the company succeeded. In the early '80s the normal noise level was reduced below the prescribed level of 53dB(A) at night (according to measurements by GE Plastic's environmental laboratory at the borders of the site), and it has been maintained at this level ever since, notwithstanding many further expansions. In the past few years several measures have been taken to reduce the noise level from specific sources.

The number of complaints addressed to the Province, used by GE Plastics as an indicator of the degree of the noise problem, shows some fluctuation over the years but was reduced to five in 1992 and two in 1993 (GEP, 1991a, 1992, 1993).

A similar result was obtained with respect to the reduction of odour nuisance. The Noryl plant used to be the major source. In 1990 a substantial improvement was achieved by the revision of the waste gas treatment installation ('Adsox'). However, this did not solve the problem completely, and a test of 'bio-filtration' was performed in 1992. The results of this test were satisfactory, because 98% of the offending substance was removed. This device was scheduled for permanent installation in 1995. A project to improve the effectiveness of the 'vent scrubber' in the BPA plant has also been scheduled (GEP, 1992, 1993; interview).

Meanwhile, the number of registered odour complaints fell sharply during the 1980s and has since then been stabilised at an acceptable level of 12 to 14 in the 1990–1993 period (GEP, 1991a, 1991b, 1992, 1993).

Energy

According to the voluntary agreement reached by the Dutch government and the chemical industry, the latter is committed to reduce its energy consumption by 20% within ten years, with 1989 as reference. GE Plastics has put much effort into its energy-saving programme and had realised the aim of 20% reduction by 1993. This achievement can be traced to implementation of an overall monitoring and adjustment approach, an increase in the amount of hot steam sold to a neighbouring company, and an improvement in the efficiency of both COGEN installations through better adjustment of the internal energy demand and supply. In the resin plant the possibility of energy savings up to 15% was detected by external consultants; some 10% had been realised already by 1993. A similar study was planned for the BPA-plant.

In general, the COGEN installations make a large contribution to the reduction of energy costs, since they enable GE Plastics to generate energy 2.5-times more efficiently as conventional power plants.

Apart from the reduction in absolute energy consumption, GE Plastics has a similar objective for relative energy consumption, or energy efficiency. In 1990 the relative energy consumption (expressed as GJ per ton production) rose from about 10 GJ/ton in the preceding years to 11 GJ/ton, due to the introduction of the COGEN-II and a fall in the production volume during the first half of that year. In the following years a reduction was effected down to 8.5 GJ/ton in 1993 (GEP, 1991a, 1992, 1993).

Polymer Recovery

In 1991 the subsidiary 'Polymer Recovery' was established, with the intent of supporting recycling projects of GE Plastics' customers on a commercial basis. Within the framework of this project, GE Plastics can negotiate with its clients conditions for retrieving its own products at the end of their life cycle, study possible applications of recycled materials in the client's products, and set up a local infrastructure for the collection of used plastics.

Agreements about take-back and recycling were concluded with major clients, such as Ford, Opel, Océ van der Grinten (copying machines), Siemens, and Nixdorf (cash registers). A major inducement for GE Plastics to undertake this project are the developments in Germany with the legal take-back duty for producers, which may be extended to apply to the whole European Union. Here GE Plastics focuses especially on the car producers among its clients, almost all of whom are involved in pilot projects for dismantling and recycling automobiles.

Although Polymer Recovery has not been very profitable yet, GE Plastics has decided to pursue this business (Milieumagazine, 1993; interviews).[3]

Polycarbonate Bottle

Since the end of the 1980s, GE Plastics Europe has been working on a polycarbonate bottle. Good market opportunities for durable bottles had been anticipated. Two types of applications were originally intended: soft drinks and milk. Polycarbonate, however, was not suited to outperform the well-known PET-bottle for soft drinks. PET, on the other hand, is not suited for milk, since it cannot withstand the heat needed for cleaning empty bottles.

When GE Plastics focused the project on milk bottles, it was opposed by the environmental movement. As a reaction to this, GE Plastics turned this project into something of a showcase, and much effort was invested in persuading the NGO involved of the advantages of polycarbonate above glass. A life-cycle analysis of alternative packaging for milk showed that polycarbonate is to be preferred over glass, as well as over cheaper plastics and cardboard, provided the polycarbonate material from old bottles is re-used two or three times. A polycarbonate bottle has a trip rate of about sixty, a glass bottle of about thirty, and a polyethylene (PE) bottle of about three. Another advantage with respect to transport and storage is the square shape of the bottles. A major advantage is the good recyclability of polycarbonate, a feature it shares with glass. Although the material costs of polycarbonate are much higher than those of PE, the impact of this on the overall costs equation is limited. The ecological movement finally had to admit that polycarbonate could be in fact an alternative to glass (FD020589; FD210391; VK041194; KIvI, 1991; interview).

Nowadays the project is very successful. GE Plastics is one of two companies in Europe producing a type of polycarbonate suited for milk bottles. Its customers are milk producers in Austria, Denmark, Germany, the Netherlands, Sweden, and Switzerland. The product sold is not just polycarbonate, but an integrated concept comprising also the design of a milk bottle that can withstand all treatments of filling, transport, consumer-handling, and cleaning, plus advice to the moulder, who actually has to fabricate the bottles (interviews; VK041194).

4   ENVIRONMENTAL MANAGEMENT

This section deals with the organisational context within which the environmental achievements were accomplished. The focus begins with a general description of the corporate environmental policy and then shifts to the environmental management of GE Plastics Europe and the environmental management system at the Bergen op Zoom site.

---

[3] In the Netherlands, GE Plastics is also involved in 'Project Goes'. More about this project is mentioned in the chapter about NedCar (Chapter 10).

Environmental Policy

One of the respondents describes the essence of GE Plastics' environmental policy –
as with its quality and safety policy – as the interlocking of technique, maintenance,
and operations. If any of these elements is missing, the policy will fail. In addition,
the company's specific corporate culture states that it should aim to be the best on
every project it undertakes, whether the effort concerns marketing, quality, safety,
or environment.

Corporate policy   This corporate culture at GE Plastics Europe, which is said to be
an important thrust behind the environmental policy, is inspired to a large extent by
the overall corporate culture of the GE Company, which is very actively advocated
by the CEO, J. Welch. Apart from 'being the best', features of this culture are
flexibility (or 'boundarylessness'), alertness (to market developments), and openness
(to good ideas). A company like GE cannot prevent things from happening in its
setting, but it can control its own perception of such developments, whether they
are seen as threats or as opportunities. One of these developments in the recent past
was public concern about the environmental impacts of companies like GE and the
resulting demands put upon them by society (interviews; GE, 1992).

The environmental policy in all parts of the GE organisation is shaped by this
perspective. Apart from the general objective of minimal environmental impacts,
GE's policy is characterised by the absolute norm of compliance with all legal
standards involved. Since legal standards in the US are – generally speaking – more
severe than those in Europe, compliance with US law is no easy matter. The opportunities perceived are not only the advantages of a green public image, but also
market opportunities stemming from clients asking for environmentally-sound
products and production processes.

The corporate policy of the GE Company overall is effected by GE Plastics
at division level. This effort has involved, for instance, a solvent reduction programme aiming at a 75% decrease during the 1987–1995 period. GE Plastics Europe
is also committed to this aim. At the end of 1992 emissions at the Bergen op Zoom
site had already been reduced to 78% of the 1987 level. This specific corporate policy
was a stronger inducement than the Dutch governmental programme stipulating a
voluntary 58% reduction of hydrocarbon emissions (KWS-2000, reference: 1981) for
the Dutch chemical industry, to be reached in 2000 (GEP, 1991b, 1992; interview).

Further, environmental protection is associated with quality management. At GE
Plastics quality means 'zero defects', that is: no accidents, no complaints, no
emissions, and perfect products. The 'zero defects' logo is portrayed all over the
Bergen op Zoom site, revealing GE Plastics' American origin.

For GE Plastics Europe, the principle of compliance with the law is translated
into the 'principle of equivalence', the guiding principle for the ecological standards
the company has imposed on itself. The idea is to use uniform standards all over
Europe, which equal the most severe legal standards prevailing in any of the
countries involved, or – in case of an absence of relevant legal or formal standards – standards which are related to 'best practices' in the chemical industry of
northwestern Europe. This principle is not an absolute directive, since legal standards based on densely populated or ecologically-vulnerable areas are not simply

transported to less critical situations, especially not if they call for large investments. Of course, the ultimate norm remains compliance with the local legislation.

Information and communication   The style of the environmental management in GE Plastics Europe, as developed in the '80s, has been praised for its openness to public authorities, the local population, the environmental movement, and the general public. In contrary to what happened in the past in Bergen op Zoom, the present style is certainly more in tune with the GE Company's culture.

The annual report on health, safety, and the environment for the Bergen op Zoom site is one of GE Plastics' major tools in this respect. Starting in 1989, GE Plastics was one of the first chemical companies to make these annual reports public. This was not initiated by top management, to contribute to the green image of the company, but was the logical result of the frequent communications with external stakeholders and the need to coordinate information on environmental issues (interview).

Some concrete actions in this respect have been:

- formally arranged consultations with the regional authorities, such as the Province and the municipalities, on environmental and safety issues at a frequency of about eight times per year;
- regular consultation and information conferences with the local population and neighbouring companies;
- a communication programme, drawn up in collaboration with the municipalities of Bergen op Zoom and Halsteren, and executed since 1989 (GEP, 1991b).

Respondents emphasise that GE Plastics is not just pursuing a green image without living up to it in reality, as some firms tend to do, but that its guiding philosophy is that such an image should result from good performance. In the publication on the Environmental Prize awarded to GE Plastics the manager HSE-Europe is quoted:

'We have learned that our external relationships prosper through openness and honesty. Everything GE Plastics does with respect to the environment is verifiable to everyone. Apart from figures on our production capacity, we don't keep anything secret.' (KIvI, 1991)

HSE-Organisation

At the level of GE Plastics Europe, environmental protection is managed by the Health, Safety, and Environment Europe group (HSE-Europe), a staff department reporting to the European manufacturing group. In relation to its tasks HSE-Europe is staffed rather leanly, with four managers and some administrative support. The head of the department is the Manager HSE-Europe (respondent for this study), the others are the Manager Safety Programmes, the Manager Industrial Hygiene and Production Safety, and the Manager Environmental Programmes (contact person for this study).

Each production site has a local HSE department headed by the HSE manager. The tasks of the local HSE departments are: measuring and registering all emissions into air, water, and soil; training and informing the staff; coordinating the relationship with the authorities on environmental affairs; and coordinating the removal of waste. Due to the increased number of legal regulations, the administrative tasks have come to dominate the informative ones (Smits, 1993).

Each local HSE department has two different supervisory relations. The first one follows the formal lines of command, where HSE management reports to the site management. The responsibility for all environmental affairs on the site level lies with site management. In addition, each local HSE department has a strong functional relationship with HSE-Europe. This concerns the (further) development, implementation and auditing of environmental protection. Although HSE-Europe sometimes has a considerable influence in this respect, its relationship to local HSE units and site managers is purely advisory.

The HSE department at the Bergen op Zoom site is staffed by the HSE manager (respondent for this study), three environmental specialists, and some analytical and administrative support staff. One of these environmental specialists is the supervisor of the environmental laboratory on the Bergen op Zoom site, which acquired the ISO-9002 Quality Certificate in early 1993.

Similar to arrangements at the European level, the HSE management at the site level holds only an advisory position *vis-à-vis* site management. Therefore, the effectiveness of the environmental policy on the site level is dependent on the personal qualities of the HSE manager. The main skills in this respect are conceived to be of a technical nature, thus enabling the HSE manager to participate in discussions about production processes and technologies, while legal expertise might be hired.

Environmental Management System

In the case of GE Plastics in Bergen op Zoom, the environmental management system has a relatively long history. According to the respondents, this system evolved gradually during the period in which GE Plastics finally got control over the contingencies that had caused such a number of incidents in the past. It was not so much a matter of implementing a specific design, but rather formalising and completing a daily practice, acquired through what one could call organisational learning. Therefore the implementation of the environmental management system in the Bergen op Zoom site can be considered a success, although the section on environmental management, encompassing the documentation of the system, was not finished and fully approved by all responsible managers until 1993. The elaboration into procedures and work documents, as well as the further implementation in the daily practice, is still in progress (interviews; GEP, 1992, 1993).

GE Plastics has an advanced measurement and registration programme. Each plant has adequate facilities to monitor its own production process. The central environmental laboratory on the site supervises these measurements at the plants and performs general emissions monitoring: the concentration of solvents in the air, the level of noise at the border of the site, the quality of the waste water, and risks of soil contamination. For all of these tasks sophisticated equipment is available.

Before 1990, environmental audits were conducted based on US regulatory standards. Nowadays, HSE-Europe has developed a system for internal environmental audits for all sites under its regime, an arrangement that is considered to be a crucial element of the environmental management system. The auditing system is based on a detailed checklist containing a huge amount of issues arranged in ten chapters. The audit is performed by a team headed by HSE-Europe's Manager for Environmental Programmes, and staffed by people from the site together with colleagues – not only environmental specialists – from other GE Plastics sites in Europe. The checklist is completed according to the results of an on-site inspection and accompanying deliberations. It takes about three to five days.

The results of an audit represent the present environmental performance of the site. The quantitative aspects are expressed as ratios, as the realised proportions of ideal situations. Such ratios are assessed for each single item, each chapter of the audit manual, and the whole situation.

The audit system was tested first at the Amsterdam site, in December 1990, and later in 1991 also at the Bergen op Zoom site. Although these were the first audits of this type conducted at GE Plastics Europe, action programmes were formulated based on the audit reports. In the 1992–1993 period all other sites with a functional relationship to HSE-Europe were also audited. The system is intended to have a cycle of two or three years.

5   THE HISTORY OF THE BERGEN OP ZOOM SITE

GE Plastics Europe was founded in 1969, and chose to start its activities in the Netherlands. The decision to come to Bergen op Zoom, between the Rotterdam and Antwerp harbours, was induced by financial support from the Dutch government through regional stimulation programmes. After GE Plastics had decided to purchase some land in the industrial area called Theodorushaven, one of the two municipalities involved, Halsteren, objected to the funding of a chemical industry near its residential areas. The other one, Bergen op Zoom, was eager to get GE Plastics, because of the opportunities for new employment. Halsteren withdrew its objections when the national government threatened to bring the entire industrial area under the dominion of Bergen op Zoom.

During the '70s the firm expanded swiftly at this site. After the start of chemical production processes in 1976, GE Plastics underwent a very dramatic period involving numerous incidents related to safety and the environment: excess acetone in the waste water drained into the ecologically-vulnerable Oosterschelde, along with continuous fugitive emissions of solvents, such as methylene chloride and benzene. Relations with the local authorities and the local population deteriorated rapidly, and each event was extensively covered by the local press.

The Phosgene Plant

The largest problem concerned the phosgene plant. A number of incidents were recorded in 1976 and early 1977. On one occasion, 18 employees had to be taken to hospital. The climax was reached when another phosgene release incident occurred in December 1977 (accounts vary regarding the size of the release: between six to seventy kilograms). Thanks to favourable weather conditions, no harm was caused to the local population. The Governor of the Province of Noord Brabant decided that production should not be resumed before adequate measures had been taken. As a consequence the entire production line of polycarbonate was stopped. In January 1978, the Governor allowed GE Plastics to resume production, after the company had agreed to reconstruct the plant as a matter of high priority. This reconstruction involved complete enclosure of the plant, to prevent any phosgene from reaching the open air. The Governor's decision – which ran counter to most of the official advice and was opposed by a minority of the Provincial Board ('Gedeputeerde Staten') – aroused national commotion. Critical comments were published in the national press. It was suggested that the provincial authorities were pressured by GE Plastics not to obstruct production by the company's threatening to leave the country immediately, despite it having made a recent announcement about planned additional hiring. It was employment, after all, which was so important to the region, and in which so much had been invested by the provincial and national authorities. It was mentioned that the design for the reconstruction of the facility was similar to what the Province had requested back in 1974, when the construction of the plant had been announced, and which had been rejected at that time by GE Plastics for technological reasons. Critics were sceptical about this argument; they were not convinced of the feasibility of such swift technological development within a few years and suggested that the issue had been a matter of money from the very beginning. The regional ecological movement contested the decision of the Governor before the Council of States (the highest authority in the Netherlands on administrative matters), but their complaint was finally rejected (April 1979). The reconstruction of the phosgene plant was eventually realised in summer 1978, when the whole production line of polycarbonate was interrupted once again for two months.

Noise Nuisance as Barrier to Expansion

In 1980, when the relationship with the local and regional authorities and the local population was still very bad, the Province of Noord Brabant made the issuing of a permit for GE Plastics' latest expansion project conditional on a substantial reduction in the overall noise level. For years there had been many complaints by the inhabitants of a residential area situated less than one kilometre from the border of the site. Violations of the prevailing permit limits had been registered quite frequently. According to GE Plastics, the preconditions demanded by the Province for the new expansions were unfeasible. Threatening to leave the Netherlands, they appealed to the national government, which in turn placed heavy pressure on the Province. In December 1980, an extensive delegation of ministers went to Noord Brabant. Two weeks later, at a conference at the prime minister's residence attended by all parties, the provincial authorities capitulated and granted GE Plastics the permit it had requested. Formally the maximum noise levels in this permit satisfied the requirements of the Province, but GE Plastics was given several years to achieve this level. Experts thought that this adjustment should be feasible, but GE Plastics doubted it. Nevertheless the company made great efforts to reach the target, and they succeeded. By the early '80s – much sooner than anyone had expected – the noise level was reduced to an acceptable level, not only in terms of permit requirements but also in the view of the local population, and this level has been maintained ever since, notwithstanding many further expansions of the site.

The Chlorine Plant

The sequence of environmental initiatives had not yet come to an end. Ever since GE Plastics began producing polycarbonate in Bergen op Zoom, the firm has needed large quantities of chlorine. When GE Plastics wanted to increase the frequencies of the transports by rail, the Province of Noord Brabant refused. Since GE Plastics needed more chlorine, the only option was to build a chlorine plant on the Bergen op Zoom site. In 1985 the Province decided to grant GE Plastics a permit for the chlorine plant. Objections to this decision from the ecological movement and the local population were rejected. They argued that the distance between the new plant and the nearest residential area was only 750 metres, while legal standards stipulated three kilometres. The Province argued that this plant would improve rather than worsen the environmental and safety situation. The only concession granted to the critics was that GE Plastics was not allowed to produce chlorine for third parties. The environmental movement appealed to the Council of States against the provincial decision. Its main argument was that the safety study conducted by GE Plastics, on which the granting of the permit had been based, was not available to the public. The verdict, issued in December 1987, was that the study should be made public, and that the decision to grant the permit should be maintained.

Toluene

The last development concerned the emission of ten kilograms of toluene gas during the repair of a clogged pipeline in September 1987. Although the incident as such was a minor one, it too caused some commotion, because GE Plastics had failed to report the incident to the Province. When the Province heard about it one month later, during regular consultations on environmental affairs, the information caused a great panic, especially at the municipality of Halsteren. A rumour had spread that the incident involved ten tons of toluene (instead of ten kilograms). When the case was cleared up, GE Plastics had to apologise for this negligence.

Environmental Prize for Dutch Industry

GE Plastics not only succeeded in ending this sequence of incidents, but also in acquiring a position among the ecological front-runners of the Dutch chemical industry.

In 1991 GE Plastics Europe received the Environmental Prize for Dutch Industry, in the environmental management category. This is a rather prestigious prize, awarded every two years by a number of national industrial and engineering organisations under the auspices of the European Union. The jury, which relied mainly on information provided by the candidate companies themselves, mentioned a number of achievements: the sizeable and demonstrable reductions of emissions of solvents and acids; open and active communication on ecological affairs; and GE Plastics' environmental management system with its unique system of environmental auditing. The modifications in the chlorine plant were also mentioned prominently (KIvI, 1991).

6 ASSESSMENT OF THE SUCCESS STORY

This section reviews the conditions explaining the success story. A distinction is made between general conditions that apply to the entire chemical industry or to its engineering thermoplastics segment, specific historical conditions that apply to the company under study, and the prevailing environmental orientation within this company, through which these conditions are perceived.

In this section the concept of 'scope of choice' is used, which refers to the strategic options left open for an organisation under consideration of situational constraints. Three perspectives are to be distinguished: general conditions applying to all firms in the same market, specific circumstances applying to the unique firm under study, and the orientation through which the firm perceives its own scope of choice.[4]

General Conditions

If one is to look for an explanation for the success of this firm, it is useful to review the market conditions under which GE Plastics is operating. Compared to more diversified chemical companies, GE Plastics Europe is relatively tightly dependent on a small number of (indirect) customers – large companies with substantial market power, particularly automobile manufacturers. Notions about sustainability from 'cradle to grave' suggest that companies are held responsible by their customers for the environmental impacts of the raw materials and production processes they add to the production chain. The automotive industry is very interested in the environmental properties of the materials applied in their own products, a concern that provides GE Plastics with a strong incentive to meet these requirements. One of the respondents commented: '*The first time customers ask for such data, you are surprised. These are among the things that used to be considered as confidential. The second time you just hand them over and the third time you provide the customer with the information without being asked for it.*' GE Plastics learned to use their environmental achievements as a marketing tool. '*We discovered that environmental performance can be used as a competitive instrument, and we are ahead of most of our competitors in this respect.*'

The prevailing environmental laws and regulations constitute another general condition. The limits imposed upon chemical companies in Europe are rather severe. Most transnational chemical companies have adopted a yardstick for their environmental standards involving, at minimum, compliance with local legislation. In the case of GE Plastics Europe, this criterion is explicitly stated in the formal corporate environmental policy, not only at division level, but also at the level of General Electric Company (this policy is, however, far more ambitious than just aiming at

---

[4] This distinction is based on the theoretical framework based on the concept of 'available scope of choice' (e.g. Schrama, 1994).

legal compliance). Such a policy is economically rational. Situations of noncompliance mean considerable trouble with the local authorities. Of course, large companies are very powerful, and it takes a lot of courage for local authorities to blame them for violating environmental standards when economic interests are at stake. On the other hand, a good public image with respect to environmental performance is a valuable, but also vulnerable, asset for chemical companies. In cases of noncompliance situations, firms cannot control the resulting publicity. Bad publicity may seriously affect a company's green image, although this result is not reflected (directly) in the balance sheet. Because the economic life of plants and installations is much longer than the terms on which new legal environmental standards are announced, chemical companies like GE Plastics have an incentive to anticipate future limits, especially in relation to large investments. The extra costs of the best available techniques at the moment of investment are often less than the costs of adjusting to new requirements after a number of years.

The main environmental permits held by GE Plastics in Bergen op Zoom are granted by the Province of Noord Brabant. The company has known some situations of noncompliance in the past. This history places a heavy burden on the relationship between the parties involved, both of which in turn invested substantial effort in improving this relation over the last decade. The aim of GE Plastics is to avoid any new noncompliance, but the firm also aims at a smooth adjustment to new legal standards, in line with the general investment programme. The Province's aim is to negotiate relatively severe standards, while avoiding confrontations, since officials know that they are going to lose if GE Plastics takes the case to the national government. According to both parties, the relationship has been satisfactory over the last few years, although annoyances occur every now and then. GE Plastics, for instance, is not pleased when specific measures taken by the company at its own initiative are included in the permit at the next revision, as happened with speed limits for driving on the site.

Specific Historical Circumstances

Although the general conditions discussed suggest that chemical companies have good reason not to neglect environmental issues, these considerations do not offer sufficient explanation for the changes observed in the present case. If one turns to the more specific situation of the company under study here, one cannot ignore the events reported in the previous section. It is not unlikely that GE Plastics in fact had no other choice than to develop a very 'pro-active' environmental strategy to stop the negative spiral of incidents and to restore the company's relationship with its external stakeholders. How this reversal was achieved is discussed in the next subsection on orientation and perceptions.

Another specific condition that should not be underestimated, mentioned by some respondents, is the age of the production site. Although GE Plastics started production in Bergen op Zoom in 1971, the first plant for genuine chemical production processes was built in 1976. From that moment the site has been developed quickly. Because the company prospered, it could afford the best available techniques in the newly built plants. When the recession of the early 1990s hit the chemical industry, including GE Plastics, the site was almost finished. At the present moment GE Plastics is quite capable of realising relatively low emission levels because of its modern installations. The respondents are aware that the time will come when legal compliance will be harder to achieve, when the installations will have grown older and the environmental standards will have become more severe.

Environmental Orientation

The general and specific conditions discussed above indicate that GE Plastics' pro-active and successful approach to environmental management was perhaps the only way to survive its profound crisis. Nevertheless, a full picture of the way in which this effort was actually accomplished has not yet been given. Such coverage would require an exploration of the way GE Plastics perceived its own situation during this period, a perspective that can be analysed in terms of the environmental orientation prevailing within the company (c.f. Arentsen *et al.*, 1994).

From the interviews with respondents at GE Plastics one gets the strong impression that at first the company's world view encompasses only attention to environmental law and regulations, a perspective that proved to be insufficient for a chemical company. By the end of the 1970s GE Plastics had come to realise that a radical shift was required. The reorientation that followed encompassed both its own performance in the field of health, safety, and environment, and the way the company communicated with external parties. The installations on the Bergen op Zoom site were considered to be modern, technically adequate, and in good condition, by company executives and also according to external experts who had judged the phosgene plant. The incidents that had occurred were attributed to human failures of the kind that could occur due to high staff turnover rates and consequent lack of experience. According to a respondent, the precautions taken in the past were considered to be 'good', but they proved to be not good enough. To avoid these kinds of incidents, the very best precautions are required.

Furthermore, GE Plastics changed its communication strategy. Earlier, the outside world had been seen as completely hostile, and GE Plastics' approach was based on this perception. The idea was: '*it doesn't matter what information you give, the papers will always write whatever they want.*' The authorities and the public, on the other hand, complained about inadequate information provided by the firm about the incidents. GE Plastics' biggest failure in this respect had been the reaction of one of its managers, who suggested on the national television network (July 1978), that the phenol found in the urine of some employees could have been caused by excessive consumption of hot pepper sauce (in Dutch: 'sambal'). Thus GE Plastics learned that perceiving the environment as hostile can easily turn into a self-fulfilling prophecy.

To survive in its present location, and to achieve the expected returns on investments, the firm concluded that the sequence of incidents had to come to an end, and that the faith of the local authorities and population had to be restored. In the summer of 1977 GE Plastics held an informational campaign for local authorities and residents concerning the problem of solvent losses. The American parent company played a major role in this transformation. First of all, a new managing director was sent to Bergen op Zoom in 1978, an individual with a good reputation for open communication. According to the respondents, the results of the reversal became manifest in a remarkably short time. The company had undergone a 'catharsis'. All employees were convinced of the necessity of change, for many of them were criticised themselves in the private sphere for their assumed contribution to environmental damage. A major event was the visit of GE Company top executive Welch (the present CEO) to the residence of the Dutch Prime Minister, Van Agt, in June 1981. In the initial period after the company decided to alter its approach, GE Plastics aimed for a low profile: no incidents and no publicity.

In the early '80s, GE Plastics gradually became the second largest foreign investor in the Netherlands, after Dow Chemicals. They became known as the 'silent investor': large investment programmes accompanied by very little publicity. Somewhat later, GE Plastics changed its attitude. In 1983, much publicity accompanied the start of the PPO plant's construction. The ceremony was conducted by the former Prime Minister, Van Agt, who had been appointed as the new Governor of the Province of Noord Brabant. For the company this was a great relief, because prominent politicians were no longer reluctant to be associated with GE Plastics. *'They were willing to be pictured with us again,'* says one of the respondents.

The respondents in this case study were asked for their explanation for the environmental achievements of GE Plastics. According to them, the *organisational culture* has been favourably disposed toward the swift implementation of an ambitious environmental policy. No time was wasted in formalising the environmental management system; it was implemented and began to function even before the manual had been completed. All respondents speak of a 'pioneering spirit'. Until recently the Bergen op Zoom site remained in its developmental phase. Investment projects had to be completed on a continual basis and new plants had to be built. This process made the company flexible, creative, and oriented towards results. Good promotion opportunities were available to the relatively young management staff. In this atmosphere new plants were established with broad discretion for plant management. And these managers had to juggle many requirements, not just environmental ones. They succeeded because they were building something new, and they could use the best available technologies in doing so.

Another factor mentioned was *the commitment of GEP's top management* to the environmental policy. The successful implementation of the environmental management system was not simply a matter of following instructions in handbooks. This commitment not only stressed the importance of the environmental management system, but also involved an emphasis on openness, an atmosphere in which problems are not settled by blaming someone but by reliance on learning and error prevention. At the same time, the American origin of the company was visible in its competitive perspective, an approach that influenced the way improvements were accomplished.

Finally, *the corporate policy of the parent company* can be mentioned. In the case of GE Plastics Europe, the influence of this factor is not completely clear. On the one hand, the corporate environmental policy of GE Company and the elaboration by the GE Plastics division are said to be rather ambitious. Every GE enterprise worldwide is expected to meet the highest demands with respect to environmental issues. On the other hand, respondents in Bergen op Zoom assign only a relatively small share of the explanation for success to corporate policy. The two examples given by the respondents seem to contradict each other. GE Plastics Europe claims to have acted rather independently with respect to the removal of heavy metals from all its products. It had already decided to ban cadmium in the early '80s, before the corporate policy on heavy metals was announced. The US parent company preferred to start with the elimination of lead (less toxic, larger amounts than cadmium), while the European branch postponed actions on lead until it had finished with cadmium. The other example concerns GE Plastics' worldwide programme for the reduction of organic solvent emissions, which imposed much more severe limits on the Bergen op Zoom site than did the Dutch national policy ('KWS-2000'). Nevertheless, branches of large companies with quite a lot of discretion, as in GE Plastics Europe, have a salient motto, quoted by one of the respondents: *'don't embarrass the parent.'*

7 CONCLUSION

This chapter tells the story of GE Plastics Europe, a firm that has acquired a leading position within the chemical industry with respect to environmental protection. Its main achievements concern considerable reductions of several types of emissions, a number of environmentally-sound investments, and involvement in promising plastics-recycling projects. These accomplishments stem from what one could call a profound internalisation of environmental management and a corporate policy objective of being the best as regards environmental care.

The intention of this case study has been to explore the conditions which may explain the environmental success story of GE Plastics. General conditions identified include market conditions along with environmental rules and regulations. These contextual features stimulate companies such as GE Plastics, which operate in the ETP-market with a limited number of powerful customers and are subject to Dutch and European law, to take at least a cooperative attitude towards demands from external stakeholders. The final explanation for GE Plastics' pro-active approach to environmental management is provided by its turbulent history, which included numerous incidents and resulted in poor relations with the regional authorities, the local population and the environmental movement. Something like a 'catharsis' was needed for the company to realise that only a pro-active environmental strategy could end this sequence of incidents and restore productive relationships with the external stakeholders. Once this shift occurred, additional conditions, such as the relative newness of the site, the specific organisational culture and the corporate policy of the parent company, all contributed to the realisation of impressive environmental achievements within a remarkably short time.

LITERATURE

Annema, J.A. (1990) *Engineering Thermoplastics (ETP's)*. Utrecht: Stichting Natuur en Milieu.
Arentsen, M.J., Klok, P.J. and Schrama, G.J.I. (1994) Milieurelevante besluitvorming in bedrijven. *Beleidswetenschap*, **8**: 240–260.

BN[ddmmjj] = *Brabants Nieuwsblad* (day, month, year) (Dutch newspaper).

FD[ddmmjj] = *Het Financieele Dagblad* (day, month, year) (Dutch newspaper).

Fortune 1994 = *Fortune*, 25 July 1994.

GE 1992 = General Electric Company, *1992 Annual Report*. Pittsfield.

GEP 1987 = General Electric Plastics B.V., July 1987. *Veiligheidsbeoordeling van de Chloorfabriek te Bergen op Zoom*. Bergen op Zoom.

GEP 1989 = General Electric Plastics B.V., March 1989. *Extern Veiligheidsrapport. Lokatie Bergen op Zoom*. Bergen op Zoom.

GEP 1991a = General Electric Plastics B.V., *Jaarverslag 1991 Veiligheid, gezondheid en milieu*. Bergen op Zoom

GEP 1991b = General Electric Plastics B.V., 1991. *Inzending Milieuprijs voor de Industrie*. Bergen op Zoom.

GEP 1992 = General Electric Plastics B.V., *Sociaal Jaarverslag 1992, Veiligheids- en milieujaarverslag 1992*. Bergen op Zoom.

GEP 1993 = General Electric Plastics B.V., *Veiligheids- en milieujaarverslag 1993*. Bergen op Zoom.

HWB 1978 = Hoogheemraadschap Westbrabant, *Aanvrage voor het verkrijgen van een vergunning tot lozing van afvalwater door General Electric Plastics B.V. Bergen op Zoom*. 5 June 1978.

KIvI 1991 = Koninklijk Instituut van Ingenieurs, *Milieuprijs voor de Industrie 1991*. Den Haag, 1991.

Milieumagazine, 1993 = *Milieumagazine*, 1993, issue 5: 24–26.

NRC[ddmmjj] = *NRC-Handelsblad* (day, month, year) (Dutch newspaper).

Schrama, G.J.I. (1994) The available scope of choice for corporate environmental strategies. Preconditions, environmental orientations, and motivations for pro-active environmental management. Paper presented at *From Greening to Sustaining: Transformational Challenges for the Firm. Third International Research Conference of the 'Greening of Industry Network', 13–15 November 1994, Copenhagen*. Enschede: Centre for Clean Technology and Environmental Policy, University of Twente.

Schrama, G.J.I. (1995) The internalization of environmental management at GE Plastics Europe. Case study report submitted to the European Commission. Enschede: Centre for Clean Technology and Environmental Policy, University of Twente.

Smits, J. (1993) Environmental Information Systems for General Electric Plastics Europe. Enschede: Master-thesis University of Twente, 1993.

STEM[ddmmjj] = *Dagblad De Stem* (day, month, year) (Dutch newspaper).

12. ENVIRONMENTAL MANAGEMENT AT THE CIBA CORPORATION

Joseph Huber, Ellen Protzmann and Ulrike C. Siegert

INTRODUCTION

The following study presents achievements in environmental protection and environmental-oriented management at the chemical corporation Ciba, since 1997 part of Novartis which has its main office in Basle, Switzerland. The study describes and clarifies the following aspects:

- measurable success and verifiable improvements in environmental and resource protection,

- the tools used and the system of ecological management in the corporation,

- motives and causes which provide the background for the further development of activities toward environmental protection.

The case study was conducted between January 1993 and February 1994, based on groundwork done between September 1991 and February 1992. The study is based on an analysis of literature about Ciba, an analysis of documents from Ciba itself, and on in-depth interviews, which were conducted with high-ranking company employees in two stages. A total of 21 managers participated in the study. They belonged to either corporate management, the Basle plant, or to division management from the Agriculture, Pharma, Textile Dyes, and Additives Divisions, and are responsible for environmental protection, production, communication, planning, research, and development. In some instances the same individuals granted us repeat interviews. Therefore, we take this opportunity to express our thanks to Ciba's personnel for their extraordinary co-operation in this project.

1 THE CIBA CORPORATION

Ciba is a multinational pharmaceutical and chemical corporation, with its main office located in Basle, Switzerland. Its product palette includes items beneficial in the areas of healthcare, agriculture, and industry, for example pharmaceuticals, herbicides, insecticides, seeds, pigments and dyes, additives, and analytical instruments. The corporation is divided into 14 divisions, which are composed of 140 companies in 60 countries. Figure 1 shows the corporate organizational chart after its last reorganization in 1992.[1] The corporation was founded in 1914 in Basle under the name

---

[1] *Facts and Figures* 1993, 5.

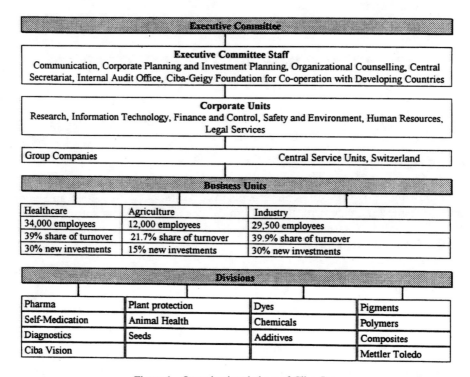

Figure 1   Organizational chart of Ciba, Inc.

'J. R. Geigy AG'. A second branch was the 'Gesellschaft für Chemische Industrie in Basel' (Ciba) which had been in existence since 1884. In 1970 the two combined to form Ciba-Geigy, Inc. Since 1992, the firm has gone by the name of 'Ciba'.

Today, Ciba employs more than 90,500 people world-wide. 17,554 of them work at the main headquarters in Basle. Seventy of the 140 locations are production and research sites, which are located not only in the classical industrialized nations, but also in countries such as Argentina, Mexico, and Brazil. With its turnover of approximately 22 billion Swiss francs (Sf), Ciba ranks among the 100 largest enterprises in Europe. With respect to turnover it is half as large as the three German chemical corporations Bayer, BASF, Hoechst, and ICI (GB), somewhat larger than the Swiss company Sandoz, and about as large as Rhone-Poulenc (F) and Enimont (I). Ciba is twice as large as Akzo(NL), Solvay (B), Henkel (D) and its Swiss counterpart Hoffmann-La Roche. Its investment of 10.6% of turnover for research and development places it in the 'high-chem sector' of the chemical industry.

The following Figure (Figure 2) provides an overview of the share of expenditures from turnover for investments, research and development, and environmental protection measures. The data for the selected chemical enterprises in Switzerland and Germany is from 1992. It is obvious that the Swiss enterprises with a higher share in speciality chemicals and pharmaceutical production have access to a much larger research budget. Ciba here, too, ranks among the top group with its share of 6.4% of turnover for environmental protection expenditures.

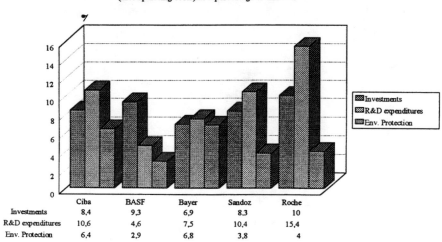

Figure 2 Enterprise comparison 1992.

2   MEASURABLE IMPROVEMENTS IN ENVIRONMENTAL PROTECTION

2.1   Resource Input

Concepts for an integrated starting point for environmental protection internal to the company are directed at optimizing production procedures, which lead to increased safety in processing and either avoidance of problematic substances and emissions or a reduction of materials used. To accomplish this, an analysis of the material flow and a description of the substance and energy balance is required. In this fashion, trouble spots become easily recognizable, thus allowing resources to be used sparingly, beginning at the 'source'. This will be verified using the resource, energy, as an example.

From 1975 to 1988, Ciba's main plant in Basle achieved a doubling of production while maintaining the primary energy usage level constant.[2] Despite this, the energy-saving trend has slowed down since the mid-1980s and separating increased production from energy use is no longer possible. As one can see in Figure 3, primary energy usage at the factory level has increased. This development is also recognizable at the corporate level.[3]

Therefore, since 1992, Ciba's corporate management has had a basic document in place establishing the entire concept behind its energy policy as a building block of Vision 2000 (see Section 3.1). According to this policy, each division, taking into account its unique situation, is individually responsible for helping the entire

---

[2] Oekotrend (1993) p. 15.
[3] Corporate Environmental Report (1992) p. 21.

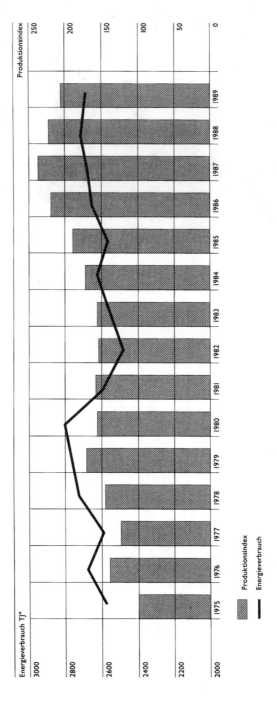

Figure 3 Energy use and productivity at the Basle plant 1975–1989.

corporation save 250 million Sf in energy costs. The focus of the energy policy lies in gaining energy from waste to save resources. The Swiss plants in Kaisten, Munchwilen, Stein, Basle, and Schweizerhalle produce 25,000 to 30,000 tons of solvent waste per year. Fifty percent of this goes to third parties to be burned. With the steam produced by the new solvent incinerator (ALV-2), 33% of the energy requirements and 80% of the steam requirements at the Schweizerhalle plant are met, leading to a yearly savings of 6,500 tons of heating oil.[4]

By introducing new technologies the usage of raw materials sank enormously, for example at the division plant protection in Schweizerhalle. Furthermore, the water usage at Ciba could be reduced by the same efforts from 1991 to 1992 by 5.8%.[5]

2.2   Emissions

Many of the improvements in environmental protection at Ciba are due to emission reductions in end-of-pipe environmental protection. From 1991 to 1992 the air-borne emissions could be reduced by 9.6%.[6] An another example of this is the 'restoration measure' towards nature, in which Ciba saw itself forced to participate after the 1986 fire in Sandoz's Schweizerhalle plant. This accident killed a consider-able amount of fish in the Rhine, and as a result the entire chemical industry in the area fell under a bad light. After the accident the affected corporations participated in 'Action Programme Rhine 2000'. The goal of this program was to have again 'salmon all the way to Basle' whereby the presence of salmon is obviously used as a bio-indicator of the purity of the water. Accomplishing this goal meant, more specifically, for the Basle plant to half the amount of waste in effluents from 1985 to 1995, which it accomplished early in 1990. This was achieved mainly by connecting it to the ARA Rhine, an effluents-purification plant. The amount of copper, for example, sank from 20,000 kg down to 500 kg per year.[7]

Organic waste in the Rhine was reduced in the same manner. One can trace the reduction back to the construction of a wet-oxidation facility in Grenzach, which was completed in 1992. The reduction of organic waste, measured in DOC (dissolved organic carbon) of the Ciba factories on the Rhine (Basle, Schweizerhalle, Grenzach, Kaisten, Stein, Muenchwilen), is illustrated in Figure 4. If one considers the fact that purifying effluent in the wet-oxidation facility devours 250 Sf per cubic meter, whereby the additional costs of its final disposal are simply delayed, it becomes obvious that the costs for the Ciba factories in the Basle region to reduce their effluents by half was quite prohibitive. The Ciba corporation must dispose of a total of 29.4 million cubic meters.

A new Swiss ordinance, in place since 1987, has strengthened Ciba's efforts to resolve the effluent problem. The ordinance defines waste more strictly, particularly hazardous waste, 10% of which in Switzerland originates from the chemical industry. Ciba decided that its directive in environmental protection would be 'to

---

[4] Corporate Environmental Report (1992) p. 6.
[5] K. Eigenmann, September (1993) p. 2.
[6] Corporate Environmental Report (1992) p. 23.
[7] Oekotrend (1993) pp. 23–25.

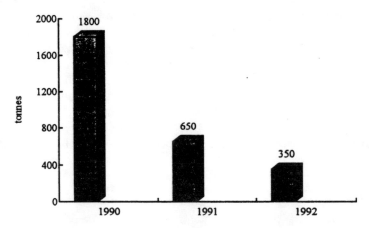

Figure 4   DOC values at the Ciba plants on the Rhine.

attack the problem right here at home. Wherever waste is created, there it must also be removed. Our goal is to install at our own initiative a waste management facility for effluents, gases, and other waste, which will comply with the existing laws and ordinances...[8] This very clearly illustrates the exponential development of waste-management costs, due to its lack possibilities for internal waste disposal.[9]

Ciba has attempted through the 1992 directive for waste management to 'pull up the problem by the roots,' giving priority to solving the waste problem in the following order:

- Avoid, Reduce,
- Recycle or Reuse,
- Treat, Dispose.

In order to obtain an overview of waste's movements, quantities, transportation, and characteristics, they developed a computer-aided 'waste accounting', which assists in documenting and supervising the path of waste from its creation to its removal. The law requires this type of documentation only after waste has left a facility. In Asia and South America companies are only required to disclose the amount of waste and its carrier.[10] With the aid of regularly performed audits (see also section 3.7) Ciba would like to ensure that the internal directives are carried out throughout the entire corporation, which also demonstrates that waste management is the central focus of the present auditing system. An employee at the Schweizerhalle plant who works in the environmental protection service describes the concept of waste management this way:

---

[8] Environmental Protection at Ciba-Geigy, p. 24.
[9] Oekotrend (1993) p. 33.
[10] Corporate Environmental Report (1992) p. 8.

'The amount of environmental damage is attributed to the product and thereby to its causal agent. The figures document the amount of harmful substances produced per product. With the help of a waste card, workers note which types of waste are produced and the information regarding the cost and means of disposal. The costs are determined on a yearly basis only from expenditures on storage, transportation, and disposal.'

The Schweizerhalle site produced 39% less waste in 1992 than in previous year. The amount of recycled waste rose from 61% to 76%. Corporation-wide, Ciba recycled 80% of the 1.5 million tons of waste produced in 1992.[11]

2.3   Changes in Products and Product Lines

2.3.1   Bio-control instead of conventional agricultural chemicals   The Plant Protection Division is the second largest division of the Ciba corporation. Its 86% share of turnover (1992), makes it the largest division after Seeds and Animal Health within the market area of agriculture. In spite of its loss of turnover, the corporation remains the number one producer of plant-protection products. Until now Ciba has been represented most strongly in industrialized nations (67%), which own only 18% of the productive land. Pesticides and insecticides sold in Europe and North America are the most important products.

Public ecological awareness of intensive farming, with its massive use of fertilizers and plant protection chemicals, lead to a decisive image loss for the chemical industry. Fertilizer usage is 15 times as high as it was in 1945.[12] At the same time, discoveries from agricultural research show that crop damage is a result of fertilization, which weakens a plant's immune system.[13] Conflicting findings regarding the length of time pesticides remain in ground water and the harmful effects to the health of farmers and workers who use them, especially in the Third World, provide further motivation for producers to search for ecologically-compatible products and technology, which will also have a promising market in the future.

Ciba calls its steps in this direction 'biological control'. Micro-organisms (viruses, bacteria), macro-organisms (natural pests), and natural products (plant extracts) fall into this category. The object is to use natural, self-limiting, regulating cycles, such as insect-growth regulation, which hinder a pest's development instead of killing it. Ciba plans to have 10 biological products of this type on the market by the year 2000.

The use of such high-tech products in Third World countries is hindered by their predominantly low educational level and strained financial situation. The representatives of the agro-chemical industry view the application of bio-products by farmers of small independent farms as more of a stumbling block than a chance for constructing a new market based on social and ecological factors. In spite of this, in 1991 dedicated employees within this division were able to make progress through a program called 'Farmer Support Team (FST)'.[14] FST is directed at helping small

---

[11] K. Eigenmann, September (1993), p. 24.
[12] D. Meadows (1991) p. 36.
[13] I. Seidl (1993) p. 208.
[14] Ciba-Magazin 1/93, 15 and interview with Bill Vorley.

farmers by allowing them to participate in an educational program aimed mostly at user safety. The group works with the local Plant Protection Division, the distributors, local and national institutions, research institutes and private organizations. Its long-term programmes, based on principles of integrated plant protection, provide education and training regarding the correct application of plant protection products and cover the related questions on user safety.

> '*The program deals with environmentally-responsible and well-directed plant protection, which creates an atmosphere of trust, and thus a long-term market share in an area where the competition in the Third World is not yet so active.*'[15]

2.3.2   Textile dyes   Environmental protection in the area of textile dyes is an especially-current topic. The two main problems are the dye content and metal content in effluent. Ciba is also researching this field to provide ecologically superior alternatives to conventional products. This wide spectrum of modern dyes is designed for cellulosic fibres and cellulose-mixed fibres. The advantage of this product family lies in the total setting of the colour on and in the fibres. These products, with their 95% transfer rate of the dye into the fibres, greatly surpass conventional products. The latter achieve only a 65% transfer rate. Above all, the resulting effluent is greatly reduced. Therefore, the energy- and water-saving dye technology, the Kalt-Verweil technology, is an advantage. It makes a reduction of water usage to 50 litres/kg of cotton possible, whereas the average to date is around 150 or more litres/kg. The disadvantage is that these dyes still contain metals and the dying process requires large amounts of salt.

2.4   An Attempt to Compare Ciba's Advances in Environmental Protection with those of Other Chemical Corporations

Ciba's success in conserving resources, reducing emissions, and substituting or eliminating environmentally toxic or dangerous substances, as described in the previous sections speaks for itself. The question which now arises is whether improvements display something special and Ciba-specific, or do other enterprises in this field present a similar picture, due to the ever-increasing emphasis on ecology. The latter would in no way minimize Ciba's successes, but it would alter their context relation.

This section must be prefaced by saying that any type of comparisons between enterprises are extremely difficult. For one thing, the data from the enterprises is barely comparable. For another, there are in some cases wide differences in each company's range of products from which the possible environmental pollution originates. The German chemical companies Hoechst, BASF, and Bayer, in contrast to Ciba, produce large quantities of base chemicals, which involves converting large amounts of materials and energy. The Swiss chemical companies, compared to their German counterparts, produce many more speciality chemicals and pharmaceuticals. In a similar fashion, the varying product portfolios result in differing types and amounts of emissions, with partially-differing environmental-measure demands. For

---

[15] Ciba-Magazin 2/1992, p. 26.

example, the chemical enterprises with a large share of pharmaceutical products, like Hoffmann-La Roche, produce a wide-spectrum of emissions in small amounts, which are complicated to dispose of or recycle. These are mostly solvents used in running chemical processes.

One must further consider that the enterprises employ different analytical methods in determining certain amounts of air or water pollution. Additionally, the companies publish their emissions data either by individual site, like Basle or for the Basle region, for the BASF plant in Ludwigshafen, or for the entire corporation.

In spite of these handicaps towards making a complete comparison of firms, one can determine a trend. BASF, Bayer, Hoechst, Hoffman-La Roche, Sandoz, and others, have made progress in environmental protection comparable with that of Ciba.[16] One must thus assume that Ciba's success in conserving resources and reducing emissions is similar in tendency and scope to that of the other firms in this line of business, although Ciba belongs to the group of chemical companies which spends relatively more money than most for environmental protection.

This brings up a series of questions, which at this point certainly cannot be answered, but should be mentioned at least once:

1. How is it possible that with varying expenditures, nearly the same results are achieved in such a highly regulated area of business?

2. Has Ciba possibly earned a respectable 'eco-image' which it does not deserve? Have its environmental achievements been categorized, by comparison, too positively or are the other chemical enterprises, along with the entire chemical industry as a whole viewed too negatively? Has the chemical industry's wide scope of efforts and improvements in environmental protection not been adequately appraised?

3. Will Ciba's seemingly higher expenditures on environmental protection turn out in the end to be a useless and competitively-damaging expense, or will Ciba reap the benefits of a decisively higher eco-performance and its related advantages only in the future?

4. How much must a company do, in the way of building new and costly environmental protection 'structures', in order to fulfil certain environmental protection functions, or could the existing structures (in the areas of planning, production, logistics, R&D, etc.) in most cases take over the required functions within the framework of a multi-purpose organization?

The principle of integrated environmental protection seems to suggest reducing extra facilities down to the necessary minimum, and tacking the major portion of environmental protection functions on to already-existing structures or even combining them. From this standpoint having a visibly large share of environmental management would be an indicator of a high, and possibly excessive, amount of end-of-pipe environmental protection. In the following sections the individual aspects and elements of Ciba's environmental-management system will be described.

---

[16] See also J. Huber, E. Protzmann, U. C. Siegert, Ciba-Case Study (1994) full version and H. Middelhoff (1994).

3   CIBA'S ENVIRONMENTAL-MANAGEMENT SYSTEM

3.1   The Corporate Philosophy

The documenting of company ideals to give employees a corporate identity is a tradition at Ciba. Ciba began publishing its fundamental corporate philosophy in its policy publications as early as 1972/1973. In 1990, in conjunction with a sweeping corporate reorganization, this policy was renewed in the form of 'Vision 2000'. During the creation of 'Vision 2000', Ciba built on its already-existing environmental protection policy. 'Environmental protection and safety in line with the most modern technology' was to be just as important a goal as product innovation and profitability. Nevertheless, Vision 2000 must be qualified by the statement that, 'at no time are the three areas of responsibility exactly equal in importance', and that during recessions 'economic goals have priority over society and environmental protection'. One must understand the company's often criticized balancing of 'ecology and economics' as an optimization problem. Only short-term profit orientation acts contrary to ecology. For the long term, the overlap between ecology and individual business goals becomes greater.

Ciba's role as an 'early follower' compared to other large European chemical enterprises[17] and its reputation in the '70s as a socially-responsible corporation have been able to sustain the company until now. From 1990–1993 the Hamburg Environmental Institute (HUI) extensively researched the world's 50 largest chemical and pharmaceutical companies. Ciba here belongs to the 'pro-active enterprises, which take environmental protection seriously' and took sixth place in the HUI ranking.[18] Its environmental directives as well as its internal and external management received an above average rating.

3.2   Organizational and Personnel Change Towards Environmental Protection

The basic idea behind the 'Organization 90' concept was to give the different sections of the corporation some independence. Communication and interactive planning were to prevent them from becoming too independent and to maintain the corporate bond. In addition to the autonomous divisions of the two-dimensional matrix organization (global and local), Ciba formed corporate units as so-called 'centres of competence' for strategically important areas such as research, information technology, safety and environmental protection, human resources, and legal services, as well as for the executive committees communication and corporate planning.

Environmental protection was institutionalized in its own executive branch and is organized according to the motto. 'Think globally, act locally'.[19] Since environmental protection presents a problem on the local level and since problems must be solved locally with local know-how, it makes sense to delegate this responsibility to

---

[17] Ciba employees nearly unanimously view the corporation, in contrast to its competitors, as an 'early follower' and completely competitive.

[18] Hamburger Umwelt-Institut (1993).

[19] Th. Dyllick (1989) p. 46.

the individual divisions. The development of generally valid principles and directives, in contrast, is a central function of the Executive Committee for Safety and Environmental Protection, since at that level all the knowledge and experience should be available for everyone. In the years following, Ciba developed for the 'three main environmental functions, i.e., production safety, safety, and the environment'[20] corresponding principles, directives, and instructions, which corporate management elevated to apply corporate-wide. The organizational responsibility for product safety is decentralized and resides in the individual divisions. The division manager is responsible for product safety. A technical section, however, is in charge of technical matters. Each plant manager is responsible for reducing emissions and resource usage in the regular production process and for the reduction of accident risks and their consequences for humans and the environment. The factory managers are entirely responsible for integrating their factories into the environment in the best possible way.

Ciba employees understand 'empowerment' to mean that there are no strict regulations made by corporate management, but that the divisions' own directives will be transformed into concrete goals and strategies. Alex Krauer, president of the administrative council personifies the vision on which the changes in the company structure were founded.[21] It is mostly due to him that environmental protection and 'sustainable development' became the 'bosses concern'.[22]

3.3   Environmental Communication

Perceiving external pressure and that from its own employees, and suffering from an extreme loss of image – especially after the burning down of the storage facility in Schweizerhalle on November 1, 1986 – , the corporate directors developed a new communications concept as an expression of Ciba's offensive communications policy. Communication and its improvement are a central theme and objective of the corporate leadership. Communication is used as a tool to attain the vision, improve the exchange of information and experience between employees, promote the leadership principle of 'interconnected independence', win employee support, and stimulate open and controversial internal discussions.[23]

Corporate and public affairs and communication and issue management are not only responsibilities at the corporate-management level. At the division level – and this is a Ciba-unique innovation – communications and policy departments have also been initiated and developed (in the Pharma Division since 1978), in order for each division to have social and environmental sensitivity regarding the effects of its own line of products.[24]

---

[20] R. Saemann (1990) p. 38.

[21] According to statements by an employee in Ciba's Corporate-Communication Division.

[22] Ciba-Magazin 3/92, p. 12.

[23] See also Roeglin/Grebmer (1988), p. 66, who report on investigations which show that the enterprise's know-how is lost down at the level of the individual employee if in-house communication does not convey it. See Ciba's goals in the Ciba-Zeitung, 12/91, p. 4.

[24] According to statements by employees at Pharma policy.

The 'AGora' dialogue forum, set up in the Plant Protection Division, in which employees, division representatives from interest groups, politicians, businessmen, and scientists take part, is one example of Ciba's activities concerning the furthering of internal communication. The dialogue forum on controversial topics from the areas of agriculture and plant protection began in 1990 and is held several times a year. The 'Corporate Environmental Report 1992'[25] verifies, for example, that effective methods of communication are also in place at Ciba. Additionally, issue-management procedures are conducted as trend monitoring. 'Issues' are things like socially-relevant topics, environmental protection, social movements, city and local planning, and experiments with animals. Furthermore, any other topic which seems relevant can become an issue which is covered systematically. The assessment of 'issue potentials' is to serve as an early warning system, used to avoid developments in the wrong direction and problems of acceptance.

3.4   Consideration of Environmental Protection in Investment Planning

Responsibility for the natural environment comes into Ciba's investment planning in two ways. First, environmental and resource protection are decision-making criteria for investment planning. Second, Ciba makes special investments, which are used mainly to promote environmental protection. Ciba uses the results of environmental-compatibility tests, besides infrastructure and business criteria, to make decisions on whether or not to change or introduce a certain processing procedure or on where to locate a new factory. Decisions are made by the corporate directors and the corporate planning staff for safety and environmental protection on matters involving more than 10 million Sf. This corporate entity, with its 30 highly qualified specialists in safety and environmental protection, ensures that the company's directives are followed. For investments under 10 million Sf the division or factory management performs the same procedure. Investments in safety and environmental protection play a large role in Ciba's investment planning, as do investments in expanding capacity or creating of a new line of products. In 1991 Ciba invested 440 million Sf world-wide, and in 1992, 380 million Sf in safety and environment protection. These investments and operating expenses for environmental protection and safety of one billion Sf in 1991 and 1.04 billion Sf in 1992 are viewed as 'important investments in the future'. Since 1988 investments in environmental protection and safety totalled approximately 15–25% of total investments, and operating expenses totalled approximately 4.5–5% of turnover.

3.5   Priority for Integrated Solutions Instead of End-of-Pipe Measures. Starting Point for Environmentally Oriented Product Development and Product Policy

Ciba would like to take a pro-active, innovative, and strategic stance, as mentioned previously. Ciba views 'the inclusion of ecological accountability in the management process..., the transition from end-of-pipe installations to process-integrated solutions, and the attainment of a competitive edge through the development of

---

[25] Corporate Environmental Report (1992).

environmentally compatible products and production processes'[26] as its greatest long-term challenges. Therefore, with the help of an 'ecological product life cycle'[27] employees thoroughly analyze ecological problems in each step of the production process. Ciba realized already in 1988 that these so-called 'curative measures'[28] helped only to a small degree if the chemical processes are not extensively developed at the same time.

Another important aspect, besides the improvement of production processes, is the topic environmentally-friendly products. The main question is: 'What are the unintentional repercussions of our products on the environment?'. Each of the affected divisions investigates this issue. An example for the so-called applications guidance is the 'Ecology and Energy Service'.[29] The customer is viewed as a partner to be assisted with knowledge gained from many years of experience, concerning not only the products and their applications, but also the disposal of effluents. For example, experts provide customers with advice on solvents tailored to fit their problems and inform them as to how to stay within the legal limits when funnelling effluents into a purification plant.

3.6   SEEP – Safety, Energy and Environmental Protection Reporting

Safety, Energy and Environmental Protection (SEEP) reporting is an internal reporting tool used at Ciba. The report contains data, documented yearly, about the ecological situation of each of Ciba's branches and divisions. Ciba began with this type of standardized reporting in 1987 at the plants in Switzerland and Grenzach, since similar legal statutes, environmental management systems, and methods of data collection and analysis existed there. Since 1990, each of the different sites has presented its data in a standardized fashion. Since 1993, Ciba has used this data as the basis for its Environmental Report covering the entire corporation, which is used as an external report. Each of the larger sites must submit a yearly SEEP report. In 1992 this involved 82 sites. These included all of the chemical plants and all of the larger production facilities, which together expel 90 percent of all emissions.[30] Ciba developed the concept behind the SEEP reports in accordance with the directives of the European Chemical Association (C.E.F.I.C) as part of the Responsible Care Program – one of the chemical industry's charters for long-term and sustainable development. Ciba, itself, was actively involved in the development of these directives. It, therefore, comes as no surprise that it, and many other enterprises like Sandoz and La Roche, published their first environmental reports that same year. The actions of the chemical industry and its associations can be traced back mainly to the lack of satisfaction with the chemical industry as a whole, which they hope to compensate for with a gradual opening up of their enterprises. Upon comparing the environmental reports of the different chemical companies, it is evident that the

---

[26] H. Kindler, in: Corporate Environmental Report (1992) p. 19.
[27] See also according to the concepts of PLA Dyllick (1989a) p. 23 and Pfriem (1989).
[28] G. Eigenmann, in: Chimia 5/78, 178.
[29] See also Ciba-Magazin 3/91, 30.
[30] P. Naish, September (1993) p. 2.

firms could only come to a minimal consensus within the C.E.F.I.C. This is true for the manner in which the reporting should be conducted as well as for the choice of environmentally-relevant data to be published (see Section 2.4).

3.7 Environmental Auditing

Ciba has performed corporate-wide environmental and safety audits since 1980. This early-warning system serves to minimize a company's risk of liability. This audit, used as an inspection tool, has moved into the centre of management's responsibilities in the past few years. However, according to reports, the audit has been around, at companies like Sandoz, since 1977 and at La Roche since 1980. This is due to the lack of standardization of the audit. Risk precautions have been in place for quite a while, but throughout the course of time, companies have switched from selective inspections, oriented towards supervising quality and safety in the workplace, to systematic and thorough inspections with emphasis on environmental protection. For this reason, Ciba's own Environmental Auditing Section has been placed on the level of executive management, within the department of Safety and Environmental Protection. The larger plants are inspected every three to five years. Many corporations perform additional audits of their own accord.

The following points are inspected:

- conformity to the enterprise's policies and internal directives,
- technical, organizational, and personnel-directed measures, which improve environmental protection,
- environmental protection measures and criteria, which are fixed by investments,
- the existence of problems and the determination of appropriate actions.[31]

Ciba uses the audit as an internal tool, as a guideline for reducing packaging, resources, and emissions, and as a means of discovering weaknesses. Ciba's directors, however, oppose using the audit in any form as an external report, which the EU is advocating. There are no standardized processes or norms in place, which could help verify the published audit results. The group of corporations is sceptical towards the EU's demands for external inspections and the publication of auditing data, since first, there are not enough trained auditors available, and second, the results are barely comparable between enterprises. Ciba, therefore, views the value of externalizing the audit as questionable.

In 1992, Ciba inspected 19 of its subsidaries in Switzerland, France, Great Britain, Argentina, Brazil, and Japan. In 1992, 37 additional audits were made, for which the corporation had responsibility.

3.8 Eco-Balances, Environmental Compatibility Tests

The enterprise follows the concept of eco-efficiency, i.e. the establishment of environmental protection measures which are profitable. One instrument used to determine profitability is the eco-balance, a process which quantitatively measures

---

[31] Corporate Environmental Report (1992), p. 17.

the environmental pollution from products, processing, and activities, thus making these items comparable. There are several workgroups at Ciba, which are involved with this topic. The divisions are responsible for performing the eco-balance and have access to a special software program which was developed for it. Ciba utilizes the eco-balance more as a subject of research than as a finished management tool. Since there are no uniform methods and standards for the eco-balance, it is actually more of a materials- and energy-flow analysis, with partial balancing, which is supposed to make the materials' ecological strengths and weaknesses visible and workable. An example is the input and output analysis of pigment production from the Pigments Division. The eco-balance showed that 90% of the pollution from energy resulted from the creation of the energy and only 10% resulted from the chemical processes themselves. In a comparison of the production sites in Switzerland and Germany, the energy balance of the latter came out clearly worse, since Germany's subsidizing policy dictates that coal must be used as an energy source. It is thus understandable why Ciba is pushing so strongly for international energy emissions reductions in its argumentation.

3.9   Environmentally Oriented Accounting, Financial Direction

In order to calculate the resource and emissions flows and assign them to the correct cause, Ciba's main office records the amount of pollution per ton of each product produced. Each production site pays for purifying the pollution from its own effluents and gases and for its own waste disposal. In this way, motivation through financial direction causes the individual factories to reduce the amount of waste they produce.

In the Polymers Division the environmental costs serve to make the producers eliminate products and methods of processing which are no longer competitive and substitute them more quickly. An unsuccessful idea from Ciba employees was to introduce 'artificial' in-house payments, for example, towards energy savings within a company as a directing tool for resource conservation.

3.10   Training the Workforce

Educating its employees with respect to environmental protection has been a central matter of concern at Ciba since the beginning of the 1980s. Every Ciba employee receives general information concerning the safety and environmental policy of the corporation. Furthermore, environmental protection and general safety are an integral part of each educational and training programme whether it be for factory workers or trainees, or for foremen, shop-floor managers, chief laboratory technicians, or junior management in their promotion courses. Education in the areas of factory and production safety, environmental safety and environmentally-compatible technology, as well as in development of safe and environmentally-acceptable processing methods, is directed not only towards every level of the headquarters, but also towards every level throughout the corporation in the form of seminar topics for the company directors as well as course material for workers in production plants.

3.11   Moving Environmental Protection Activities to Company Level

Employees can submit suggestions and put ecological innovations into practice through the company's suggestion programme called 'Cigenius'. Many workers participate in energy-conservation contests or other competitive events instituted to promote the corporate resource-conservation goal, such as the contest for 'AgriNova', an innovation prize. Suggestions which are carried out receive corporate-wide acknowledgement and are rewarded with gifts and cash awards for up to several thousand Swiss francs. At the Basle plant, for example, important suggestions are published in 'Oekotrend' with the details concerning costs, cost savings, resource conservation, and environmental advantages. Furthermore, an important effect is for corporate management to receive with feedback regarding how much the employees identify with environmental protection. In this way, the programmes serve as a type of progress check.

4   CONCLUSIONS CONCERNING EFFECTIVE FACTORS AND CONDITIONS

In Figure 5 the significant environmental protection improvements related to the case study are listed again with factors which Ciba has determined to have caused each improvement. There are 22 examples of varying complexity and importance. These cannot be discussed right away, but simply adding them up provides one with a basic orientation, which agrees with the findings of other studies in their basic structure.

Of the 22 examples, the following were the deciding factors (taking repeat causes into account):

- laws, ordinances, conditions laid down by authorities   17
- image reasons, public pressure   10
- direct cost issues, cost control reasons   9
- market share acquisition, strategic marketing position   3
- general prevention   2
- do what others do   2
- pro-active, self-imposed mandates   1
- self-imposed mandates due to public pressure   1

The factors named are interconnected in an effects structure. It is especially notable that each factor eventually leads to an increase in the costs for environmental protection. Thus, in actuality, the cost factor, in contrast to the others, always plays a role, because each factor is joined to an increased-cost coefficient.

Even at Ciba it was necessary for the environmental protection activities to first originate from pressures outside the enterprise, i.e. from the legal authorities and the government, which mostly react to public pressure, from the ecology movement, the nature movement, and from citizens' environmental protection initiatives, which are

Example	Cause	Time Frame
Switching to energy sources with low emissions	New air-pollution ordinance	mid 80s
Ceasing Galecron production	User accidents with resulting image damage and loss of market share	1980s
Environment-oriented Accounting	Cost control, quicker substitution of non-competitive products	1980s
Environmental auditing, Life Cycle Assessments	Minimizing risk liabilities, general prevention methods	1980s
Bio-control, integrated pest management	Drinking-water ordinance, user safety and image matters, important for success in 3rd world market	late 80s
Discontinuing Merfen production	Self-imposed mandate based on internal directives	late 80s
Action program Rhine, reduction of heavy metals in effluents	Image crisis after the fire in Schweizerhalle and the ensuing attempt to make reparations	1986
DOC and TOC reductions in Schweizerhalle	Mandate from the authorities	1986
Reduction of volatile organic compounds	National air-pollution ordinance	1986
Waste management	Waste ordinance, waste-disposal costs, image damage from waste-disposal accidents	1987
Moving atrazin production elsewhere	Drinking-water ordinance, public protests	1987
SEEP reporting, Oekotrend, Corporate Environmental Report	Comprehensive initiation of environmental-cost control, pressure from EU, competitors and environmental movement	1987-1993
Substitution of cadmium in PVC stabilizers	Pressure from EU, competitors, environment movement	1980 & 1990
Plastics waste management and Venture Group Additive	Packaging ordinance, acquiring a market share in additives	1990s
Substituting heavy metals in foils	Self-imposed mandate based on internal directives (failure due to consumer behaviour)	1992
Energy concept of Ciba, Inc.	Rising environ. protection costs, increasing energy consumption & air-borne emissions, new air-pollution ordinance	1992
New usage instructions for plant-protection products	Image problems due to severe injuries, resulting public pressure, other companies move in the same direction	1992
Expanding air purification to a multi-levelsystem	Sensitive Swiss/French/German public	1992
Substitution of heavy metals in pigments	Problems with users surpassing allowable amounts	1992
Eco-balances	Cost-control, prevention is a weak point, helpful for argumentation against bureaucratic orders	1992
Effluent-Free Dye Standard in Toms River, U.S.A.	Order imposed by local authorities for reducing effluent levels	1993
Construction of a Toxic-waste incinerator	Waste-management costs, export to the GDR no longer an option	start in 1994

Figure 5   Exemplary compilation of effective factors.

strengthened by the media and have increasing influence directly on the corporation. The more widely ecological thinking is diffused in the population and in the economy, the more obvious become the limits of end-of-pipe environmental protection which is mandated by law. Therefore, environmental protection and resource conservation in the corporation become increasingly more important. Systematic environmental protection management emerges, which can in no way be separated from environmental legislation and the government's environmental policy instruments, and which remains further tied to them. Nevertheless, it begins to take on a shape of its own, in that it contains continually more enterprise-specific, situational, and pro-active elements outside of government intervention. Last, but not least, it contains elements of preventative and integrated environmental protection and resource conservation, which are impossible for the government to dictate. Ciba experienced its worst ecological catastrophe to date in 1986. However, the Schweizerhalle accident was mainly fuel for the fire of the ecology movement and the media. In the enterprise itself it caused nothing fundamentally new; developments already in process were simply sped up and intensified. The Schweizerhalle accident essentially caused the Swiss and German chemical industry from that time on to openly recognize their ecological responsibility. It also forced the chemical associations and their enterprises to declare environmental policy an institutionalized matter of concern. In this sense, Schweizerhalle in 1986 marked the official turnround of the chemical industry's behaviour from ignorant defensive to one that gradually became pro-active. However, it did not achieve this transformation on its own.[32] In view of the driving forces and motives for converting to ecological management, Ciba is not set apart from other firms. The three most significant reasons for ecological adaptation in the industry are listed below:

1. Reactions to financial pressures, to maintain competitiveness;

2. Political loyalty to the government (compliance with the law) to guarantee stable conditions for corporate and investment planning;

3. Efforts to integrate into society, to unite the corporation with its business environment in the interest of ensuring the necessary transfers of personnel, knowledge, and capital, and to acquire the needed political legal, and administrative co-operation, imperative for an enterprise's long-term reproduction.

These three items affect the industry in general. To explain why they lead to earlier and more developed consequences at Ciba than at other firms, one must rely on a series of intermediate factors of medium influence. The following, and others, belong to this category:

• Ciba's tradition and reputation as a 'socially'-focused enterprise;

• an especially intense willingness on the part of the Swiss to dedicate themselves to environmental protection and to do so in an accurate and responsible manner;

• the situation that the economy during the '80s, up until 1988/90, was generally booming and brought in high profits, especially for the chemical industry; in

---

[32] See also Longdius (1993).

return this made hefty investments in end-of-pipe and integrated environmental protection possible.

If the chemical industry is open today to ecological matters, it is not due to a Rousseau-like, romantic, and vital love of Nature. A much more plausible reason is the utilitarian attitude of homo oeconomicus, supplemented by the insight that the ecological expansion of the production model lies in the enlightened self-interest of the economic participants. Such an attitude may be criticized from a cultural standpoint; however, it brings the desired results of an increasingly-environmental orientation of the industry. One can permanently build on the ecological modernization of industry only if the positive sentiments for Brother Tree and Mother Earth can stand up to the sharp wind of competition and the cool climate of rational calculation.

LITERATURE

Dyllick, Thomas (1989a) *Ökologisch bewußtes Management*, Bern.
Dyllick, Thomas (1989b) *Management der Umweltbeziehungen—Öffentliche Auseinandersetzung als Herausforderung*, Wiesbaden.
Eigenmann, Gottfried (1978) Umweltschutz: Eine Herausforderung für die chemische Industrie. *Chimia* **5/178**.
Eigenmann, Kasper (1993) Der erste Konzernumweltbericht von Ciba, 23 September, 1993.
Hamburger Umwelt Institut (HUI) (1994) The TOP 50 Ranking, January, 1994.
Hamburger Umwelt Institut (HUI) (1993) *Manager Magazin* **1/94**, p. 68ff.
Huber, J., Protzmann, E. and Siegert, U. C. (1994) Fallstudie Ciba AG. Report. University Halle.
ICC (International Chamber of Commerce) (1989).
Konzern-Report 1992 der Hoffmann-La Roche AG Basel: Sicherheit und Umweltschutz.
Longelius, St. (1993) *Eine Branche lernt Umweltschutz*, Berlin.
Meadows, D. H., Meadows, D. L. and Rauchers, J. (1992) *Die neuen Grenzen des Wachstums*, Stuttgart.
Middelhoff, H. (1994) Die Organisation des betrieblichen Umweltschutzes in der schweizerischen und der chemischen Indiustrie. Dissertation, St. Gallen.
Naish, Peter (1993) SEEP—The standardised report on Safety, Energy and Environmental Protection, Konzernweite Erfassung von Umweltschutzdaten, September, 1993.
Saemann, Ralph (1990) Ansatzpunkte für eine effektive Umweltschutzorganisation im Unternehmen.
Seidl, Irmi (1992) Unternehmenskultur-ein Einflußfaktor auf die Ökologieorientierung von Produktionsinnovationen (Dissertationsschrift), St. Gallen.

Publications of Ciba, Inc.:

AGROinfo, 13/September 1993; *AgriNova-Broschüre*, 1992/1993.
Auf einen Blick, Broschüre der Division Pflanzenschutz, 1992.
Case Study der Ciba AG: Manufacturing Dyestuffs with Minimum Pollution Candra/Sari, Indonesia.
Case-Study der Ciba AG: Environmental Protection—Treating Waste Water.
Case Study der Ciba AG: Incineration: Treating Special Wastes.
Cigenius-Broschüre; Core Values of Communications; Corporate visual identity, 1992.
Das neue Logo, 1992.
Facts and Figures, 1993.
EC Vade Mecum, No 3/91.
Führung und Zusammenarbeit, 1991.
Annual Report, 1992.
Grundsätze von Ciba-Geigy der Informations- und Kommunikationspolitik, 1989.

Konzept und Programm für die Stammhauswerke, 1989.

Corporate Environmental Report, 1992.

Ökotrend, Taschenausgabe, 1992; 1993.

Oekotrend Ciba Werk Basel, 1993.

Organisation 90 (new edition 1992).

Pharmaceutical research and development, 1990.

Sicherheit und Umweltschutz in der Produktion, 1990.

Umweltbericht Werk Lampertheim, 1993.

Umweltschutz bei Ciba-Geigy: Konzept und Programm fürr die Stammhauswerke.

Umweltschutz und Sicherheit bei Ciba-Geigy, 1991; 1993.

Umweltschutz in der Produktion, 1982; 1984; 1987.

Vision 2000, 1990.

Ciba-Magazin, 3/4/87; 4/88; 3/91; 4/91; 2/92; 3/92; 4/92; 1/93; 2/93; 3/93; 4/93.

Quartierzeitung, 1/90; 2/92; 4/92; 3/93.

Ciba-Zeitung Beilage, 21.11.89; 12/91; 2/93; 3/93; 8/93; 11/93.

III. COMPARATIVE EVALUATION

13. TYPICAL AND VARYING ACTORS AND STRUCTURES

Jobst Conrad

1 INTRODUCTION

This chapter concentrates on the specific results of the case studies presented in Part II, Chapters 4 to 12,[1] describing their features on the levels of action and structure, whereas subsequent chapters embed these results in the broader context of other investigations in environmental management. However, the focus on actors and structures of successful environmental management is accompanied by treatment of main features of the success stories in general. Table 1 summarizes the principal characteristics of the companies and their environmental management performance.

In looking for typical and varying actors and structures of the success stories, it is important to distinguish between the three types of evaluative analysis pursued (outcome, process, and institutional evaluation) because they will necessarily tend to relate to different actor constellations and structural arrangements. In particular, successful environmental protection measures as part of an environmental management system already well established within a company will typically exhibit other action and structural characteristics than early success cases as yet without such socio-organizational embedding. Thus, the comparative analysis of actors and structures underlying the case studies does not allow their systematic comparison in view of the differing areas of successful environmental management under investigation.

With this caveat the present chapter first addresses the evidence of typical actor constellations as well as the variance of actor involvement underlying the success stories; second looks for the features essential to environmentally-oriented behaviour by these actors; and third points to the structural similarities and dissimilarities among the case studies. The final objective of this analysis is to find out if and how any general pattern of successful environmental management can be hypothesized from the case studies, as discussed in Chapter 17.

2 ACTOR CONSTELLATIONS

Key aspects of actor constellations are the number of actors involved, the role of key individuals, the in-house pattern of actors, the type of actors external to the company involved in the success story, the absence or exclusion of actors likely to

---

[1] For their extensive description the reader is again referred to the corresponding case study reports cited in the literature.

Table 1   Main features of companies studied.

company feature	Amecke Fruchtsaft	Diessner	ABC Coating	Glasuld	Hertie	Kunert	NedCar	GEP Europe	Ciba
type of evaluative analysis	outcome	outcome	outcome	process	process	process	institutional	institutional	institutional
company size	medium	medium	medium	medium	large	medium/large	medium/large	medium/large	large
turnover/year (Mio DM)	30		5	90	6000	700	1700	2000	26500
independent company	(yes)	yes	no	no	(yes)	yes	no	no	yes
industry branche	beverage	chemistry	surface refinement	construction materials	retail trade	textile	automobile	chemistry	chemistry
environmental management strategy existent	no	(no)	no	in progress	in progress	in progress	partly	yes	yes
internally/ externally determined project	cooperative	internal	cooperative	cooperative	cooperative	internal and consultancy	cooperative	mainly internal	mainly internal
favourable economics	yes	yes and no	yes	yes	yes and no	yes	yes and no	yes	yes and no
public environmental policy significant	no	partly	yes	yes	partly	no	hardly	partly	partly
diffusion of environmental improvements	yes	no	small	yes	partly	no	small	small	small
controversy important	no	no	in the beginning	in the beginning	partly	partly	no	in the beginning	hardly
many actors involved	no	no	yes	yes	no	no	no	medium	medium

participate, the cooperative, neutral or confrontational relations among actors, and the patterns of interest, power and problem perception among actors that facilitate or impede successful environmental management.

Overall, few (3 to 5) macro-actors, i.e. corporations, governmental agencies, and other civil society organizations (environmental groups, trade unions, etc.) were involved in the success stories investigated, except for the Danish case studies with up to 10 macro-actors participating. On the level of organizational subunits, typically around 10 to 15 actors were involved, mainly various divisions of the particular company. On the level of participating individuals, there is greater variation among the case studies, although only a few (5 to 10) individuals formed the core actor constellation in nearly every specific case of substantive environmental action and protection.[2] Thus, 10 to 20 individuals had to be considered in the outcome-analysis-oriented case studies (Amecke Fruchtsaft, Diessner, ABC Coating), 20 to 30 individuals played some role in the process-analysis-oriented studies (Glasuld, Hertie, Kunert), and 50 to several hundred individuals were significantly involved in company environmental management efforts in the case studies oriented on institutional analysis (NedCar, GEP Europe, Ciba).

On average, the actor constellation in the success stories typically consisted – within the company – of few highly committed (technical and power) promoters, top management, environmental quality circle(s), the division leaders concerned with the environmental improvements investigated, and – from outside the company – one or two key consultants, cooperating public agencies or firms, public authorities. Only in the two Danish case studies (ABC Coating, Glasuld) were comparatively many actors substantially involved, due to the cooperation-oriented national culture as well as the corresponding arrangement of the success cases, such as workers' council, county council, health authorities, unions, industrial (standard-setting) associations. Interestingly, other actors that might well have participated, like environmental groups (except for Hertie), environmental policy administration (except for ABC Coating, Glasuld), or financial institutions, particularly banks, were hardly directly involved in the success stories. This is not so unexpected, however, because the installation and practice of an environmental management system is mainly an inhouse affair, and pioneering eco-technical developments will usually occur largely inside a company, even a small one, like Amecke Fruchtsaft, Diessner or ABC Coating. However, the non-participation of such company-external actors and frequently of competitors, suppliers (except for ABC Coating, Amecke Fruchtsaft) and customers of the firm does not imply that these actors do not play an important role on the structural level (Conrad, 1996). Table 2 gives an overview of the actual (direct or indirect, favourable or unfavourable) influence of all potentially relevant 26 social actors divided into six groups. Existing environmental awareness in society has an impact via market acceptance or rejection of ecological

---

[2] Distinguishing outcome, process and institution oriented case studies, an increase in the number of participating actors can, as expected, be observed, but this increase remains rather limited, mainly due to the growing number of environmental management activities and areas under consideration, but hardly produces actor constellations involving additional, new types of actor. The figure is certainly higher for environmental management systems in big companies as a whole (Ciba, GEP Europe).

Table 2   Influence of social actors on corporate environmental management (projects).

social actor	Amecke Fruchtsaft	Diessner	ABC Coating	Glasuld	Hertie	Kunert	NedCar	GEP Europe	Ciba
board/top management	d+	d+	d+	d+	d+	d+	(d+)	d+	d+
workforce	i+	i+	d+	d+	i+	i+	d+	i+	i+
key persons (promoters)	d+	d+	d+	d+	d+	d+	n	n	n.
mother company			i+	i+			i+	d+	
investors	n	n	n	n	n	n	n	n	n
suppliers	d+	(i-)	d+	i+	i+	(i+)	(i+)	(i+)	(n)
competitors	n	(i-)	(i-)	n	(io)	(i-)	(io)	(io)	io
buyers/customers	n		io	io	io	io	(i-)	io	io
trade	i+	i-		i-		n	n	n	n
waste management/environmental service firms	do	n	(n)	(n)	(i+)	(i+)	(n)	(i+)	(n)
cooperating enterprises	d+		(d+)	(n)	(n)				(n)
external consultants		n	i+	d+	d+	d+	n	n	(n)
industry association	io	n	i+	i+	i-	n	(i+)	n	(io)
trade unions	n	n	i+	i+	n	n	n	n	n
environmental groups	n	n	(n)	i+	d+	n	n	i+	(n)
consumer groups	n	n	n	i+	n	n	n	n	n
other lobby groups	n	n	(n)	(n)	(n)		n	n	(n)
politicians/government	n	i+	i+	i+	i+	i+	(i+)	i+	i+
parliament	n	(n)	n	n	(n)	(n)	n	n	(n)
local politics	i+	i+	i+	d+	n	i+	n	i+	(n)
(environmental) authorities	io	i+	(d+)	d+	i+	i+	i+	i+	i+
judiciary	n	n	n	i+	n	n	n	n	n
(external) science	n	n	d+	i+	(n)	i+	(i+)	n	n
neighbours	n	n	i+	i+	n	(n)	n	(i+)	(i+)
media	i+	n	i+	i+	i+	i+	n	i+	i+
the public	n	n	i+	i+	(n)	i+		i+	(d+)

d: direct participation, i: indirect influence, n: no influence, (): only partly correct qualification,
+: favourable influence, o: ambivalent influence, -: unfavourable influence

products, intensifying public environmental regulations, or production-relevant pro-tests by environmental action groups. Thus, non-interacting but structurally signifi-cant actors were not absent from the actor constellation due to deliberate exclusion or intentional absence, but for case-specific structural reasons. Overall, pragmatic or enhanced cooperation due to eco-functional tasks assumed and shared, and neutral, formal relations predominated among the actors involved in the success stories, whereas antagonistic relationships were much less in evidence and due far more to contrasting duties (sales and marketing in Diessner, Hertie, Kunert, availability of financial resources in Glasuld, Ciba, NedCar) than to conflicting personal or group interests as far as traceable. So the case studies represent examples of environmental cooperation rather than environmental conflict, or of social processes of learning from environmental conflict to cooperative environmental management. This is probably due largely to the selection of genuine success stories for the case studies.

The main features of interest patterns, power relations, and problem perception within actor constellations investigated are as follows:

1. The pattern of (environmental) problem perception is typically characterized by a gradually-broadening understanding of ecologically relevant facts, interrela-tions and the scope of the production process, e.g. the progressive inclusion of the whole value chain (Porter, 1985) in a (continuous) practical learning process around actively sought environmental improvements, sometimes driven by eco-logically detrimental hazards or events (Glasuld, GEP Europe, Ciba, ABC Coating).

2. The pattern of interests significant for the success stories can be characterized by the combination of environmentally positive attitudes – or at least a change of mind in this direction – with corresponding interests in environmental improve-ments, the interest of securing and strengthening one's own position within the organization by the environmental commitments undertaken, and the interest in economic savings and gains by pollution prevention or by efficient use and thus reduction of material input resources. Interests opposed to environmental im-provements, for instance because of additional costs, lack of market demand, un-favourable earning conditions for field staff, the provision of company-internal information to the public, or scepticism towards environmental cooperation with outside (environmental) groups (and vice versa), hampered but rarely blocked efforts to improve environmental management and protection. More extensive counteractive (strategic) interests, for instance company or competitor attempts to achieve competitive advantage through environmentally detrimental low pro-duction costs, or public authority imposition of time-consuming and/or expensive environmental regulatory prescriptions in retaliation to past ecological failings by the company, were of only minor importance. Rather, public environmental ordinances and requirements (indirectly) contributed to quite a number of envi-ronmental management efforts investigated (ABC Coating, Glasuld, Diessner, Hertie, GEP Europe).

3. Power distribution among actors was to some extent rather equal and partly hierarchical. As a result, successful environmental management was the outcome of mutual dependence among major actors and of willingness on the part of

subordinate actors to follow the main line agreed upon by the major actors. So a multi-layered, but still relatively clear-cut (power) game within the actor constellation was characteristic of the success stories: top management and power promoter have greater directive power in installing and performing environmental management than division leaders, technical promoters, environmental quality circles, (hired) consultants, public authorities or environmental action groups, but are nevertheless (crucially) dependent on their cooperation or at least consent. Further participating actors, such as the (company) workforce, professional associations, health authorities, still have some power to facilitate or to retard, but usually cannot obstruct effective environmental management and protection. Overall, the configuration of actors as such was not a qualifier *structurally* hindering the evolution of the success stories.

After this rather abstract general description of actor constellations, we list a few observations of more specific interest:

1. Under circumstances of (radical) change, competent motivated individuals play a key role in initiating and effecting this change, whereas institutionalized provisions may substitute this precondition at a later date.[3]

2. As would be expected, the prominent role of key individuals is especially significant in the smaller companies studied.

3. Typically, the bigger companies maintaining continuous working relations with public authorities emphasize communication policy in environmental affairs.

4. The main actors involved in successful environmental management tend to belong to upper management levels in a company and cooperating organizations, such as suppliers or public authorities. Although the shop-floor finally has to implement environmental management concepts in everyday working practice, employees on this level at least receive training in environmental management practices, but have hardly any influence on general decision-making in this respect.[4]

5. Cooperation matters in environmental management, but the type of actors collaborating to achieve substantive environmental improvements varies considerably from essentially in-house actors only (Diessner) via common projects with one or two other companies (Amecke Fruchtsaft, NedCar) to joint ventures with public institutions and environmental consultants, including environmental organizations (Hertie, ABC Coating, Glasuld).

3   ACTOR BEHAVIOUR

Against the background of the prevalent configuration of actors favouring the success achieved, it is important to analyse the impacts of the following factors on

---

[3] In Ciba, for instance, employees nowadays largely (have to) act in line with the established environmental management system, whereas the establishment of the corresponding basic corporate philosophy occurred only because of the commitment of key persons in top management.

[4] The two Danish case studies differ somewhat in this respect, partly due to their conscious coupling of environmental with health and safety concerns.

the development, change, and stabilization of actors' (environmental) behaviour (see WBGU, 1993: 256):

- The perception and evaluation of environmental conditions,
- Environmentally relevant knowledge and information processing,
- Attitudes and values,
- Incentives for action (motivations, reinforcement),
- Offers and opportunities for action,
- Perceivable consequences of action (feedback).

As indicated above, the perception and evaluation of environmental conditions is typically characterized by a social (learning) process in which genuine ecological concerns gain increasing relevance. This process is due mainly to overlapping forces of (initiating) non-ecological pressure and interest, diffusion of environmental awareness among company staff, improved and more comprehensive understanding of the environmental aspects of production and products, strategic (environmental) planning and action by core actors, and experience of substantial (environmental) success and company internal and/or external recognition.

Non-ecological pressure and interest are emphasized as significant driving forces because, on a general abstract level, they point to the need to couple environmental objectives with existing structures and interests and they indicate motivating company internal or external boundary conditions. These trigger environmental management efforts and thus provide incentives for action that are due to competitive strategy considerations (cost reduction, environmentally oriented consumer demand), or to the negative consequences and scandals resulting from (illegal) environmental burdens from production or products (ABC Coating, Glasuld, GEP Europe). The diffusion of environmental awareness among company staff points to the necessary sociopsychological internalization process of (genuine) ecological attitudes and values in order to increase environmental concern in a social learning process. On the cognitive level, this learning process necessarily implies the acquisition of environmentally relevant knowledge and information processing leading to an improved understanding of ecologically relevant facts and interrelations not yet conceived of in its starting phase. Experience of substantial (environmental) success, e.g. cost reduction, economic gains, significantly reduced environmental pollution, appears to be a necessary condition for generating motivational self-dynamics within a company in its pioneering environmental efforts to install environmental management on a continuous institutionalized basis and strategy. This tends to be further supported by company internal and/or external recognition.[5] Furthermore, on average there appear to be plenty of offers and opportunities for environmental action with highly visible consequences during the first decade of installing systematic

---

[5] The motivational self-dynamics generated by the new commitment to environmental management followed by rather immediate experience of environmental success gave important momentum to the continuation and further expansion of environmental management efforts in most cases. Particularly in the case of Kunert, public interest and recognition probably played a decisive role since they gave impetus to the continuation of its eco-accounting efforts and for repeating them on an annual basis.

environmental management within a company which do not contradict its economic objectives. Finally, strategic planning and action in the direction of environmental management was essential for the success stories investigated. A conscious, explicit corporate environmental policy was either the basis for specific substantive measures in environmental management or it was induced by and developed in parallel with such measures.

Some further observations common to all case studies may sum up company behaviour and strategy in this respect.

1. In all cases the main efforts towards environmental management occurred from about 1990, so that the companies despite certain pioneering improvements in environmental protection may be considered early followers in general (socio-economic) terms, but not genuine overall pioneers in environmental management.

2. When good profits are being made, available financial resources allowed some attempts to sell environmentally friendly products or additional environmental investments which were not profitable and involved economic losses.

3. In the success cases investigated, economically beneficial steps in environmental management plus some attempts at pro-active environmental protection and ecological product innovation and marketing predominated, but no risk-taking strategy in the portfolio of market chances and environmental risks (Steger, 1993) was followed which relied on potential, but improbable market opportunities for ecologically oriented investments and products.[6]

4. On a general abstract level, the initiating events and starting points for improving company environmental management were always negative ones aiming at avoiding certain effects, but they differed with respect to their internal or external causation: sometimes the conditions triggering environmental management efforts are due to competitive strategy considerations (cost reduction, environmentally oriented consumer demand), sometimes they are due to the negative consequences and scandals resulting from (illegal) environmental burdens from production or products (ABC Coating, Glasuld, GEP).

5. The environmental improvements to be observed in the case studies are mainly production oriented improvements, whereas innovative, environmentally friendly products are of only secondary importance, though by their very nature some changes in the production process also lead to environmentally more favourable products (NedCar, ABC Coating, Diessner).

4   STRUCTURAL FEATURES

On the structural level, and thus independent of actors' (current) articulation of interests and modes of procedure,[7] special account is typically to be taken of the

---

[6] Only Ciba's strategy contains components which aim at integrated process technologies of environmental protection and at service instead of product orientation in the future agrochemicals market.

[7] For further discussion of the distinctive impacts of structure and action on politics see cf. Conrad, 1990, 1992.

following relevant structural elements and conditions: ecology (1), economics (2), corporate organization and culture (3), law, politics and administration (4), general sociocultural conditions (5), and contextual know-how (6):

1. In an ecological perspective, one may ask if the company was confronted by severe environmental problems, and publicly and politically perceived as such; if their treatment involved conflicts with other, especially economic, company objectives; if the solutions pursued might lead to other environmental burdens, i.e. involved ecological trade-offs; and if significant potential existed for (economically favourable) savings and efficiency gains in resource utilization and environmental protection.

   Typically, severe environmental problems plus corresponding public and/or political pressure were in some cases the events triggering serious installation of, and investment in, environmental management practices and development into becoming an environmental front-runner. This was especially true of Glasuld and GEP Europe, whereas in other case studies, like Amecke Fruchtsaft or Hertie, the environmental impacts of production or products did not pose a particularly severe problem and could well have been dealt with in a routine manner without extraordinary environmental management efforts. Similarly, in some case studies the potential for economically favourable savings and efficiency gains were considerable, but much less so in others. So in the cases investigated there is, contrary to expectation, no significant correlation between the pursuance of clearly available options of savings or efficiency gains in resource utilisation and environmental protection, on the one hand, and the absence of severe environmental problems and corresponding political and public pressure on the other. That points to the only limited unequivocal impact of the size and significance of an environmental problem and concern on actual environmental management efforts. In addition, mutual ecological or ecology–economy trade-offs, the handling of which constitutes a major task for a fully established environmental management system, did not play an important role, with the partial exception of NedCar. Increased production costs due to environmental protection measures are frequently considered acceptable to a quite considerable degree. However, environmentally favourable but economically unviable products seem to be abandoned in the longer run (Ciba, Hertie). Furthermore, if concern about environmental protection falls together with health protection concern, their viability typically increases strongly (ABC Coating, Glasuld).

2. Major economic impact on a company's environmental management efforts can be expected from the following structural features: available economic resources, essentially due to the profitability of its activities, its competitive situation (or its degree of market dominance, respectively), the (general) business outlook for the industry, and the pursuance of an economically sound (competitive) company strategy.

   With some variation and few exceptions (meagre business outlook in the textile industry in the case of Kunert, partial market dominance of Glasuld/Saint Gobain in glass-wool and of GEP Europe in thermoplastics for cars) these structural features predominated in the case studies. The presence of economic resources, (strong) market competition, economic prosperity, and a sound corporate strategy cannot be interpreted, however, as strictly necessary conditions for successful environmental management because no counter examples have been investigated. Taking into account other case studies (see Chapters 2 and 14) it appears reasonable to assume that these structural characteristics are supportive of, but are not necessary conditions for, successful environmental management.

3. Corporate organization and culture, i.e. company-internal structural conditions of environmental management, refer particularly to the degree of (in)formality of its decision-making and working process, general working atmosphere, low or high employee turnover, cooperative or hierarchic style of leadership and corporate action, company size/number of employees, external contacts and cooperation, long or short-term time horizon for company strategy and expected returns on investment, and environmental awareness among employees.

Since most of these conditions were not explicitly ascertained in an operationalized and validated manner, only a number of plausible conclusions can be drawn with respect to the influence of corporate organization and culture. As would be expected, growing company and staff size tends to lessen the significance of informal communication, cooperation and working processes, of cooperative leadership, of good working atmosphere, and of low employee turnover for viable environmental management systems, because large corporations are able to institutionalize corresponding task routines in environmental management, thus making it relatively independent of personal intrigues or occasional high turnover rates. And they need more formal and hierarchic arrangements, even if a higher degree of informality, cooperative organization, satisfying working atmosphere etc. probably facilitate successful environmental management. So, as far as can be ascertained, working atmosphere was relatively positive in the companies investigated with the limited amount of intrigue and controversy facilitating effective cooperation. This appears to be correlated with low employee turnover as well as management confidence in the workforce allowing for effective ecological learning processes in most companies investigated. Furthermore, such a constellation appears to lead to environmental management practices covering more or less all divisions and levels of the company whereas in other cases frequently only sectional/partial environmental management efforts can be observed. In addition, external contacts and cooperation and longer time horizons in corporate strategy appear to be favourable for successful environmental management. Finally, environmental awareness among company employees is certainly a rather necessary condition for successful environmental management. It typically develops on a larger scale, however, only during the process of developing and implementing environmental management within a company. Therefore environmental consciousness and effective environmental management presuppose each other, each being both the condition for, and the result of, the other. Overall, the existence of the above structural characteristics of corporate organization and culture, namely considerable informality, cooperative leadership, satisfying working atmosphere, low staff turnover, external contacts and cooperation, long time-horizon in corporate strategy, and significant environmental awareness among the employees, tend to favour successful environmental management.

4. The structural relevance of law, politics and administration for environmental management can be traced by the (non)existence of clear environmental laws, of substantial conditions set by environmental policy, of case-specific administrative prescriptions, of the authorities' capacity to implement and enforce public environmental policies and regulations, and by the homogeneity or fragmentation of the (environmental, economic, labour, health, technology etc.) policies relevant for (successful) environmental management.

There is usually a relatively clear legal framework of environmental regulations in West European countries within which the investigated cases occurred: Denmark, Germany, The Netherlands and Switzerland. Apart from these legal boundary conditions, however, substantial environmental policy norms and programmes, specified administrative prescriptions, and strong public enforcement capabilities were relatively unimportant except in those cases where administrative refusals and requests initiated environmental management efforts due to (foreseeable) serious violations of environmental standards (GEP Europe, ABC Coating, Glasuld, Diessner), or where public authorities collaborated with the companies investigated to advance environmental management in general (Glasuld). This does not come unexpectedly because by definition the cases of successful environmental management imply that the company has achieved (pioneering) successful environmental improvements beyond average industrial standards required by environmental law. Therefore the relative insignificance of environmental policy regulations for the specific environmental advances under review is rather likely. Typically, in all cases compliance with legal environmental standards and regulations is meanwhile established practice. More important, however, some companies, on the one hand, even became active in promoting the (legal) generalization of the more advanced environmental standards with which they already complied (ABC Coating, Glasuld, Ciba, Hertie). On the other hand, some companies did not consider public authorities as helpful actors in their environmental management efforts and strategy (Amecke Fruchtsaft, Diessner), or experienced the increasing counter-productivity of public end-of-pipe oriented environmental prescriptions (Ciba). Particularly in such cases, inconsistency and fragmentation of relevant policy segments and programmes can be observed, easily leading to opposing policy demands and opening up considerable latitude for action to the company. Thus, the parameter policy homogeneity versus policy fragmentation influenced the development process of environmental management efforts but hardly the substantive results of the success stories investigated.

5. General sociocultural conditions which may well influence the development of successful environmental management may be listed as the modernization capacity of a society; the importance of the state and public policy; the importance of public debate and the equivalent strength of civil society; the extent of division into different social classes or strata; the degree of public participation and socio-structurally entrenched substantive democracy; the significance of self-responsibility and liability of social actors; the degree of legalism; decentralized versus

centralized (political) culture and decision-making procedures; the importance of postmaterialistic value orientations; environmental awareness of, and behaviour by, main actors and the population in general; and the significance and social influence of the environmental movement and corresponding groups.

Since these general sociocultural conditions are usually assumed to refer to a society or country at large, and since these conditions do not differ considerably among the countries where the success stories occurred, no distinctive differences in the impacts of these sociocultural conditions on them can be expected. National culture may have some influence on the development of successful environmental management, but at best seems to play the role of a background variable in the case studies, undertaken in four neighbouring countries only (Swiss social responsibility and environmental concern in Ciba, American business culture in GEP Europe, workers' participation in Glasuld).

One may only assume with some plausibility, but cannot demonstrate, that the prevalent modernization capacity, the important role of the state, the available opportunities for public participation, the minor role of class conflicts, the prominence of legalistic traditions, the federal structure of policy-making, the considerable spread of postmaterialistic values, the comparatively high degree of environmental awareness and concern, and the non-negligible political influence of environmental action groups, will all tend to support (corporate) environmental management.

6. Contextual know-how has an impact on successful environmental management due to the availability of the corresponding (scientific) information and corporate manpower to acquire it; to the availability of environmentally sound technologies; to the knowledge about environmental impacts of company action and behaviour; to the feasibility of (ecological) modernization strategies; to the monitoring efforts of environmental impacts of company production process and products; to access to information and knowledge sought about corresponding environmentally significant programmes and measures in other social spheres and systems; and to the embedding of environmental management in strategic corporate planning.

Although these structural conditions referring to the availability and processing of information tend to be prevalent in the success stories, there is still quite some variation in this respect. Knowledge about existing know-how and environmental management programmes in other social spheres and systems, and even about available environmental technologies, is only partly available, particularly in smaller companies, and stringent comprehensive environmental impact assessments are still more the exception than the rule in spite of the broadening of the ecological perspective within many companies to include the whole product lifecycle. Necessary expenditures in time and manpower set limits on corresponding information-gathering and -processing efforts. Nevertheless, the companies investigated seem to have more and better knowledge than the average about environmental aspects and alternative, ecologically favourable options for their production and/or products and the ability to embed their environmental management activities more clearly in an explicit corporate strategy.

In sum, the structural features of the case studies still show considerable similarity on the level of general abstract structural requirements for successful environmental management, but differ to quite some degree in their specific expression. The existence of typical structural features can be described (in analytical terms) and assessed as favourable conditions of the success stories with quite some plausibility, but cannot be deduced as necessary preconditions of successful environmental management because, in accordance with the preference for hermeneutic understanding of the success stories, they were neither operationalized in precise terms nor tested systematically.

5 SOCIAL NETWORKS

As indicated in Chapter 1, sufficient understanding of social processes and projects needs to take into account the level of social networks with their own internal logic of action on the basis of a communicative rationality, a level beyond the actions by (individual) actors and of (systemic) structures not simply open to alteration by the actors involved (cf. Scharpf, 1993; Weyer, 1993). Social (policy) networks are able to generate institutional arrangements and order beyond the mere rationales of individual actors and of systemic structures. Social networks evolve by a process of mutual production of behavioural expectations. Their stability is based on the capability of the participating actors not only to address demands to others but also to cope with demands imposed on them by others. Via self-commitment to interactive relations structured in such a way, the parties' room for manoeuvre is thus not only expanded but also restricted at the same time. For rules of behaviour develop in social networks which no party can control exclusively but which have to be complied with if one wants to participate in the network. Furthermore, if one's own action potential depends strongly on the (financial, legitimation, argumentation) resources mobilised by social networks, to question the network implies a questioning of oneself, too. Different rationales of action involving different time horizons, namely (individual) instrumental rationality based on situational opportunity structures, more stable, though terminable communicative rationales within collective structures, and hardly alterable system rationales, will typically be experienced as conflicting objectives which can only be reconciled temporarily (Weyer, 1993).

Looking now at the various case studies, their actor constellations tend to exhibit characteristics of social networks to a higher or lower degree. This is particularly true for the two Danish case studies, but the important role of environmental cooperation in successful environmental management implies the significance of binding social rules and the structural consequences of social networks in the other cases, too. Without the mutual production and acceptance of behavioural expectations among the actors involved, successful environmental management would hardly be feasible. So the specific interests and behaviour of actors within the actor constellation of the success stories are to be understood in terms of the combination of individual intentions and strategies, existing structural boundary conditions, and the requirements of participation in a social network providing benefits.[8] It appears worth mentioning, however, that with partial exception of the Danish case studies, crucial actors are members of differing (overlapping), partly loose social networks relevant to the success stories. Although some (central) actors clearly disposed of more resources and power than others they were clearly unable to fully control the behaviour of other participants or even the rules of behaviour within the social networks. Obviously, the social networks described in the case studies were able to generate cognitive, financial, political and manpower resources, which usually could not have been generated by one (corporate) actor alone, such as the cooperation between Amecke Fruchtsaft and KHS, the progress in eco-auditing around Glasuld on the county level, or the impetus to develop a chromium-free black colouring agent by the social interaction of different departments as well as individual actors – an environmental consultant and the supplying dye manufac-turer – in the case of Kunert.

6 SIMILARITIES AND VARIATIONS

If we now look at the differences between, and the homogeneity within, the three types of evaluative analysis carried out, similarities between outcome, process and institution-oriented investigations of company eco-performance appear to be more important on a general level than differences, especially when compared with similarities and dissimilarities among companies in the same type of evaluative analysis. Basic features of actor constellations, of structural boundary conditions, and of social networks characteristics in each case do not differ significantly. What differs is company size, the number of individuals involved in the success story, and the existence of an environmental management strategy and system in those com-panies only where their institutional entrenchment was investigated (see Table 1). However, these differences are partly due to the, in this respect arbitrary, selection of companies for investigation, and to the type of evaluative analysis applied, because in-stitutional (and process) evaluations simply presuppose an environmental manage-ment system to exist or at least to be projected, and will tend to cover a broader sphere of company action with more individuals concerned than does outcomes evaluation focusing on a substantive individual case of environmental management.

The predominance of basic structural similarities between the success stories does not come as a surprise, however. If actor constellations, structural boundary conditions, and cooperation in social networks are favourable and typical for successful environmental management, these features should prevail irrespective of the evaluative perspective applied. Differences are then to be expected on a lower level of abstraction, e.g. in the role of industrial associations, company organization, specific actors involved in a success story, prominence of good housekeeping measures versus innovation orientation in environmental management strategies (cf. Dyllick et al., 1994; Meffert/Kirchgeorg, 1993; Post/Altman, 1992; Steger, 1993; Wicke et al., 1992). However, such differences are found as much among different

---

[8] The case-specific description of the concrete elements contributing to this combination explaining the actor constellation and the behaviour of participating actors is not spelled out here in any greater detail, but can be explicitly or implicitly found in Part II presenting the individual success stories.

companies studied by the same type as by differing types of evaluative analysis. Thus, the initial caveat of this chapter does not play a significant role when comparing typical actors and structures of the case studies carried out.

To sum up, typical and varying actors and structures can be observed in the investigated cases of successful environmental management such as to invite the conclusion that there are typical actor constellations, structural boundary conditions, and (cooperative) social networks on an abstract level of analysis, as described above, even if not all of the individual characteristic features identified are likely to be present in each case, whereas, on a more concrete level of analysis, the setting of participating actors, the type of social networking, and the structural features of the success stories vary to a considerable degree.

Whereas one may reasonably query the explanatory value of this statement, because it probably holds for most instances of social analysis, the statement may well be the maximum of generalizable social science explanation feasible for this type of question. The question of a generalizable pattern of successful environmental management is taken up again in Chapter 17.

LITERATURE

Conrad, J. (1990) *Nitratdiskussion und Nitratpolitik in der Bundesrepublik Deutschland*. Berlin: edition sigma.
Conrad, J. (1992) *Nitratpolitik im internationalen Vergleich*. Berlin: edition sigma.
Conrad, J. (1994) Ökonomischer Wegweiser ins ökologische Mehrweg-Optimum. FFU-Report 94-4, Berlin.
Conrad, J. (1996) Unternehmensexterne Determinanten betrieblichen Umweltmanagements. In: (eds.) L. Mez and M. Jänicke, *Sektorale Umweltpolitik*. Berlin: edition sigma.
Dyllick, Th. *et al.* (1994) *Ökologischer Wandel in Schweizer Branchen*. Bern: Haupt.
Holm, J. *et al.* (1994) Two Cases of Environmental Front Runners in Relation to Regulation, Market and Innovation Network. Report, Roskilde University.
Huber, J. Protzmann, E. and Siegert, U. C. (1994) Fallstudie Ciba AG. Report, University Halle.
Kirschten, U. (1995) Ökobilanzierung und Aufbau eines Öko-Controlling bei dem Kunert-Konzern. Report, Free University Berlin.
Meffert, H. and Kirchgeorg, M. (1993) *Marktorientiertes Umweltmanagement*. Stuttgart: Poeschel.
Porter, M.E. (1985) *Competitive Advantage. Creating and Sustaining Superior Performance*. New York: The Free Press.
Post, J. and Altman, B. (1992) Models of Corporate Greening. In: (ed.) J. Post, *Markets, Politics and Social Performance*, Vol. 13. Preston.
Scharpf, F.W. (ed.) (1993) *Games in Hierarchies and Networks*. Frankfurt: Campus.
Schrama, G. (1994) The Internalization of Environmental Management at GE Plastics Europe. Report, University of Twente.
Schrama, G. (1995) The Environmental Factor in the Designing of Passenger Cars by NedCar. Report, University of Twente.
Steger, U. (1993) *Umweltmanagement*. Wiesbaden: Gabler.
Taneja, M. (1994) Erfolgsbedingungen betrieblicher Umweltpolitik am Beispiel der Firma Diessner Farben und Lacke GmbH und Co KG. FFU-Report 94-7, Berlin.
WBGU (Federal Government scientific advisory council 'Global Environmental Changes') (1993) Welt im Wandel: Grundstruktur globaler Mensch-Umwelt-Beziehungen. Jahresgutachten 1993. Bonn: Economica.
Weyer, J. (1993) System und Akteur. *Kölner Zeitschrift für Soziologie und Sozialpsychologie* **45**, 1–22.
Wicke, L. *et al.* (1992) *Betriebliche Umweltökonomie*. München: Vahlen.
Will, S. (1994) Hertie und BUND—eine Kooperation im Spannungsfeld zwischen Ökonomie und Ökologie. FFU-Report 94-6, Berlin.

14. COMPARISON WITH OTHER CASE STUDIES OF ENVIRONMENTAL MANAGEMENT

Jobst Conrad

In this chapter the major results obtained from the case studies are put into perspective by comparing them with other studies of environmental management presented in the literature (see Dillon/Fischer, 1992; Dyllick, 1989, 1991; Dyllick et al., 1994; Fischer/Schot, 1993; Freimann, 1994; Hildebrandt et al., 1994, 1995; Hopfenbeck/Jasch, 1993; Huisingh et al., 1986; Kirchgeorg, 1990, 1995; Longolius, 1993; Meffert/Kirchgeorg, 1993, 1995; Middelhoff, 1991; Oberholz, 1989; Ostmeier, 1990; Prisma-Industrie-Kommunikation, 1992, 1993; Rappaport/Flaherty, 1992; Schmidheiny, 1992; Steger, 1992; UBA, 1991; Vietor/Reinhardt, 1995; Wild/Held, 1993). The procedure is as follows. First, the differing research strategies and methods followed in the various studies on environmental management are compared in order to arrive at an evaluative assessment of the theoretical and methodological quality of the case studies undertaken. Second, the main substantive results of other studies of environmental management are summarized. Third, these results are compared with those of the project underlying this book.

Since major results of environmental management research have already been presented in Chapter 2, the focus of this comparison is on other case studies and less on questionnaire-based surveys of environmental management practices. The selection of these case studies shows a German bias due to their easy availability to the author. However, taking into account corresponding Anglo-Saxon literature (cf. Davis, 1991; Dillon/Fischer, 1992; Fischer/Schot, 1993; Groenewagen et al., 1995; Huisingh et al., 1986; McKinsey & Company, 1991; Morrison, 1991; Piasecki/Asmus, 1990; Rappaport/Flaherty, 1992; Shen, 1995; Smith, 1992; Vietor/Reinhardt, 1995; Ytterhus et al., 1995), this bias should not disqualify the results of this particular comparison because of the overall similarity of environmental management features and development in western industrialized countries in spite of considerable differences in corporate culture and environmental policy style and because of the national location of the project's case studies.

1 CONCEPTUAL AND METHODOLOGICAL APPROACHES

In the literature on case studies of (successful) environmental management four methodological approaches and thus levels of increasing (scientific) validity, can initially be reasonably distinguished:

1. Many case studies outline the main substantial and institutional results of a company's or industry's environmental management efforts, partly supplemented

by narratives of key persons about the history of this process, without further cross-checks of the validity of this reconstruction (Hopfenbeck/Jasch, 1993; Oberholz, 1989; Prisma-Industrie-Kommunikation, 1992, 1993; partly Schmidheiny, 1992). These case studies hence remain relatively superficial with respect to the valid reconstruction of social processes leading to (successful) environmental management and cannot assert sufficient claim to scientific method.

2. Other case studies represent a more detailed (critical) reconstruction of a company's environmental management efforts and their results by insiders who are themselves more or less involved in the case presented (Dyllick, 1991; partly Prisma-Industrie-Kommunikation, 1992, 1993; Schmidheiny, 1992; Steger, 1992; UBA, 1991; Wild/Held, 1993). These stories can probably claim higher (scientific) validity, but necessarily have a subjective bias due to the personal and hindsight nature of reconstruction.

3. Case studies of this type reconstruct a company's environmental management efforts and success on the basis of documents, questionnaires, and possibly very few interviews (Dillon/Fischer, 1992; Dyllick *et al.*, 1994; Fischer/Schot, 1993; Huisingh *et al.*, 1986; Kirchgeorg, 1990; Longolius, 1993; Middelhoff, 1991; Ostmeier, 1990; partly Rappaport/Flaherty, 1992; Steger, 1992; UBA, 1991; Vietor/Reinhardt, 1995; Wild/Held, 1993). Here the data-collection method allows at least partial monitoring of the reconstruction so that it acquires a more 'objective' character, although it still is too spare to claim scientific validity for the portrayed underlying social processes.

4. Finally, only few case studies are based on the extensive gathering of all potentially relevant information and data and subsequent cross-checking by document analysis, by a large number of interviews with diverse resource persons, and eventually by longitudinal analysis (Dyllick, 1989; Hildebrandt *et al.*, 1995; Kirchgeorg, 1995; Kühleis *et al.*, 1994; Zimpelmann *et al.*, 1992; partly Meffert/Kirchgeorg, 1993; Rappaport/Flaherty, 1992). Only this type of case study represents a social-science case study in the strict sense of the term so that results can claim considerable scientific reliability and validity. They require, however, both sufficient manpower investment (around 6 person-months) and reasonable information access.

Although the literature cited and investigated represents only a portion of the many case studies reported, it is probably reasonably representative. Unfortunately, the sources of information are quite frequently not clearly indicated and therefore can only be guessed at. Most of them belong to the first three classes of investigation. Thus, only very few, i.e. around 50 out of 5,000 cases have a methodically comparative case study design which permits, strictly speaking, more rigorous comparison with the project's nine case studies. Since, however, the substantive results of the different classes of case studies point overall in a quite similar direction, the general rather than detailed comparison presented in this chapter concerns the results of all case studies. Nevertheless, this analysis of the investigative method of case studies on environmental management clearly demonstrates the limits to scientific knowledge concerning the complex social processes underlying (successful) environmental management.

There are a number of further formal features of environmental management case studies:

1. The unit of analysis is usually a company (or an organizationally autonomous sub-company), and sometimes an industry, e.g. chemical industry, detergent industry, or the banking sector.

2. The case studies typically refer either to individual environmental management projects, or to the various management efforts and organizational processes installing and improving environmental management, or to the description of the structure of, and experience with, the environmental management system of a company, and thus parallel the three distinctive types of outcome, process, and institutional evaluation of the project's case studies.

3. Two main reference points of many case studies are exemplary cases of successful environmental management and of distinct environmental management efforts initiated by (public) environmental scandal and subsequent environmental conflict, whereas practically no case studies of unsuccessful environmental management exist.

4. The case studies relate either to specific elements or to major corporate segments of a comprehensive ideal-type environmental management, for instance development and marketing of the returnable PET bottle by Coca-Cola (Meffert/ Kirchgeorg, 1993), or the installation of environmental auditing in Norsk Hydro (Schmidheiny, 1992), as well as Elida Gibbs' turn from ecological crisis management to preventative environmental management strategy (Gerhardt/Kühleis, 1994), or the development and installation of a waste-water-free textile mill by Brinkhaus (Oxenfarth, 1994), but not to such a comprehensive environmental management system that transcends company boundaries.

5. Few case studies focus on the process of interaction between (private) environmental management and (public) environmental policy to identify the interactive social and system dynamics of corresponding actor strategies, behaviour, and substantive ecological results.

6. Only few case studies present a differentiated analysis of the complex social processes leading to the environmental management efforts and successes described. Most concentrate on the description of technical features, organizational arrangements and processes, and strategic considerations of environmental management efforts.

7. To my knowledge there have been no genuine (real-time and not *ex post facto*) longitudinal studies covering long-term development processes and changes within a company over more than a decade, including the durable effects of its environmental management measures as well as the feedback of environmental improvements on the further shaping of the environmental management system. Such studies could avoid the inevitably somewhat distorted reconstruction of social development processes by investigations with a more short-term and *ex post facto* case study design.

8. As already indicated in Chapter 2, predominant categories of investigation in environmental management favour a micro-economic perspective

(e.g. profitability, financing, marketing) and are more descriptive-taxonomic than theoretical (e.g. differentiation among industrial branches, company functions, organizational units, (environmental) technologies applied, corporate (environmental management) strategies), which are frequently connected in a rather additive manner for purposes of synthesis. Attempts towards a more elaborated theoretical-analytical perspective on a case study basis can be found in Dyllick, 1989; Fischer/Schot, 1993; Freimann, 1994; Hildebrandt *et al.*, 1995; Kirchgeorg, 1990; Ostmeier, 1990; Rappaport/Flaherty, 1992.

9. Few case studies are organized (also) to test specific (theoretical) hypotheses (cf. Dyllick, 1989; Rappaport/Flaherty, 1992), whereas the majority aims at presenting illustrative examples of (successful) environmental management (cf. Hopfenbeck/Jasch, 1993; Schmidheiny, 1992).

10. In order to substantiate empirically based hypotheses about the determinants of successful environmental management, it would be necessary systematically to compare successful and unsuccessful cases of environmental management. No such rigorous comparative investigations are available, besides the lack of case studies of unsuccessful environmental management which occurs quite well too.

Altogether, from a methodological point of view the project's nine case studies belong to the relatively rare fourth class of case study in environmental management, aiming at the rather detailed analytical reconstruction of corresponding success stories going beyond a merely economic perspective, including all relevant actors involved, though not yet offering real-time longitudinal analysis and not testing specified theoretical hypotheses.

As far as the analytical categories and frameworks employed in the theoretically more elaborate case studies are concerned, a combination of the following key components – presented with partly divergent classificatory notions – predominates in the endeavour to reach an adequate understanding of corporate environmental management, namely reference to:

1. external pressures and demands on companies:

   • the quantitative and qualitative growth of environmental problems due to industrialism;

   • rising environmental public awareness and pressure;

   • more stringent environmental policy regulations and the limitations of bureaucratic environmental protection;

   • growing (non-economic) demands on firms (politicization of production; Conrad, 1990; Kitschelt, 1985) by different social groups such as employees, investors, suppliers, customers, competitors, service providers, collaborating companies, public authorities, societal institutions and groups (citizen groups, media, churches, cultural institutions, consumer groups, the general public, future generations);

   • consequently, the market, politics, and ethics as external control systems for corporations (Dyllick, 1989);

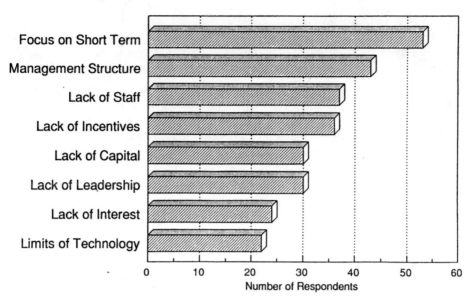

Figure 1   Factors preventing better EHS. (Source: Rappaport/Flaherty, 1992: 141)

2. differentiated substantive (technical) environmental management requirements:
   • production, products, or environmental services as points of departure for environmental management;
   • (additive) end-of-pipe, recycling, integrated environmental technologies;
   • the corresponding need for an integrated, systemic-evolutionary, supra-functional approach to environmental management on all levels, with particular reference to life-cycle assessment and the diverse classes of environmental burden (consumption of resources, consumption of energy, air, water and soil pollution, noise, waste volumes, synergetic impacts), the environmental compatibility of normative, strategic, and operative management, as well as to the various corporate functional domains;
3. formal considerations of corporate (environmental) strategy:
   • the need for environmental protection as a genuine company objective within the corporate goal system;
   • regulatory, economic, organizational, conceptual, informational, technical, product acceptability barriers and obstacles to strategic environmental management (Huisingh et al., 1986; Dieleman/de Hoo, 1993);[1]
   • different formally feasible environmental management objectives and strategies for sufficiency, efficiency, or consistency: efficiency gains, (technological) innovation and substitution, ecologically beneficial structural change, limitation of

---

[1] According to Rappaport/Flaherty (1992) management orientations and corporate culture are the main factors preventing better environment, health, and safety provisions, as indicated in Figure 1.

volumes, avoiding disasters, technological problem management (Conrad, 1994; Huber, 1995), on different substantive levels, such as housekeeping changes, materials' substitution, changes in the design and operation of processing equipment, reuse and recycling of material (Huisingh, 1988);

- strategic environmental management portfolio, based on the analysis of the company-internal ecological strengths and weaknesses, and of the company-external environmental opportunities and risks, relating to environmental protection, risk management and competition: the reduction of environmental costs, the opening up of new business fields, the exploitation of opportunities with new products and production processes, the activation of performance reserves, the minimization of liability risks, the avoidance of confrontation; competence, credibility, commitment, cooperation; cost leadership, quality leadership, orientation towards the whole market or towards market niches, timing strategy; resistance, passivity, retreat, adaptation, selectivity, innovation (Meffert/ Kirchgeorg, 1993; Steger, 1993).

Altogether, empirical investigations of environmental management practices may be distinguished in terms of the following modes of (theoretical) interpretation of their empirical findings:

1. *de facto* renunciation of theoretical explanation by illustrative explanatory reference to diverse conceptual models without stringent hypothesis testing (cf. Morrison, 1991; Oberholz, 1989; Winter, 1993);

2. explanation by one overarching concept, such as economic profitability (Huisingh *et al.*, 1986: pollution prevention pays), decision theory (Terhart, 1986: micro-economic analysis), or biocybernetics (Vester, 1992; Gorsler, 1991);

3. addition of segmented theoretical concepts addressing different analytical aspects of environmental management (cf. Hopfenbeck/Jasch, 1993; Meffert/Kirchgeorg, 1993; Steger, 1992, 1993);

4. heuristic integration of different analytical dimensions of environmental management in a historically reconstructed hermeneutic understanding of empirical cases (cf. Dyllick, 1989, 1991; Hildebrandt *et al.*, 1995; Huber, 1991);

5. testing of specified theoretical hypotheses about different forms of environmental management by causal-analytical methods of explanation (e.g. by LISREL (Linear Structural Relations System)) with the aim of identifying the relative influence of and mutual links between diverse variables on successful environmental management on the basis of statistical analysis (cf. Kirchgeorg, 1990; Ostmeier, 1990).

Looking at both the methodological and theoretical case study approaches, one clearly detects significant correlations between the methodological and theoretical solidity of the case studies. Theoretically more elaborated ones tend to exhibit stronger methodological meticulousness. Certainly, the studies that go furthest in the quantitative testing of theoretical (causal) hypotheses can rely only on extensive questionnaires and very few interviews because of the impracticability of a hundred or so in-depth case studies.

2   RESULTS OF OTHER CASE STUDIES

Turning once again to the survey given in Chapter 2, major substantive results of the various case studies of environmental management can be summarized on a meta-level[2] as follows:

1. Looking at the past decade as a transitional period towards developing and installing environmental management in corporate practice, it is no surprise that the many case studies reflect the whole spectrum of potentially feasible environmental management measures and arrangements concerning: underlying motivation and incentives (public scandals, public regulations, economic savings, competitive advantage, market opportunity, bandwagon effect, public image and credibility); organizational arrangements (location and distribution of responsibilities for environmental affairs, task force groups, environmental accounting, auditing and controlling systems, training of the workforce, environmental communication); complexity of environmental management system design (manpower and resources for environmental management tasks, environmental management activities on the levels of corporate philosophy, strategy, information processing, financial accounting and budgeting, research and development, personnel selection and education, substantive cooperation with external actors, such as consultants, suppliers, public authorities, environmental groups); inclusion and/or exclusion of certain components of the value chain (raw materials extraction, transportation and storage, upstream production, mainstream production, distribution, use, maintenance and repair, recycling and waste products); and of corporate functions (product development, purchasing/provision of inputs, production, marketing and sales, logistics, waste management, installations, organization, management systems and instruments, environmental information systems, finance and accounting, personnel, public relations).

2. Case studies aiming at an appropriate reconstruction of the complex and multi-layered social (learning) processes when installing effective environmental management systems on a broad basis emphasize the need for the positive interaction of determinants within and without the company on various levels leading to the technical, organizational, socio-economic and cultural entrenchment of (innovative) environmental management efforts, thus generating a self-supporting interaction dynamics towards substantive environmental protection. Strategic environmental management has to be pursued systematically, and only strategic environmental management tends to generate competitive advantages (Wieselhuber/Stadlbauer, 1992).

3. Typically, the case studies present environmental management efforts and successes of companies which are often front-runners in their attempts and measures better to protect the environment, but rarely genuine pioneers gaining outstanding competitive advantages by innovative environmental management moves. On

---

[2] This summary focuses on general results on the meta-level in order to avoid endless listings of individual findings and, essentially, merely repeating the results of environmental management research already listed in Chapter 2.

the contrary, environmental improvements are usually soon imitated and adopted by other (competing) companies. Examples are AEG, Ciba, Dow Chemical, Henkel, IBM, Kunert, Migros, Neumarkter Lammsbräu, Opel, Swiss Air, Tengelmann, Werner & Mertz.

4. As can be theoretically deduced, the case studies sometimes indicate that neither does the setting up of a (sophisticated) environmental management system automatically guarantee good ecological performance, nor do good environmental practices and/or a good environmental management system necessarily guarantee good business practices and market success.[3]

5. It appears rather to be the other way round: companies pursuing a successful corporate strategy in general are likely to realize a successful environmental management strategy, too.

6. There are no specific best environmental management practices because their success depends on the context (cf. Cebon, 1993). Successful environmental management practices will thus vary with contextual conditions (cf. Kirchgeorg, 1990), something that could, again, be easily derived from general social theory.

7. In highly contingent situations of social change and corresponding transitional periods, situational settings and eminent individuals play a particularly decisive role at branching points and diverging future development paths because of the significance of coincidental Cournot effects (Boudon, 1984; Mayntz, 1996). Thus, social theory can only formulate theoretical modules identifying individual causal mechanisms and, eventually, replicable constellation effects permitting insight into the conditions and hazards of potential or likely environmental-management evolution options, but not predict a single, unequivocal development path.

8. The greening of technology processes 'will necessarily be rather slow, despite the strong public call for it, due to the dominance of prevailing technological trajectories. These technologies are favoured by the existing selection environment and benefit from dynamic scale and learning efforts. New, cleaner technologies sometimes fail in the marketplace unless these barriers have been removed.' (Schot/Fischer, 1993a: 17; Kemp, 1993) This points to the probable slow diffusion rate of new sets of environmental technologies substituting for old technical assets even if companies are keen on developing new environmental technologies, which they usually are not.[4]

9. Prominent in most case studies is the emphasis placed by corporate actors on the economic viability or profitability of environmental management. Although there is frequently some willingness to spend extra money (resources) on environmental protection measures, these measures are deemed to have to pay off in the longer run.[5]

---

[3] Though this may often be the case, as assumed by the title of a paper by Huisingh (1988).

[4] 'Development of technological innovations in both product and process is a time-consuming and costly process and is difficult to justify if the primary motivation is environmental. Consequently, companies are responding to outside pressure from regulators and the public rather than developing a long-term strategy for environment.' (Rappaport/Flaherty, 1992: 140)

[5] This is only a comparative requirement: environmental protection may be costly as long as the future costs of neglecting environmental protection are likely to be even more costly.

10. Successful environmental management requires congruence of image and reality in both directions. It turns out to be counterproductive if a company's environmental practice does not match its proclaimed positive environmental image because nowadays this easily leads to subsequent loss of credibility which is hard to regain; and without advertising its environmental management competence and results, they may well tend not to be (economically) acknowledged by clients, the authorities, and the public. Consequently, Meffert/Kirchgeorg (1993) plead for the combination of competence, credibility, commitment and cooperation as preconditions for successful market-oriented environmental management.

11. Companies, especially large ones, that practise relatively comprehensive environmental management have established an overall environment, health, and safety (EHS) system, since environmental, health, and safety concerns are perceived as belonging closely together (cf. Rappaport/Flaherty, 1992; Kasperson et al., 1988; Hildebrandt et al., 1994, 1995; Middelhoff, 1991). In fact, the usually already well-established health and safety concerns and entities facilitate the incorporation of genuine environmental protection measures in daily corporate routines.

12. From a general perspective, an evolution of corporate environmental policy from passive opposition via active opposition to organized adaptation and self-organization can be observed, with a corresponding shift in (strategic) orientation from environmental conflict among (traditional) corporate managers and outside environmental critics and lobby groups as well as in-house (environmental) reformers to (partial) environmental cooperation among these groups since the, 1980s or at least 1990s. Now conflicting economic, environmental and other objectives are dealt with within such cooperative settings, as a result of social and organizational learning processes, often initiated by public debate and pressure relating to public rules and standards, social debate on environmental requirements, and public environmental scandals (cf. Dyllick, 1989; Hildebrandt et al., 1994).

Distinguishing with Wiesenthal (1995) between (path-dependent) conventional and unconventional organizational learning, where conventional learning refers to the three modes of simple (or single-loop or adaptive), complex (or double-loop or assumption-sharing) and reflexive (or deutero or problem-solving) learning (cf. Argyris/Schön, 1978; Pawlowsky, 1992), and unconventional learning refers to processes of intrusion, such as invasion, dissidence and intersection, allowing abrupt changes in cognition and strategy, too[6], the development of environmental management efforts and systems in corporate organization,

---

[6] It is important to distinguish between individuals and organizations as subjects of organizational learning, which need not coincide in time or substance. Theories of conventional learning (cf. Argyris/Schön, 1978; Dodgson, 1993; Duncan/Weiss, 1979; March/Olson, 1975; Morgan, 1986; Shrivastava, 1983; Simon, 1991) relate gains in competence to the adequacy of rule systems and the linkability of core beliefs, i.e. to potentials of learning organizations, which are capable of reliably controlling their boundary to the environment. In contrast are processes of unconventional learning based on diffuse borders between organizations and their environment, allowing for intrusion processes, but they cannot

culture and daily routines can be described as juxtaposition of simple and complex as well as unconventional learning via processes of invasion (e.g. by external environmental consultants), dissidence and intersection (e.g. by the BUND's involvement in Hertie's ecological reorientation). However, in accordance with theoretical analysis reflexive learning of organizations, as opposed to individuals (Dodgson, 1993; Wiesenthal, 1995), cannot be observed, also due to the complicated and contradictory rationale of organizational development and despite the partially reflexive self-dynamics of organizational change, including the installation of innovative and anticipatory structures.

13. Such corporate development and learning processes furthermore appear to depend on early initial success for environmental management efforts, which generates sufficient socio-organizational momentum to continue these efforts. Initial potentials of environmental and economic benefits and thus of successful environmental management are mostly considerable because the reducibility of environmental pollution by means of savings in input resources, integrated process technologies, or the minimization of production and product wastes prevails especially in the early stages, whereas at a later stage further progress can usually be achieved only with greater difficulty.

14. Although feedback loops and corresponding (corporate) learning processes from past experience and that of other companies certainly influence the process of installing, implementing and enlarging environmental management, there exist in many cases bifurcation points in this development dynamics where situational decisions are taken in favour of or against entering further new development phases of corporate environmental management.

---

be explained by organization analysis because they refer to influencing factors beyond organizational control, such as non-members, illoyalty, anomy and diffuse borders to the environment, which therefore cannot constitute elements of intentional learning strategies of an organization (Wiesenthal, 1995). Simple learning concerns operative skills of the correct application of existing rules and thus relates to routine controlled trial-and-error behaviour according to permanent incremental processes of adaptating systems of rules to a changing environment, where the company reacts to deviations from existing standards by error identification and correction within a given behavioural framework. Complex learning refers to cognitive skills evaluating the adequacy of rule systems in the light of alternatives and thus relates to changes in (organizational) strategies and premises, possible even without external stimuli, which implies the development and change of organization-specific theories of action, where organizational hypotheses, norms and action routines are integrated in the organization's knowledge system by feedback from observation of the environment, so that behavioural reactions to perceived dissonances between responses of the environment and own theories of action can lead to these theories changing and thus to changes in the behavioural framework itself. Reflexive learning refers to second-order cognitive skills aiming at improving the learning capabilities of the organization itself in order to enlarge its problem-solving potential and options for action so that the corresponding transformation process includes, beyond mere reactive adaptation to the outside environment, the change in the goals and meanings of organizational theories of action. Reflexive learning on the level of organizations, however, can only be analytically reconstructed but not found in the world of real organizations. Investigations of the impacts of organizational learning point to gains in experience, the availability and instructiveness of which are frequently overestimated, to exchange of (key) personnel as a practical but risky instrument to effect complex change, to changes in orientation as a prototype result of complex learning, and to multiple identity expressing the capability to deal with fragmented orientations (Wiesenthal, 1995).

15. Differences between American and West European companies in environmental management mainly reflect different corporate cultures and differing (environmental) policy styles. Many U.S. firms seem to be more concerned with the vagaries of environmental regulation enforcement than with the regulations themselves, whereas European and Canadian firms tend to have greater confidence in, and respect for, environmental regulators. Concerning (environmental) credibility with the customer, the customer comes first in the United States; less so in Europe and Canada. 'With surprising uniformity, companies are putting most of their resources into new pollution control devices, improving existing pollution control devices, and complying with federal and local environmental regulations. Only in the United States is there an emphasis on source reduction, attempting to control pollution as a first rather than a last step.' (Morrison, 1991: 25) Protection of workers' health, both in the workplace and the overall environment, toxic and non-toxic waste disposal and recycling, strengthening organizational awareness, and dealing with the weight of regulation are key concerns of most companies. Frequently, the costs of achieving technically feasible environmental standards are seen as excessive when compared to the benefits derived and therefore as a misallocation of technical resources, compared with other environmental as well as non-ecological goals. Furthermore, less-industrialized countries and their respective domestic companies often can ill-afford pollution control. (Dillon/Fischer, 1992; Fischer/Schot, 1993; James/ Stewart, 1994; McKinsey & Company, 1991; Morrison, 1991; Rappaport/ Flaherty, 1992; Schmidheiny, 1992)

We list further interesting, slightly more specific empirical results:

1. On average, outside social pressures are still mainly responsible for raising environmental awareness in industry and for stimulating industry responses, which, according to Williams *et al.*, 1993, include:

   - 'Increasingly stringent environmental legislation and enforcement;
   - Increasing costs associated with pollution control, waste disposal, and effluent disposal;
   - Increasing commercial pressure from the supply, consumption, and disposal of both intermediate and final products;
   - Increasing awareness on the part of investors of companies' environmental performance in view of the cost implications associated with liability and the 'polluter-pays' principle;
   - Increasing training and personnel requirements, together with additional information requirements;
   - Increasing expectations on the part of the local community and the work force of the environmental performance of firms and their impact on the environment.' (Schot/Fischer, 1993a: 17)

2. Environmental protection is thus still perceived more as restriction than as a market and innovation opportunity by the majority of firms, although they will probably be forced to reduce environmental burden on the input, production,

and output side in the future, where they could acquire a leadership position today through voluntary environmental management strategies (Coenenberg *et al.*, 1994; Meffert/Kirchgeorg, 1993; Steger, 1993).

3. The need for systematic environmental management is thus recognized and proclaimed by industry, but only a few companies are already considerably advanced in their environmental management efforts (Coenenberg *et al.*, 1994; Fischer/Schot, 1993; Freimann, 1994; Freimann/Hildebrandt, 1995; Hildebrandt *et al.*, 1995; UBA, 1991; Wieselhuber/Stadlbauer, 1992).

4. Thus, environmental protection improvements are still often due primarily to public regulations than to other causal factors, such as strategic marketing position, general prevention, do-what-others-do, self-imposed mandates (cf. Huber *et al.*, 1994; Jänicke/Weidner, 1995).

5. As far as hypotheses on correlation between effective environmental management and other influencing variables are concerned, 'one can say, for instance, that regardless of company size, industrial sector, consumer product recognition, profitability, or histories of well-publicized environmental incidents, EHS programs of MNCs (multinational corporations) are more likely to be effective when policies have the backing of top management, including division management; when management recognizes good business reasons for installing and implementing good programs; and when there are management systems in place that recognize the importance of cultural and political differences in overseas operations and are consistent with overall company quality goals. This general proposition emphasizes the relative importance of management over other factors in determining EHS effectiveness and has the advantage of recognizing that programme improvements can be wrought out of actions of corporations themselves and, in some cases, out of the actions of various public policy bodies.' (Rappaport/Flaherty, 1992: 136f)

6. Therefore, 'improved EHS will come through eliminating the institutional and managerial barriers to use of existing technology rather than from the development of better technology ... We see motivating people, communicating effectively, establishing influence through networks, and creating mechanisms for changing corporate culture as being absolutely critical to making significant progress in corporate environmental management in the years ahead.' (Rappaport/Flaherty, 1992: 140)

7. In this context, the role of trade as an ecological gate-keeper (and environmental adviser at the point of sale), and thus as a key agent for the diffusion of environmentally friendly products is likely to become much more important in the future (cf. Meffert/Kirchgeorg, 1993, 1995; Nielsen, 1992; Will, 1994).

8. If environmental management efforts and public environmental policy programmes and regulations complement one another, progress in substantial effective environmental protection is likely, in accordance with the necessary positive interplay of environmental management determinants emphasized above. In practice, however, corporate and public environmental policies evolve largely side-by-side in a relatively uncoordinated manner; though taking notice of one

another, they only partly complement and sometimes contradict one other. Structurally, public command-and-control policies, which may well be appropriate *vis-à-vis* industries basically opposing environmental protection by their corporate culture and strategy and organizational arrangements, are out of place in effecting continuous substantive environmental improvements which require active commitment and cooperative policy settings among the main actors involved. For such purposes, public environmental policy has to rely primarily on meta-policies of providing favourable boundary conditions for developing and installing environmental management in given social settings, on prospective intervention, and on political will and skill flexibly utilizing situational windows of opportunity (Jänicke, 1996; Jänicke/Weidner, 1995).

3 COMPARISON

Turning now to a comparison of the project's case studies with other studies of environmental management, the following more general conclusions can be drawn:

1. Compared with other investigations of environmental management (cf. Dyllick, 1991; Fischer/Schot, 1993; Hildebrandt *et al.*, 1995; Hopfenbeck, 1990; Hopfenbeck/Jasch, 1993; Meffert/Kirchgeorg, 1993; Morrison, 1991; Rappaport/Flaherty, 1992; Steger, 1992; Vietor/Reinhardt, 1995), there are no marked characteristics distinguishing these cases from others. By the selection criteria chosen, the nine cases are by definition success stories and therefore mark a certain terrain of favourable factor constellations, for instance an environmentally oriented corporate policy and committed management. However, the development path followed by these companies is in principle open to most other companies, too. Neither examples of active or passive resistance towards further environmental protection nor substantive failures of deliberate environmental management efforts have been studied, which will certainly differ in the determinants of insufficient or unviable environmental management in such stories.

2. Counteractive factors played some role in some case studies but were never strong enough to offset or even get the better of the supportive factors of influence, and thus to block successful environmental protection efforts. Such factors were thus typically in a minority position and able only to generate delays and to prevent environmental management systems from extending to many companies contributing to a whole value chain.

3. In general, successful (innovative) environmental management depends on the positive interplay of determinants within and without the company (concerning for instance the interaction between firms and stakeholders, technological development, cost-structures, perception of new market options, ensuring back-up to corporate profile and strategy) which leads to the technical, organizational, socio-economic and cultural entrenchment of environmental management efforts, thus generating a self-supporting interaction dynamics towards substantive environmental protection.

4. The specific configuration of substantive influencing factors explaining individual success stories is particular for each case, allowing little generalization.

5. Some common ground is given by the preponderance of economically profitable environmental management, by the limited number of individual (10–20)[7] and organization (1–5) actors usually involved, and – with the exception of the Danish case studies in our sample – by the frequently low significance of public environmental policy for innovative environmental management.

6. Apart from agenda-setting, environmental policy hence appears to be more important for the subsequent diffusion of examples of innovative environmental management than for their occurrence itself.

7. Furthermore, perceived pressure for change (for varying reasons), initial experience of success, the availability of resources, individuals who champion environmental management ideas within the organization, strategic action and cooperation among actors, and only limited counteractive influences, frequently seem to be a necessary condition for successful environmental management.

8. Using the push-pull model, common in innovation theory, where ecology-push signifies external pressure, such as by critical media reports, citizen group actions, regulatory directives, and ecology-pull signifies genuine market demand by customers or trade or company orientation and action towards the environment, then the joint impact of ecology push and pull factors on the degree of a company's (successful) environmental management efforts appears to be higher than would be expected by the mere addition of both types of influencing factors (Ostmeier, 1990), which again confirms the importance of the positive interaction of influencing factors.

9. As known from general social and organization theory, conceiving a company as an open system is a more challenging, more promising but more risky corporate policy than relying on the relatively comfortable and secure course of corporate action prestructured and delimited by bureaucratic environmental policy, because then it has to be innovative to develop a high degree of flexibility in its situational actions necessary for its adaptation of learning capabilities in order to cope with increasingly varying contextual conditions generating ever more uncertainty. However, taking an innovative stance is not simple and easy, since 'due to growing competitive pressures that are bringing about all kind of changes, firms are already confronted with a lot of uncertainty. The inclusion of environmental criteria in strategic considerations adds more uncertainty and complexity, and thus, firms are reluctant to do so. Moreover, some global business trends, such as shorter production life-cycles and smaller production series, are not easily compatible with environmental requirements.' (Schot/Fischer, 1993b: 370) Therefore, successful environmental management requires rather comprehensive, systematic, resource consumption and long-term concerted corporate action

---

[7] This number is certainly larger for environmental management systems of big companies as a whole (Ciba, GEP Europe, Nestlé, General Motors/Opel) as opposed to project-specific environmental management investigations.

addressing the various levels, functions, and groups of a company and taking into account the whole life-cycle of its products (Huber, 1991; Kirchgeorg, 1990; Wieselhuber/Stadlbauer, 1992).

10. Although the project's case studies point in this direction, they do not yet represent prototype examples in this respect because they neither include all of these relevant dimensions of full-scale analysis nor have the companies investigated already extended their environmental management systems and efforts to comprehensive product life-cycle assessment and to all spheres of corporate action and strategy.

11. Concerning theory and method, the project's case studies belong to the more rare fourth type of in-depth case studies with a reasonably well specified research design. They are, however, only *ex post facto* investigations without longitudinal analysis, were not organized to develop and test a specified theoretical approach comparatively, and differ in their specific unit of analysis by evaluating outcomes, processes, or institutions of environmental management.

12. In sum, successful innovative environmental management demands a systemic-evolutionary reticulate perspective, which avoids single unilinear causal models and interpretations.

Altogether, in spite of the quite interesting characteristics of the case-specific success stories, the case studies showed features of successful innovative environmental management which are to be expected by those familiar with investigations and theories of innovation, organization, political economics, and environmental management.

LITERATURE

Argyris, C. and Schön, D.A. (1978) *Organizational Learning. A Theory of Action Perspective.* Reading, Mass. Addison-Wesley.

Boudon, R. (1984) *La place du désordre. Critique des théories du changement social.* Paris: PUF.

Cebon, P.B. (1993) The Myth of Best Practices: The Context Dependence of Two High-performing Waste Reduction Programs. In: (eds.) K. Fischer and J. Schot, *Environmental Strategies for Industry.* Washington D.C.: Island Press.

Coenenberg, A.G. *et al.* (1994) Unternehmenspolitik und Umweltschutz. *Zeitschrift für betriebswirtschaftliche Forschung* **46**, 81–100.

Conrad, J. (1990) Technological Protest in West Germany: Signs of a Politizisation of Production? *Industrial Crisis Quarterly* **4**, 175–191.

Conrad, J. (1994) Überlegungen zu nachhaltiger Entwicklung und qualitativem Wachstum in Baden-Württemberg und dem diesbezüglichen TA-Akademie-Projekt. Ms. Berlin.

Davis, J. (1991) *Greening Business—Managing for Sustainable Development.* Oxford: Basil Blackwell.

Dieleman, H. and de Hoo, S. (1993) Toward a Tailor-made process of Pollution Prevention and Cleaner Production: Results and Implications of the PRISMA Project. In: (eds.) K. Fischer and J. Schot, *Environmental Strategies for Industry.* Washington D.C.: Island Press.

Dillon, P.S. and Fischer, K. (1992) *Environmental Management in Corporations: Methods and Motivations.* Medford, Mass.: Tufts University.

Dodgson, M. (1993) Organizational Learning: A Review of Some Literature. *Organization Studies* **14**, 375–394.

Duncan, R. and Weiss, A. (1979) Organizational Learning. Implications of Organizational Design. *Research in Organizational Behavior* **1**, 75–123.

Dyllick, Th. (1989) *Management der Umweltbeziehungen*. Wiesbaden: Gabler.

Dyllick, Th. (ed.) (1991) *Ökologische Lernprozesse in Unternehmungen*. Bern: Haupt.

Dyllick, Th. *et al.* (1994) *Ökologischer Wandel in Schweizer Branchen*. Bern: Haupt.

Fischer, K. and Schot, J. (eds.) (1993) *Environmental Strategies for Industry*. Washington D.C.: Island Press.

Freimann, J. (1994) Umweltorientierte Unternehmenspolitik in Deutschland In: (eds.) E. Schmidt and S. Spelthahn, *Umweltpolitik in der Defensive—Umweltschutz trotz Ökonomischer Krise*. Frankfurt: Fischer.

Freimann, J. and Hildebrandt, E. (eds.) (1995) *Praxis der betrieblichen Umweltpolitik*. Wiesbaden: Gabler.

Gerhardt, U. and Kühleis, Ch. (1994) Vom ökologischen Krisenmanagement zur präventiven Umweltschutzstrategie. Die Fallstudie Elida Gibbs. WZB FS II 94–203, Berlin.

Gorsler, B. (1991) *Umsetzung ökologisch bewußten Denkens*. Bern: Haupt.

Groenewagen, P. *et al.* (eds.) (1995) *The Greening of Industry Resource Guide and Bibliography*. Washington D.C.: Island Press.

Hildebrandt, E. *et al.* (1994) Politisierung und Entgrenzung—am Beispiel ökologisch erweiterter Arbeitspolitik. In: (eds.) N. Beckenbach and W. van Treeck, *Umbrüche gesellschaftlicher Arbeit*. Soziale Welt, Sonderband 9. Göttingen: Schwartz.

Hildebrandt, E. *et al.* (1995) Industrielle Beziehungen und ökologische Unternehmenspolitik. Final Report, Volume, 1 and 2, Science Center: Berlin.

Hopfenbeck, W. (1990) *Umweltorientiertes Management und Marketing*. Landsberg/Lech: moderne industries.

Hopfenbeck, W. and Jasch, C. (1993) *Öko-Controlling*. Landsberg/Lech: moderne industrie.

Huber, J. (1995) Nachhaltige Entwicklung durch Suffizienz, Effizienz und Konsistenz. In: (eds.) P. Fritz *et al.*, *Nachhaltigkeit in naturwissenschaftlicher und sozialwissenschaftlicher Perspektive*. Stuttgart: Wissenschaftliche Verlagsgesellschaft.

Huber, J. *et al.* (1994) Fallstudie Ciba AG. Report, University Halle.

Huisingh, D. (1988) Good Environmental Practices—Good Business Practices. WZB FS II 88–409, Berlin.

Huisingh, D. *et al.* (1986) Proven Profits from Pollution Prevention. Case Studies in Resource Conservation and Waste Reduction. Institute for Local Self-Reliance, Washington D.C.

James, P. and Stewart, S. (1994) The European Environmental Executive. Report, Ashridge Management Research Group, Berkhamsted.

Jänicke, M. (ed.) (1996) *Umweltpolitik der Industrieländer. Entwicklung—Bilanz—Erfolgsbedingungen*. Berlin: edition sigma.

Jänicke, M. and Weidner, H. (eds.) (1995) *Successful Environmental Policy*. A Critical Evaluation of 24 Case Studies. Berlin: edition sigma.

Kasperson, R.E. *et al.* (1988) *Corporate Management of Health and Safety Hazards. A Comparison of Current Practice*. Boulder: Westview Press.

Kemp, R. (1993) An Economic Analysis of Cleaner Technology: Theory and Evidence. In: (eds.) K. Fischer and J. Schot, *Environmental Strategies for Industry*. Washington D.C.: Island Press.

Kirchgeorg, M. (1990) *Ökologieorientiertes Unternehmensverhalten*. Wiesbaden: Gabler.

Kirchgeorg, M. (1995) Umweltorientierte Unternehmensstrategien im Längsschnittvergleich von 1988 und 1994. In: (eds.) J. Freimann and E. Hildebrandt, *Praxis der betrieblichen Umweltpolitik*. Wiesbaden: Gabler.

Kitschelt, H. (1985) Materiale Politisierung der Produktion, *Zeitschrift für Soziologie* **14**, 188–208.

Kühleis, Ch. *et al.* (1994) Von ökologischer Ignoranz zum integrierten Umweltschutz? Der Fall Boehringer. WZB FS II 94–201, Berlin.

Longolius, S. (1993) *Eine Branche lernt Umweltschutz. Motive und Verhaltensmuster der deutschen chemischen Industrie*. Berlin: edition sigma.

March, J. and Olson, J. (1975) The Uncertainty of the Past. Organizational Learning Under Ambiguity. *European Journal of Political Research* **3**, 147–171.

Mayntz, R. (1996) Gesellschaftliche Umbrüche als Testfall soziologischer Theorie. In: (ed.) L. Clausen, *Gesellschaften im Umbruch*. Frankfurt: Campus.

McKinsey & Company (1991) *The Corporate Response to the Environmental Challenge*. Amsterdam: McKinsey and Company.

Meffert, H. and Kirchgeorg, M. (1993) *Marktorientiertes Umweltmanagement.* Stuttgart: Poeschel.

Meffert, H. and Kirchgeorg, M. (1995) Green Marketing. In: (ed.) M.J. Baker, *Companion Encyclopedia of Marketing.* London/New York: Routledge.

Middelhoff, H. (1991) Die Organisation des betrieblichen Umweltschutzes in der schweizerischen und deutschen chemischen Industrie. Diss. St. Gallen.

Morgan, G. (1986) *Images of Organization.* Newbury, CA: Sage.

Morrison, C. (1991) Managing Environmental Affairs. The Conference Board. Report 961, New York.

Nielsen (ed.) (1992) Umweltschutzstrategien im Spannungsfeld zwischen Handel und Hersteller. Report, Frankfurt.

Oberholz, A. (1989) *Umweltorientierte Unternehmensführung.* Frankfurt: Frankfurter Allgemeine Zeitung.

Ostmeier, H. (1990) *Ökologieorientierte Produktinnovationen.* Frankfurt: Peter Lang.

Oxenfarth, A. (1994) The 'waste-water free textile mill' ecological project of Brinkhaus GmbH & Co. KG as an example for successful environmental management. Ms. Free University, Berlin.

Pawlowsky, P. (1992) Betriebliche Qualifikationsstrategien und organisationales Lernen. In: (eds.) W.H. Staehle and P. Conrad, *Managementforschung 2.* Berlin: de Gruyter.

Piasecki, B. and Asmus, P. (1990) *In Search of Environmental Excellence—Moving beyond Blame.* New York: Simon & Schuster Inc.

Prisma-Industrie-Kommunikation (ed.) (1992) *Neue Wege im Umweltmanagement* (1. SZ-Umweltsymposium). Dießen.

Prisma-Industrie-Kommunikation (ed.) (1993) *Aufbruch in die Kreislaufwirtschaft* (2. SZ-Umweltsymposium). Dießen.

Rappaport, A. and Flaherty, M.F. (1992) *Corporate Responses to Environmental Challenges.* New York: Quorum Books.

Schmidheiny, S. (1992) *Changing Course: A Global Business Perspective on Development and the Environment.* Cambridge: MIT Press.

Schot, J. and Fischer, K. (1993a) Introduction: The Greening of the Industrial Firm. In: (eds.) K. Fischer and J. Schot, *Environmental Strategies for Industry.* Washington D.C.: Island Press.

Schot, J. and Fischer, K. (1993b) Conclusion: Research Needs and Policy Implications. In: (eds.) K. Fischer and J. Schot, *Environmental Strategies for Industry.* Washington D.C.: Island Press.

Shen, Th.T. (1995) *Industrial Pollution Prevention.* Berlin: Springer.

Shrivastava, P. (1983) A Typology of Organizational Learning Systems. *Journal of Management Studies* **20,I,** 7–28.

Simon, H. (1991) Bounded Rationality and Organizational Learning. *Organization Science* **2**, 125–134.

Smith, D. (ed.) (1992) *Business and the Environment.* London: Chapman.

Steger, U. (ed.) (1992) *Handbuch des Umweltmanagements.* München: Beck.

Terhart, K. (1986) *Die Befolgung von Umweltschutzauflagen als betriebswirtschaftliches Problem.* Berlin: Dunker & Humblodt.

UBA (ed.) (1991) Umweltorientierte Unternehmensführung. Report 10/91. Berlin: Erich Schmidt.

Vester, F. (1992) *Leitmotiv vernetztes Denken.* München: Heyne.

Vietor, R. and Reinhardt, F. (1995) *Business Management and the Natural Environment.* Cincinnati: Southwestern Publishing Company.

Wicke, L. *et al.* (1992) *Betriebliche Umweltökonomie.* München: Vahlen.

Wieselhuber, N. and Stadlbauer, W.J. (1992) Ökologie-Management als strategischer Erfolgsfaktor. Dr. Wieselhuber & Partner Unternehmensberatung, München.

Wiesenthal, H. (1995) Konventionelles und unkonventionelles Organisationslernen: Literaturreport und Ergänzungsvorschlag. *Zeitschrift für Soziologie* **24**, 137–155.

Wild, W. and Held, M. (eds.) (1993) Umweltorientierte Unternehmenspolitik—Erfahrungen und Perspektiven. Tutzinger Materialie Nr. 72, Tutzing.

Will, S. (1994) Hertie und BUND—eine Kooperation im Spannungsfeld zwischen Ökonomie und Ökologie. FFU-Report 94-6, Berlin.

Williams, H.E. *et al.* (1993) Corporate Strategies for a Sustainable Future. In: (eds.) K. Fischer and J. Schot, *Environmental Strategies for Industry.* Washington D.C.: Island Press.

Winter, G. (1993) *Das umweltbewußte Unternehmen.* München: Beck.

Ytterhus, B.E. *et al.* (1995) The Nordic Business Environmental Barometer. Report, Norwegian School of Management, Oslo.

15. DETERMINANTS AND DYNAMICS OF SUCCESSFUL ENVIRONMENTAL MANAGEMENT

Jobst Conrad

This chapter applies the five-fold analytical framework (see Figure 1) described in Chapter 1 to the project case studies and – in connection with the results presented in the two preceding chapters – draws some conclusions about its usefulness as a tool for understanding successful environmental management in general.

1 EXTENDED SOCIOSPHERE

In the model of the extended sociosphere (developed from Huber, 1989, 1991) most of its components are more or less important in the environmental management efforts and systems of the companies investigated.

The physical environment is clearly important as the major – almost by defini-tion – concern of environmental management, both systemically and also as primary experience in actual working situations.

Technology and professional work are the inevitable means by which (industrial) firms practise environmental management. Although mainly determined by sys-temic-structural needs, actual environmental management practice is influenced by the (informal) lifeworld of the working sphere, too. However, in the institutional evaluations of environmental management, in particular, but also in the other case studies, lifeworld considerations were neither investigated nor do they appear to play a significant role in corporate environmental management.

Systemic economic considerations and activities play a major role in environ-mental management, but mainly as a limiting boundary condition (e.g. by preven-ting further expansion of traditional production schemes for ABC Coating, Ciba, Diessner, Glasuld, GEP Europe) and much less as a distinctive device for deciding the admissibility of specific environmental management measures. As long as sufficient economic resources are available, the (at least initially) additional costs of environmental management were considered acceptable. With the partial exception of NedCar, where the product design & engineering department promotes environ-mentally-conscious design as a strategy for survival, the companies investigated were always in an economically relatively comfortable situation when furthering their environmental management efforts.

In the regulative dimension (politics), public policy played only a background role of setting minimum standards, except in the two Danish case studies. However, many regulative decisions have been taken on the corporate level, which in some cases were influenced by personal relationships, too (Amecke Fruchtsaft, Hertie, Kunert), so that genuine politics played a role in environmental management mainly

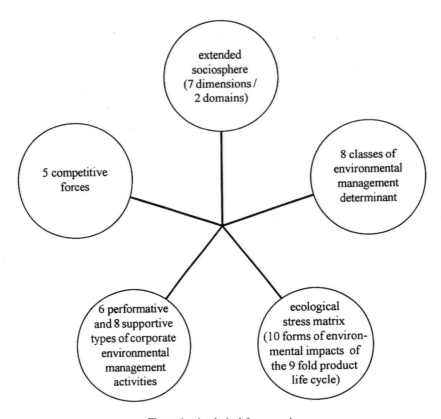

Figure 1 Analytical framework.

on the level of the organization and the individual, and somewhat less on the level of the differentiated political system.

In the normative dimension, formal legal standards and provisions (i.e. policy outputs of the past) are significant boundary conditions of corporate environmental management activities. However, these were mainly determined by self-imposed normative environmental standards and objectives, not (yet) required by environmental legislation. Furthermore, personal environmental values and norms with only limited or even without formal obligation for compliance significantly influenced individual commitment in corporate environmental management. So, as in the regulative dimension, primary and organizational domains are more prominent than the legal system with respect to ethics and the law relating to environmental management. However, environmental standards and norms sociohistorically often induce corporate environmental management efforts, as confirmed by other case studies and by the still predominant role of command-and-control oriented environmental policy.

In the semiotic-symbolic dimension, environmental management has at least threefold prominence in both domains of the sociosphere. It is conceptualized in scientific and technical language, it incorporates the case-specific experience and knowledge of employees, and it is typically based on a corresponding corporate culture. These findings are endorsed by those of other case studies, too.

Clearly, without sufficient commitment on the part of a company's management and of staff concerned with environmental management, corresponding success is very unlikely, particularly for innovative environmental management. Apart from training courses in environmental management and ecological marketing, however, professional psychological management does not play a significant role in environmental management activities and systems.

Nearly all 14 components of the extended sociosphere model are thus significant in the 9 project case studies, with the partial exception of its allocative and operative lifeworld components and the quite limited role of its regulative, normative and psychic system components. This feature is, however, hardly peculiar to these case studies. Only the regulative and normative system components will tend to play a stronger role in cases of less innovative environmental management.

The relevance of nearly all components becomes evident by mental experiments, where some components are missing, or by cases of environmental management failure, which are known of but could hardly be studied yet. Without going into detail, it is at least highly probable that a lack of commitment, of know-how, or of environmental technology, or environmental management measures insensitive to cost considerations, will undermine successful environmental management.

2   CLASSES OF DETERMINANT

Regarding different classes of environmental management success-story determinants (developed from Conrad, 1992, 1994) addressing more specific influencing factors, and formally distinguished in terms of a different segmentation of the extended sociosphere, it is again the positive interplay between these qualifying factors that largely explains the specific development path of the success stories.

The conditions for substantive ecological problem-structure vary considerably from case to case, quite apart from the question of substantively differential aptness of a given environmental problem to ecological solution. Whereas Amecke Fruchtsaft was able to effect considerable water savings in a clearly delimited project without significant negative side-effects and without embedding it in an overall, comprehensive environmental management system, the saving potentials for Hertie or NedCar were relatively minor, the general environmental management strategies followed by Glasuld, GEP Europe or Ciba embraced many different projects and aspects of environmental protection and were thus less clear-cut, and some environmental improvements achieved by NedCar, Diessner or Kunert involved negative (environmental or economic) side-effects, too.

Techno-economic context refers to the (essentially company-internal) technical and economic circumstances of an environmental management project or system, such as its technical size, design, organization, and monitoring, its embedding in a particular category of corporate strategy and management, its economic profitability, and the economic and manpower resources available for and devoted to it over time. Such techno-economic determinants are clearly crucial for the success or failure of environmental management efforts, since the way in which they are planned, the availability of sufficient resources, the technical know-how and competence of the personnel involved, sufficient monitoring of their actual effects, economic profitability, or the linkage with other corporate strategy components, such as marketing, communication, distribution or price policy have a decisive impact in this respect. Overall, the techno-economic context was favourable in the project case studies: environmental management projects tended to be embedded in corporate strategy, even where an environmental management system had not already been institutionalized; the necessary financial means were made available by top management; considerable time was invested in acquiring and utilizing the necessary technical information and competence; conscious organization of environmental management activities occurred formally or informally; their results were at least partly monitored; they were to a large extent economically profitable with the partial exception of Diessner, Hertie and NedCar. Thus, despite substantial variation in the specific techno-economic arrangements in the different success stories, these determinants were always helpful to a greater or lesser degree in ensuring the success of the companies' environmental management efforts.

The wider social context has been described in Chapter 13 and does not significantly differ for the case studies investigated, if one considers, in particular, the modernization capacity of a society, the importance of the state and public policy, the importance of public debate and the corresponding vigour of civil society, the extent of division into different social classes or strata, the degree of public participation and sociostructurally anchored substantive democracy, the significance of social actors' sense of responsibility and their liability, the degree of legalism, decentralized versus centralized (political) culture and decision-making procedures, the importance of postmaterialistic value orientations, environmental awareness and behaviour of the main actors and the general public, and the significance and social influence of the environmental movement and relevant groups, as major, indirectly operative determinants. Again, according to empirical evidence in international comparison, these general social determinants tend to favour corporate environmental management in the countries concerned, though they may occasionally have an unfavourable effect, too, for instance if bureaucratic legalism discourages a company's willingness to undertake unconventional innovative environmental management.

The economic boundary conditions, i.e. general economic situation and development of the national and global economy, of the industry in question, and of the company, in particular, including cash flow, market position, turnover, financial reserves, long-term returns on investment, competitive position, and productivity enhancement, are noted separately because they are certainly of crucial relevance in the economic system of industrial corporations. Whereas the particular economic boundary conditions of each company shape its specific environmental management options and activities, they can be characterized in all as more or less positive for Amecke Fruchtsaft, ABC Coating, Glasuld, Hertie, GEP Europe, and Ciba, and more or less neutral for Diessner, Kunert, and NedCar. In no case were they negative. Thus, the overall economic situation of the companies investigated did not (indirectly) block environmental management efforts over this period.

Psychological context refers to sociopsychological features, such as the company's working atmosphere, its (external) credibility, environmentally favourable attitudes among company external actors (e.g. shareholders, suppliers, competitors, customers, public authorities, environmental groups, media), actors' cooperativeness (and competence), staff commitment, the leadership qualities and ecological orientation of (top) management, viable and functioning division of labour and communication among different divisions and hierarchy levels of the company, motivation through initial successes and motivational momentum. These psychological determinants can be said to have been, or at least to have become partly favourable for the companies studied. Whereas an externally triggered, but internally realized ecological reorientation first had to alter the public image of some companies (ABC Coating, Glasuld, GEP Europe) from environmental villain to ecological good guy, psychologically favourable conditions finally preponderated for all case studies. In the majority of cases working atmosphere, actor cooperativeness, staff commitment, management ecological orientation, the division of labour, and motivation through initial success supported environmental management efforts, whereas external credibility, external actor attitudes, leadership qualities, and communication among hierarchy levels were usually more of less neutral.

Situational structure relates to situation-specific, non-general social conditions framing the pattern of perception, interest, power and conflict within actor constellations. Situational determinants are relatively stable (short-term) social structures which influence environmental management activities and systems of a single company because situational structure will usually be different for other companies or organizations. One may mention for instance company size, favourable or unfavourable operating conditions, informal versus formal company organization, distribution of competencies, number of divisions strongly involved, short-term availability of financial resources, and additional market opportunities by vertical cooperation. By definition, situational structure differs from case study to case study. Therefore, no general conclusions can be derived on situational determinants. Nevertheless, some situational determinants appeared typically to support the development of successful environmental management. Among these were informal company organization for Amecke Fruchtsaft and additional market opportunities for the collaborating supplier, the strategic siting decision and the cooperative arrangements on the county level for Glasuld, or the interest of various actors in the survival of a Dutch automobile company facilitating the environmental strategy of the NedCar product design & engineering department.

In contrast to situational determinants, historiographic events and conditions refer to historically and situationally contingent chance qualifiers which may well be (causally) explainable by certain (social) regularities and strategies, but which are not structurally related to the object of investigation, namely environmental management. Again, chance events repeatedly furthered environmental management in the companies investigated. In the Amecke Fruchtsaft case, the vacancy of the position of technical manager in 1988 obliged Amecke-Mönnighof to pay closer attention to the question of technical savings potential, the economic and corporate policy to cease Coca Cola bottling in 1990 and the 1985 fundamental corporate-strategy decision to stake the future of the company on fruit-beverage production in returnable bottles made a commitment towards water saving in the bottle-washing machine very reasonable. In the Kunert case the interest in practical realization of the eco-balance concept of the later external consultant of the university of Augsburg, his past acquaintance with the assistant to Kunert's chairman of the board, their chance meeting at a Nuremberg convention, and the resulting opportunity for the chairman's interest in using environmental protection to enhance the company profile, were together necessary conditions to initiate Kunert's eco-accounting project at that point in time.

Whereas psychological context determinants refer to sociopsychological features significant for certain social groups and constellations, the role and personality structure of key individuals concerns individual psychological characteristics. As mentioned in Chapter 13, the will and skill of committed individuals is particularly important under the circumstances of change that characterize the installment of (innovative) environmental management. Such key individuals, acting as technical or power promoters were especially prominent in the success stories of Amecke Fruchtsaft, Diessner, ABC Coating, Glasuld and Hertie. They have at least several of the following qualities: technical skill, leadership role and qualities, forcefulness, mediation capability, zeal and working intensity, and strategic timing. Key individuals with favourable personality traits thus played a decisive role in the case studies oriented towards outcome or process evaluation.

Although other case studies have not been systematically evaluated concerning the interplay of the formally always relevant eight classes of determinant, plausibility checks confirm the prevalence of similar features for them. In particular, the realization of corporate environmental management can be attributed to neither company-external pressure nor to company-internal engagement and strategy alone, but to the positive interaction of company-internal and external qualifying factors.

3   COMPETITIVE FORCES

The concept of competitive advantage and related company strategies, as developed by Porter (1985, 1990), provides an analytical frame for assessing a company's relative competitive position and success, and eventually its environmental management strategy. The scheme of determining (industrial) competition by the rivalry among existing competitors, the bargaining power of suppliers as well as of buyers, the threat of new entrants, and the threat of substitute products or services, and of achieving competitive advantage by generic strategies of cost leadership, cost focus, differentiation, and focused differentiation, puts knowledge about the economic sphere and boundary conditions into perspective by providing information for the analytical comparison of the competitive strategies and advantages of (industrial) companies as decisive context and functional points of reference for assessing the (relative) economic success of their environmental management efforts and strategy.

As indicated in Table 1, the companies investigated were mainly in competition with existing competitors, with GEP Europe probably in a relatively more comfortable situation. Demanders' bargaining power is usually stronger than that of suppliers. Threat of new entrants is significant only for Kunert (additional textile

Table 1 Assessed intensity of competitive forces and generic strategies.

	Amecke Fruchtsaft	Diessner	ABC Coating	Glasuld	Hertie	Kunert	NedCar	GEP Europe	Ciba
Rivalry among existing competitors	++	++	++	++	++	++	++	0	+
Bargaining power of suppliers	0	0	0	0	–	0	–	0	–
Bargaining power of buyers	+	+	+	+	0	+	+	+	0
Threat of new entrants	–	0	+	0	–	+	–	–	–
Threat of substitute products or services	–	0	–	0	–	–	0	0	–
Generic strategy	Differentiation	Focused differentiation	(Focused) differentiation	Differentiation	Differentiation	Differentiation (+cost leadership)	(Focused) differentiation	Focused differentiation (+cost focus)	(Focused) differentiation

++, very strong; +, strong; 0, medium; –, weak.

producers in developing countries) and for ABC Coating (more coating being done by steel manufacturing firms themselves), whereas the threat of substitute products or services is actually rather insignificant.[1] Although quite a few companies, particularly Kunert, made considerable efforts to reduce costs by transferring production to countries with lower labour costs, corporate strategies of differentiation or focused differentiation predominate, as far as they can be attributed to one of the major generic strategies.

These qualitative assessments certainly depend on the product segments selected, e.g. high quality fruit juice and not fruit beverages and Coca Cola in general for Amecke Fruchtsaft, cars and not just plastic car components for NedCar, engineering thermoplastics and not polymers in general for GEP Europe.

Whereas this emerging configuration of competitive forces is probably due to the accidental selection of companies, so that no conclusion can be made that successful environmental management will be favoured by strong market competition in particular, the priority of differentiation over cost leadership strategies may well correlate with pronounced environmental management efforts, because both represent a corporate orientation towards qualitative product and/or production characteristics rather than towards quantitative cost-related aspects, and because market-oriented environmental management strategies are sometimes pursued as a genuine differentiation strategy (Hertie, and to some extent Ciba, Kunert).[2] Thus, whereas competitive advantage based on effective corporate strategy appears to be a major contextual determinant of successful environmental management, the companies studied are not under extraordinary pressure from all potential competitive forces, apart from intense market competition, that could induce a corporate climate preventing innovative environmental management activities, and embed their environmental management strategy primarily in generic strategies of (focused) differentiation. No further general conclusions can be drawn on (successful) environmental management as part of corporate strategy to achieve competitive advantage.

4   PERFORMATIVE AND SUPPORTIVE ENVIRONMENTAL MANAGEMENT ACTIVITIES

Looking now at performative and supportive environmental management activities[3] in the companies investigated (cf. Meffert/Kirchgeorg, 1993; Steger, 1993; Wicke

---

[1] In principle, a shift in transportation modes from private automobiles towards public transport, or market penetration by new types of insulation materials could be such a threat for NedCar or Glasuld, respectively.

[2] Gaining an environmental profile may be the first (symbolic) step of a company's differentiation strategy followed either by substantial eco-performance and competitive green accounting or – without such substantive environmental management efforts – by loss of credibility and competitiveness.

[3] The first type relates to product development, purchasing/provision of inputs, production, marketing and sales, logistics, and waste management; the second type to installations, organization, management systems and instruments, environmental information systems, finance and accounting, personnel, public relations, and corporate culture.

et al., 1992), it is important to distinguish between the different types of evaluative analysis undertaken because it would be reasonable to expect a broader coverage of a company's functional/organizational domains for process and institution-oriented environmental-management case studies than for outcome-oriented (project-specific) studies. In addition, the chosen focus of case study analysis does not necessarily reflect a company's actual spread of environmental management activities.

With its comparatively advanced environmental management system (Hamburger Umwelt-Institut, 1994) Ciba covers, in varying degrees, more or less all functional domains. To a somewhat lesser extent the same can be said for GEP Europe. Since the NedCar case study focuses on the Product Design & Engineering unit, it necessarily describes only the environmental orientation of the unit's respective functions/tasks in product development, waste management, and to some extent installations, environmental information systems, corporate culture, particularly concerning materials selection and recycling and environmentally conscious design. As far as the process-oriented evaluations of environmental management are concerned, their focus was on establishing and implementing eco-accounting and auditing schemes in Glasuld and Kunert, resulting in considerable production improvements as well as product-oriented environmental amelioration. Most company functions are thus affected in some way, but no systematic check of the degree of penetration by environmental concerns can be made. The primary fields tackled by Hertie in environmental management were the environmental compatibility of its product range, packaging, product marketing and sales, but also logistics and waste management; and, as secondary fields, organization, environmental information systems, personnel, public relations, and corporate culture. For the outcome-oriented evaluations of environmental management by corresponding projects of Amecke Fruchtsaft, Diessner and ABC Coating, it is interesting to note that these projects, mainly concerning environmental improvements to the production process, although largely developed by company directors and a few key people from in-house and, in two cases, external technical units, led to the inclusion of environmental considerations in quite a number of the company's functional domains (e.g. product development, purchasing of inputs, production, marketing and sales, waste management, installations, environmental information system, personnel, corporate culture) and served to pursue further systematic environmental management efforts. In sum, environmental management efforts by the companies studied extended to many functional corporate domains concerning both performative and supportive activities. This observation, which tends to be confirmed by other case studies, supports the thesis about the need for environmental-management organization to be relatively systematic and comprehensive if it is to be successful, although there is no proof that successful environmental management is impossible without such arrangements.

5 ECOLOGICAL STRESS MATRIX

We turn finally to the ecological stress matrix (Dyllick *et al.*, 1994) used to identify the contents and extent of the diverse environmental impacts of the whole cradle-to-grave value chain.[4] At least where the company is concerned, their various elements are more or less, but not completely, addressed by the environmental information and accounting systems of ABC Coating, Ciba, GEP Europe, Glasuld, Hertie and Kunert; although this does not imply that all ecological weaknesses identified have already been reduced or eliminated. As described in the respective case studies, the other companies tackled more specific environmental problems but also tried to deal with the environmental impacts of their production and/or products more systematically.

As described in the various case studies, major environmental impacts identified and tackled stem – with varying salience from company to company – from resource consumption, energy consumption, air, water and soil pollution, packaging; and only occasionally construction, noise and synergetic impacts. They mainly concern mainstream production, recycling and waste products, disposal of end products, although to some extent upstream production, distribution and use, as well. This distribution of environmental management activities indicates that most types of environmental impact are diagnosed and partly dealt with, but only the company's subdivision of the value chain is substantially addressed. This pattern of company environmental management efforts shows that comprehensive approaches to life-cycle assessment are still largely lacking, but is in accordance with the socially realistic assumption that companies can and should first of all address the environmental problems they are able to and for which they are directly responsible. In this perspective the companies' environmental management coverage of relevant components of the product life-cycle and of environmental burden by their efforts can be judged as not unreasonable.

However, the pattern indicated does not necessarily imply that it reflects the companies' most urgent environmental problems, although their favourable and vigorous attitude towards environmental management makes mere symbolic activities less likely than were to be observed in many companies in the past. Furthermore, there exists no objective scale for measuring the relative (socio-ecological) importance of different environmental hazards, particularly determined by current environmental discourse, existing regulatory requirements, available options of cost reduction, given (corporate) technical trajectories, skills and know-how, a company's actual perceptions of and attitudes towards different environmental risks. So the strategic utilization of situational windows of opportunity will usually tend to lead to more substantive environmental improvements than will taking a path towards effective environmental protection that is more stringent by ecological criteria but socially less viable. Systematic assessment of, and environmental improvements within an industry's or a company's ecological stress matrix are largely still the exception and not the rule in both corporate environmental management practice and scientific analysis thereof (cf. Dyllick *et al.*, 1994; Hallay/Pfriem, 1992).

---

[4] As enumerated in Chapter 1, the life-cycle can be divided into raw materials extraction, transportation and storage, upstream production, mainstream production, distribution, use, maintenance and repair, recycling and waste products, disposal of end products; with the environmental impacts of each life-cycle phase relative to the consumption of resources, energy consumption, air, water and soil pollution, noise, construction, packaging, waste and synergetic impacts.

6   EVALUATION

Passing in final review the usefulness of applying the five-fold analytical framework[5] (extended sociosphere, classes of determinant, competitive forces, primary and secondary environmental management fields, life-cycle assessment) to the case studies, one may conclude that, as a relatively comprehensive analytical framework it allows one to place and thereby to assess a company's specific environmental management activities, strategies and system in a broad, multidimensional setting; and that it indicates the need to combine most of the various aspects of environmental management so as to make it substantively more effective, and thus consciously and strategically to embed successful environmental management in such a relatively broad perspective. This is indicated in these and other case studies by the circumstance that different dimensions of the extended sociosphere are simultaneously affected – the mutually-reinforcing interplay of determinants of successful environmental management from various contextual types, the involvement of a company's most performative as well as supportive functional activities in environmental matters, and the addressing of diverse types of environmental impact; whereas the (cooperative) integrative ecological treatment of the whole cradle-to-grave value chain and the strategic embedding of environmental management in a competitive strategy taking into account the differing competitive forces are still more of a desirable (future) goal than current practice.

LITERATURE

Conrad, J. (1992) *Nitratpolitik im internationalen Vergleich*. Berlin: edition sigma.

Conrad, J. (1994) Ökonomischer Wegweiser ins ökologische Mehrweg-Optimum. FFU-Report 94-4, Berlin.

Dyllick, Th. *et al.* (1994) *Ökologischer Wandel in Schweizer Branchen*. Bern: Haupt.

Hallay, H. and Pfriem, R. (1992) *Öko-Controlling*. Frankfurt: Campus.

Hamburger Umwelt-Institut, 1994: The TOP 50 Ranking. Eine Untersuchung zur Umweltverträglichkeit der 50 graößten Chemie- und Pharmaunternehmen weltweit. Report, Hamburg.

Huber, J. (1989) *Herrschen und Sehnen. Kulturdynamik des Westens*. Weinheim: Beltz.

Huber, J. (1991) *Unternehmen Umwelt*. Frankfurt: Fischer.

Meffert, H. and Kirchgeorg, M. (1993) *Marktorientiertes Umweltmanagement*. Stuttgart: Poeschel.

Porter, M.E. (1985) *Competitive Advantage: Creating and Sustaining Superior Performance*. New York: The Free Press.

Porter, M.E. (1990) *The Competitive Advantage of Nations*. London: Macmillan.

Steger, U. (1993) *Umweltmanagement*. Wiesbaden: Gabler.

Wicke, L. *et al.* (1992) *Betriebliche Umweltökonomie*. München: Vahlen.

---

[5] As stated in Chapter 1, these five different perspectives on environmental management cannot be simply combined to form one theoretical model or reduced to less explanatory dimensions, but have their own independent theoretical-analytical legimacy each and rather complement one another for a better overall (hermeneutic) understanding of environmental management structures and development.

16. PERSPECTIVES FOR ENVIRONMENTAL MANAGEMENT

Jobst Conrad

1 ANALYTICAL PERSPECTIVES

To some degree, seeking environmental management perspectives beyond those peculiar to the specific case studies implies prognostic statements outside the ambit of the social sciences. It is possible, however, to point out relatively probable as well as – due to theoretically known social regularities – certainly impossible developments in environmental management (cf. Mayntz, 1996; Mayntz/Scharpf, 1995). Thus one cannot predict the precise future development path of environmental management, which can vary considerably according to changing contextual conditions, e.g. the occurrence of trade wars as opposed to continuous, fairly well regulated economic competition among Western industrialized countries. But one can derive likely development trends on the basis of certain (explicated) premises and social and economic regularities. In this respect one may well refer to the five-fold analytical framework utilized in the preceding chapter.

Three basic assumptions are major explanatory structural factors concerning the (future) development and possible spread and deepening of environmental management over time. First, a *social environment pressing* for enhanced environmental management as common (industrial) practice urges companies to act accordingly in order to survive. Second, the *self-dynamics of a field of social activity*, once installed, on both the industry and individual company level, generates the momentum and self-interest of corresponding social groups to continue the expansion of environmental management activities. Third, only on the basis of *time-consuming continuous social learning, experiments and setbacks*, any complex long-term cross-sectional programme as a systems approach (including cultural aspects) needed for effective environmental management can become common social (industrial) practice. Thus, under the condition of (mainly company-external) continuous and multifarious social pressure, the self-dynamics of environmental management activities, once initiated, will gradually and slowly lead to the diffusion of systematic and comprehensive environmental management. Under these general assumptions, such a development path is *very likely* but not yet certain, and it would *probably* effectuate significantly improved environmental protection.

Before sketching the social context of environmental management significant for its future development, few distinct (analytical) observations will be made:

1. Frequently, there still exists an enormous degree to which economically profitable conservation potential and possibilities for enhancing efficiency are likely to be lying fallow as a result of sheer ignorance in industrial and service enterprises.

2. On the other hand it is important not (implicitly) to generalize the existence of such economically beneficial potential for economizing, in so far as the limitation or solution of environmental problems elsewhere in the production and consumption of goods can indeed generate higher additional costs. Bearing in mind that continuous commitment to corporate environmental management depends in the last resort on the combination of ecological as well as economic gains, there thus exist limits to profitable environmental management (including the environmental protection industry itself). As Palmer *et al.* (1995) pointed out, on the whole, environmental protection and corresponding regulation amount to a significant net cost to most companies.

3. For the purpose of achieving development that is sustainable from an ecological perspective (in the sense of a production-consumption cycle as closed as possible and an adequate supply of energy for recycling resources), gains in efficiency represent only one – in itself not sufficient – among six instrumentally rational, formal sustainability strategies, namely gains in efficiency; scientific/technical innovation and substitution; ecologically beneficial structural change; limitations in volume; disaster avoidance; and technological supervisory and control capacities (see Conrad, 1993; Jänicke *et al.*, 1992).

4. If ecological sustainability consequently demands in the last resort changes in deeply rooted, globally increasingly prevailing western civilizational patterns also implying limits to growth, relatively closed production-consumption cycles, and an immaterialization of demand in goods and thus of production, such social change may be supported but cannot be effected by broadly established environmental management practices in industry alone. Overall ecological productivity and ecological rationality in society can only be achieved via a general, and not only a system-specific process of social change, combining sufficiency, efficiency and consistency of production and lifestyle (Huber, 1991, 1995).

5. If environmental management affects a number of areas at the same time, perhaps also outside the company, and has unwanted consequences elsewhere, for example socially regrettable distribution effects, making a success of it will be more costly and difficult, and an unwanted logic of failure (Dörner, 1992) is more likely to develop.

6. The strategic task of promising environmental management especially in such more complex situations is so as to encourage and organize the positive dynamic interplay of various pull and push factors, as demonstrated in our comparative analysis of the case studies, so that actors, in exercising their functions in each social area or subsystem, reinforce every other socio-functional system in the same direction as continually and consensually as possible (Huber, 1991).

7. This implies a development towards strategic environmental management, actively tackling environmental issues through a mode of operation going beyond corporate bounds and holistic in nature that operatively anchors environmental protection as cross-section function in the various functions and spheres of the enterprise (Conrad, 1994). This corresponds to a corporate strategy that is economically viable and long-term, provides guidance for action, is flexible

in detail, is technically and organizationally capable of implementation, and which takes refuge neither in environmental goals of a general corporate philosophy and ethics that are of little practical relevance, nor exhausts itself in isolated operative measures representing an eco-activism, unrelated to an overall concept.

2 ENVIRONMENTAL MANAGEMENT IN THE GLOBAL SOCIAL CONTEXT

Taking an initial, more detached, view of the general (global) social context of environmental management, the development and the general features of environmental management not unexpectedly exhibit few distinctive structural characteristics and offer few new insights in comparison with other fields of application in economics, organization theory and innovation theory. The understanding of this social context, which is crucial in explaining the origin, evolution and impact of environmental management, may be outlined with respect to overall global societal development trends and, more specifically, with respect to environment-related social processes.[1]

With regard to the first overall sphere, a number of keywords will serve primarily to summarize the range of consequences provoked by an increasingly prevalent modernity project:[2]

1. Increase in contingency, increasing complexity and interdependence, growing differentiation and manifold conflict potential;
2. Globalization, commercialization, informatization, the technology race;
3. Loss of legitimacy for public policy and state power, the spread of 'chaos power' and turbulence in world politics, a growing demand for public participation and more effective models of (interactive) democracy;
4. The disaster potential of modern technologies, and organized irresponsibility in a risk society (Beck, 1986, 1988);
5. Heightened structural and cultural violence (Galtung, 1990) and a structurally entrenched North-South conflict (due to the unequal distribution of resources, competence and power) with reinforced structural dilemmas of security, development, ecology and coordination (Senghaas, 1994);
6. Processes of individualization, standardization and role segmentation of life strata, the ongoing destruction of humane social relations, the growing disintegration of values, norms and traditions, identity problems and fundamentalism.

Thus, as Giddens (1990: 53) has pointed out when addressing the fundamental ambivalence of the project of modernity, 'living in the modern world is more like being aboard a careering juggernaut rather than being in a carefully controlled and well-driven motor car.'

---

[1] Reference to the existing extensive literature is deliberately omitted from this section.
[2] Attempting a plausible résumé, both subsequent listings, taken from Conrad (1995) are neither arbitrary nor unequivocal and exclusive.

With reference to the second environment-related sphere, the following deserve mention:

1. Grave (global) environmental problems (ecological crisis, the question of the sustainability of modern industrial capitalist society, estimated annual environmental damage of around DM 200 billion in Germany (see Meffert/Kirchgeorg, 1993));

2. Partial ecological improvements due to structural change rather than environmental policy, often compensated for by the growth of production and consumption and by the transfer of environmental problems to other forms/media and locations;

3. The collective ecological 'ignorance' and 'organized irresponsibility' (Beck, 1988) attributable to the system mechanisms of modern society impeding due attention to, and integration of environmental issues in, e.g. industrial production, agriculture, tourism, and traffic and transportation systems;

4. Growing and extensive environmental awareness and concern: the gradual greening of all social spheres since the early 1970s, resulting in a relatively strong environmental movement, green parties, a public environmental debate, the evolution of environmental policy and regulation, a change in consumer behaviour, and finally the (partial) acceptance of environmental problems and responsibility by industry, too, and a corresponding diffusion and penetration of environmental concerns in all social systems inducing a gradually broadening management perspective;

5. Gradual improvements in environmental management and policy regulation: internalizing environmental costs, environmental self-regulation replacing bureaucratic environmental regulation, environmental (business) ethics and an ecologically-oriented corporate culture, slow recognition and testing of ecologically oriented innovations and market opportunities.

6. Enormous diversity of both ecological problem structures and social settings ruling out generalizable successful environmental management strategies on a relatively concrete level.

In view of the social context referred to, it appears reasonable to conclude that *environmental management as a task of (industrial) corporations* had and has to develop in the end because of the severity of environmental and resource problems, of the growing awareness and concern in this regard, of a relatively favourable economic situation in the 1980s, and in response to changes in consumer preferences and to the threat of (public) over-regulation.

One could also anticipate that such a development process would occur mainly in a hesitant, spasmodic, troublesome, and roundabout manner with the typical phases of problem denial and denunciation by critics, counter-argument and problem-shifting, defensive and obstructive bargaining, and willingness to find a compromise in business strategy, even if in retrospect these strategies tend to result only in the postponement of substantial environmental management at the expense of a considerable loss of credibility and escalating public regulation (Steger, 1993).

Furthermore, it could be expected that actual environmental management practice is still largely a (culturally entrenched) technocratic approach involving rationalistic control and power-based management to solve the more basic problem of ecological sustainability in accordance with the culturally, socially and economically predominant attitudes and mechanisms in western capitalist societies, such as profitability, control, competition, technological innovation, hierarchic mentality and organization, manifold structural asymmetries, even if one is justified in doubting the viability of such an approach, or at least has reasonable grounds to infer its limitations.

However, the gradual evolution of environmental management within business also shows that the economic system is not hermetic and has, in principle, an open future, which may lead to the successful establishment and integration of environmental management in business, or to the failure of these efforts, e.g. in the event of economic decline, but most probably to something in between. For the first development path, this would mean further gradual, more or less continuous penetration and expansion by already existing environmental management concepts and practices into business and industry, including a shift from mainly defensive towards more strategic, offensive and partly innovative environmental management. Conceptual innovations may primarily involve greater attention being paid to the complexity and uncertainties of relevant social processes. The major threat to this diffusion process has to be seen in the occurrence and persistence of critical economic (and socio-political) situations. This reflects the contextual influence of and embedding in the overall societal development trends mentioned, like increasing interdependence of company objectives and dependencies, global competition, enhanced self-regulation and communication policy by industry, socio-economic (industrial) coping strategies with technological risks and disaster potential, attempts at life-politics (Giddens, 1991).[3]

Future development trends in environmental management may thus be as follows. One can expect the continuation, pragmatic integration and gradual expansion of (holistic) environmental management in business, though in an incongruous diffusion process. This may result in a trend towards daily routine in environmental management with partly relaxed standards. However, even to stick to these standards of environmental management once they have been introduced will produce severe tensions in a worse economic climate with doubtful profitability. In view of the social context indicated, all options of including or excluding serious environmental management will thus probably remain open, always involving the chance and risk of considerable profit or failure. Aside from companies exercising advanced (strategic and offensive) environmental management and making a profit by selling environmentally sound products, there are also companies that make a profit even by exploiting environmental concern and regulations for environmentally

---

[3] As Willke (1989) pointed out, highly complex social systems cannot be controlled and steered in a strong (traditional) sense, but require more than mere muddling through, which is also true for environmental management. A solution may be seen in combining the natural evolution of autopoietic (modern) social systems with improved decentralized self-regulation and the civilizing of their self-referentiality and use of power by means of reflection, contextual intervention, and systemic discourse.

detrimental business (cheap regional environmental benefits via cost externalization), such as companies exporting (hazardous) waste declared as an economic good to other (developing) countries. In particular, environmental management will probably face substantial problems from the implicit export of environmental hazards by business strategies, on the one hand, and by the lack of (economically) viable options in less developed, environmentally more polluted regions on the other.

3   PERSPECTIVES OBTAINED BY THE ANALYTICAL FRAMEWORK

Examining these general development trends within the five-fold analytical framework, one may arrive at the following conclusions:

1. Only where interaction favourable to environmental management among various dimensions of the extended sociosphere is mutually supportive is its development likely under the three general assumptions mentioned above. Personal commitment, conceptual penetration, economic profitability, the availability of environmental technology, practical examples of success, and unequivocal environmental improvements are typical elements. This indicates the mutually reinforcing dynamics of ecological push-and-pull factors in different sociofunctional systems and spheres as a necessary condition for the further spread and implementation of (comprehensive) environmental management.

2. Similarly, the further spread and implementation of (comprehensive) environmental management can be expected only if, on balance, ecological problem structure, techno-economic context, the wider social context, economic boundary conditions, the psychological context, and situational structure all are more or less favourable to environmental management, which can – though not necessarily – be further supported by historiographic events and conditions, and the personality structure of key individuals. However, the presence of favourable determinants cannot be taken for granted, so that this diffusion process will at best be hesitant, arduous and inconsistent.

3. The process of (systematic) diffusing environmental management will (possibly) be driven by all types of generic corporate strategies: cost leadership, cost focus, differentiation, focused differentiation (see Porter, 1985), because systematic (comprehensive) environmental management strategies offer opportunities for (specialized) cost leadership through (economic) savings in input resources and increasingly costly emissions and wastes and of (focused) differentiation towards a growing market of products or services with a good environmental reputation. However, there are inherent limits to such a strategy towards competitive advantage because further (even strategic) environmental improvements typically become very expensive beyond a certain stage of ecological quality if practically all competitors have reached a similar stage.

   Environmental management strategies can well play an important role with regard to competitive forces, not only by fomenting rivalry among existing competitors, but also by increasing the bargaining power of certain suppliers, e.g. the prospect of higher profits on environmentally advanced technological equipment, by increasing the bargaining power of certain buyers, e.g. large retail companies imposing special environmental standards on their suppliers, by threatening to become a new entrant, e.g. by transferring ecological competence developed in one business field to other ones not previously engaged in without the newly acquired ecological superiority, or by threatening existing markets with ecologically superior products or services in growing demand but hitherto unavailable.

So systematic (comprehensive) environmental management may well turn out to be a spur to competition for some decades covering all kinds of generic strategies and competitive forces, though with its own (genuine) risks up to and including the innovation trap (Steger, 1993), where early high (investment) costs, insufficient demand at consequently high prices, a lack of practical experience, the lack of information among buyers and insufficient knowledge about potentially equivalent efforts undertaken by competitors, involve a severe risk of market failure, particularly for companies with quite limited resources and, thus, staying power.

4. If environmental management efforts proliferate in the coming decade(s), they will increasingly cover all performative and supportive activities, though probably with some initial preference for the first type. But there is some truth to the contrary assertion: only if environmental management activities extend to more or less all spheres of corporate action and strategy, will they be sufficiently systematic and comprehensive to pay off in comparison with other options and therefore to find acceptance in most companies. Again, such a development path is quite likely but far from certain within the social context and under the assumptions sketched above.

5. More or less the same can be said of the inclusion of the whole cradle-to-grave value chain in (standard) environmental management practice. Whereas conceptualizing environmental improvements in a total life-cycle perspective is to some extent likely to become common practice, comprehensive life-cycle assessments and the corresponding product and production process design, requiring immense data collection and processing, sufficient time and resources, intense cooperation among all or most actors involved, and giving greater salience to ecological objectives in corporate decision-making, may well become more frequent but for various reasons will probably remain the exception rather than the rule.

4   FUTURE DEVELOPMENT TRENDS

After this general abstract discussion, these probable (average) development trends in environmental management may be summarized and exemplified in somewhat more specific terms as follows, without further elaboration on individual statements:

1. An ongoing, slow and gradual process of greening in industry can be expected with occasional setbacks because industry has not yet advanced very far in substance, so that at this stage an actual rollback in environmental management is hardly possible. The practical implementation of verbally proclaimed ideas and concepts with corresponding learning processes and practical experience requires

time spans of least a decade, which has scarcely elapsed since the majority of industrial firms, apart from few pioneers, became actively, and not just defensively, concerned in environmental affairs. There has thus been much more talk than substantive action in (systematic) environmental management, and even more so with regard to sustainable development (Ashford/Meima, 1994; Clarke/Susse, 1995; Fischer/Schot, 1993; Freimann, 1994; Freimann/Hildebrandt, 1995; Wieselhuber/Stadlbauer, 1992).[4]

2. In this context, the role of (large) transnational corporations in developing countries is of special importance, where enforcement of environmental regulations, if they formally exist at all, is particularly weak. Whereas transnational corporations normally do not but sometimes may indeed use their siting policy to escape strict environmental regulations in some (industrialized) countries (cf. Blazejczak et al., 1993; Knödgen, 1982; Low/Yeats, 1992; Ullmann, 1982) and differ locally in their environmental management practices, for various reasons they still tend to have a better record in developing countries than local enterprises (cf. Rappaport/Flaherty, 1992). Accordingly, they are nevertheless a probable, though less than optimal (straightforward) medium for installing environmental management systems in developing countries.

3. In accordance with economic and political boundary conditions it is likely that environmental management activities will on average expand along the lines of (eventually costly) adherence to (public) environmental standards and regulations, of measures that save costs, do not affect costs, and finally which give rise to costs. However, because of the limited feasibility of prior cost-benefit calculation for investments in environmental management, the classification of environmental management strategies in a portfolio of market opportunities and environmental risks, containing indifferent, opportunity-oriented, risk-oriented, and innovation-oriented strategies (Steger, 1993) is more informative than Winter's model of political and economic costs (Winter, 1993). Again, the distribution among empirically observable strategies will probably give opportunity and innovation-oriented environmental management strategies greater salience compared to the indifferent and risk-oriented strategies that used to predominate.

4. With environmental protection becoming not only a restricting boundary condition but also a (fundamental) global challenge for corporate strategy and survival, one will in the coming decade observe a broad spectrum of different, varyingly comprehensive, more or less cooperative approaches towards environmental management, with few holistic transcorporate examples.

5. The gradual process of diffusing systematic environmental management is likely to show structural similarities with the historical diffusion processes for equally cross-sectional health and safety concerns. The political, social and organizational coupling and embedding of environmental management with/in existing

---

[4] Similarly Rappaport/Flaherty (1992: 149) diagnose, in some contrast to Schmidheiny (1992): 'Although some MNCs (multinational corporations) have publicly subscribed to sustainable development goals, none to our knowledge, including those in this study, have comprehensively undertaken the specific steps called for in the UNCTC document.'

health and safety institutions and routines (EHS-system) will usually help the implementation process.

6. As with health and safety concerns, environmental protection is unlikely to become a (separate) genuine goal of corporate strategy but rather a self-evident concern integrated socially, organizationally and procedurally in corporate strategy, planning, and operation, as empirical research has indicated (cf. Coenenberg *et al.*, 1994; Kirchgeorg, 1990; Piasecki/Asmus, 1990; Vietor/Reinhardt, 1995; Wieselhuber/Stadlbauer, 1992).

7. Frequently, the establishment of environmental information systems and eco-auditing procedures will tend to predominate first of all in environmental management efforts, which permit the necessary information to be acquired, without the undertaking of substantive changes in corporate practices. Elaborate (professional) ecological accounting is a necessary condition but is by no means a guarantee for effective environmental improvements. In addition, substantively desirable total life-cycle assessments imply the risk of information overload counterproductive for effective environmental management.

8. And even (within a company) well elaborated and accepted formal environmental-management standards, routines and systems do not always imply effective practical application with or without appropriate monitoring procedures. Just as public rules and norms of any (bureaucratic) environmental policy are often only partly implemented and controlled in spite of pertinent company management acceptance, corporate environmental management systems can only to a limited degree be expected to become routinized.

9. Champions of the environment – on the global as well as on the local level – will therefore remain crucial as highly visible individuals with the resources to implement effective environmental management programmes in corporations.

10. Given a trend towards increasingly professional environmental management, outside consultants and also industrial associations will probably play an important role in this respect, just as in other fields of business consulting and policy.

11. Generally, based upon the concurrence of actors' own judgement and external social pressure, the need for cooperation in effective and efficient environmental management – with corresponding performance in logistics and recycling – will probably induce voluntary agreements and cooperative arrangements among companies, vertically and horizontally, as well as among industrial associations, unions, environmental organizations, engineering associations, and public organizations involved in environmental management.

12. Given sufficient momentum in the broad installation of environmental management from the mutually reinforcing interaction among ecological push-and-pull factors, necessary (structural) requirements in various dimensions of the sociosphere are likely to gain ground and become institutionalized. Among these are environmental risk liability, limited reversal of the burden of proof, the specified attribution and internalization of environmental costs, corresponding environmental policy framework conditions, the development of intelligent clean

environmental technologies, enhanced product reuse and recycling, and avoidance of generating further uncontrolled (hazardous) wastes (cf. Huber, 1991).

13. Successful environmental management – within an offensive corporate strategy seeking among other things to defend cost leadership – will increasingly be based on an integrated cradle-to-grave perspective on, and solution of, environmental problems, an appropriate corporate culture, its entrenchment in internal and external corporate and management structures, its economic viability, the need and willingness to pay for the environmental benefits of products, packaging avoidance in the distribution of increasingly returnable products, appropriate ecological communication and adequate price policy (cf. Meffert/Kirchgeorg, 1993, 1995).

14. Assuming the social persistence of sustainable development as a guiding principle, environmental management strategies based on competence, credibility, commitment and cooperation will increasingly be perceived as an (industrial) contribution to this goal, for which appropriate environmental policy programmes and regulations, ecologically and economically efficient environmental management, growing pressure from adverse experience and the assumption of environmental responsibility are major necessary driving forces.

15. Companies in an uncertain and changing global environment have to develop generic strategies for their survival taking into account globalization, technology development, ecology, changing (sociocultural) values, systems complexity, information processing needs, and continuous learning as strategic reference points beyond traditional competitive strategy. Environmental management may well serve as a pilot function and agent of change in the development of such corporate competencies in future management, so that its establishment serves also the general non-ecological purpose of developing corporate capabilities to cope with all kinds of changes in a company's environment, such as changes in consumer values and demand, technology push, political risks, uncertainties on turbulent global markets. It also serves to combine various strategic management functions by intelligent information systems and processing, strategic controlling as conscious risk management, and coordinated strategy implementation, where company-specific chances and risks define the requirements for innovative (environmental) strategies (Steger, 1993). In this perspective, the propagation of environmental management may well be supported by general corporate strategy development and survival requirements, too.

16. Corresponding complex and unconventional organizational learning processes under conditions of prevailing uncertainty typically require, besides organization dissidents on the micro-level and epistemic communities on the macro-level, multiple actor identities on the meso-level.[5] So propagating and installing

---

[5] Phenomena described as multiple self and multiple identity dispose of polycentric characteristics, such as pluralism of interpretations induced by a heterogeneous environment, parallel information processing and action, organizational slack via abundant resources and loose coupling, and individual ambiguity tolerance. Organizations with multiple-actor identities are able to tolerate and need not explain and integrate different phenomena by a consistent integrated world view, to act on the basis of problem-specific orientations and particularistic causal hypotheses, and thus gain additional competence under

environmental management in companies may be easier in corporations with multiple-actor identities, which frequently have competitive advantage over others.

17. Since in the future, firms will very likely be confronted with a wider range of pressures from many more sources than in the past, innovative (environmental management) strategies may well become a necessity in dealing with these pressures and growing uncertainties, despite the continued existence of obstacles, such as an external incentive structure encouraging defensive strategy.[6] However, 'implementing an innovative strategy is hard and challenging work, partly because it always involves changing power structures within the company and among companies.' (Schot/Fischer, 1993: 371) So – in line with previous assertions – one can conclude that innovative strategies of environmental management will be in demand, and some companies will try to develop them but will still largely remain exceptional.

Altogether, perspectives for environmental management point in the direction of gradual formal and substantive expansion in breadth and depth in a hesitant, spasmodic, troublesome, and roundabout manner. Although environmental management will slowly become integrated into everyday corporate practice, including enhanced and new forms of transcorporate organization, and thus gradually become routinized, its general development towards systematic comprehensive performance (even within a sustainable development approach) remains an open question and at least for the coming decade or two will be the exception rather than the rule.

LITERATURE

Ashford, N.A. and Meima, R. (1994) Designing the Sustainable Enterprise. Conference Summary Report. The Greening of Industry Network. ERP Environment, Wetherby.

Beck, U. (1986) *Risikogesellschaft. Auf dem Weg in eine andere Moderne.* Frankfurt: Suhrkamp.

Beck, U. (1988) *Gegengifte. Die organisierte Unverantwortlichkeit.* Frankfurt: Suhrkamp.

Blazejczak, J. *et al.* (1993) Umweltschutz und Industriestandort. UBA-Berichte 1/93. Berlin: Erich Schmidt Verlag.

Clarke, S. and Susse, G. (1995) From Greening to Sustaining: Transformational Challenges for the Firm. Conference Summary Report. The Greening of Industry Network. ERP Environment, Wetherby.

---

the uncertain conditions of modern risk society, compared with monoreferential actors. Successful multiple-self strategies presuppose, however, the practical superfluity of an integrated frame of orientation and the existence of concrete spheres of experience and action for particularistic orientations. Organizations with multiple identity can then be expected to be more innovative in their activity spectrum and their self conception than others, because they regret harmonization of diverging particularistic orientations (Wiesenthal, 1990, 1994, 1995).

[6] Government regulation is often still geared to the implementation of end-of-pipe technologies. Even environmentally aware customers are reluctant to pay higher prices for additional environmental benefits. Legitimation by public debate and consensus for certain environmental technologies and approaches is developing only with difficulty. World market turbulences, including considerably varying exchange rates, do not offer an environment with stable profits facilitating long-term investments in environmental management and technology.

Coenenberg, A.G. *et al.* (1994) Unternehmenspolitik und Umweltschutz. *Zeitschrift für betriebswirtschaftliche Forschung* **46**, 81–100.

Conrad, J. (1993) Social Significance, Preconditions and Operationalisation of the Concept Sustainable Development. In: (ed.) F. Moser. *Sustainable Development—Where Do We Stand?* Technical University Graz.

Conrad, J. (1994) Was müssen und was können Unternehmen zur Umweltentlastung beitragen?. In: Fortbildungszentrum Gesundheits und Umweltschutz Berlin (FGU Berlin) (ed.), *Ökologische Modernisierung und industrieller Strukturwandel.* UTECH BERLIN '94, Berlin.

Conrad, J. (1995) Development and Results of Research on Environmental Management in Germany. *Business Strategy and the Environment* **4**, 51–61.

Dörner, D. (1992) *Die Logik des Mißlingens. Strategisches Denken in komplexen Situationen.* Hamburg: Rowohlt.

Fischer, K. and Schot, J. (eds) (1993) *Environmental Strategies for Industry.* Washington D.C.: Island Press.

Freimann, J. (1994) Umweltorientierte Unternehmenspolitik in Deutschland. In: (eds) E. Schmidt and S. Spelthahn. *Umweltpolitik in der Defensive – Umweltschutz trotz ökonomischer Krise.* Frankfurt: Fischer.

Freimann, J. and Hildebrandt, E. (eds) (1995) *Praxis der betrieblichen Umweltpolitik.* Wiesbaden: Gabler.

Galtung, J. (1990) Cultural Violence. *Journal of Peace Research* **27**, 291–305.

Giddens, A. (1990) *The Consequences of Modernity.* Cambridge: Polity Press.

Giddens, A. (1991) *Modernity and Self-Identity.* Cambridge: Polity Press.

Hildebrandt, E. *et al.* (1995) Industrielle Beziehungen und ökologische Unternehmenspolitik. Final Report, Volume 1 and 2, Science Center, Berlin.

Huber, J. (1991) *Unternehmen Umwelt.* Frankfurt: Fischer.

Huber, J. (1995) *Nachhaltige Entwicklung. Strategien für eine ökologische und soziale Erdpolitik.* Berlin: edition sigma.

Jänicke, M. *et al.* (1992) *Umweltentlastung durch industriellen Strukturwandel?* Berlin: edition sigma.

Kirchgeorg, M. (1990) *Ökologieorientiertes Unternehmensverhalten.* Wiesbaden: Gabler.

Knödgen, G. (1982) *Umweltschutz und industrielle Standortentscheidung.* Frankfurt: Campus.

Low, P. and Yeats, A. (1992) Do 'Dirty' Industries Migrate? In: (ed.) P. Low. *International Trade and the Environment.* World Bank, Washington D.C.

Mayntz, R. (1996) Gesellschaftliche Umbrüche als Testfall soziologischer Theorie. In: (ed.) L. Clausen. *Gesellschaften im Umbruch.* Frankfurt: Campus.

Mayntz, R. and Scharpf, F.W. (eds) (1995) *Gesellschaftliche Selbstorganisation und politische Steuerung.* Frankfurt: Campus.

Meffert, H. and Kirchgeorg, M. (1993) *Marktorientiertes Umweltmanagement.* Stuttgart: Poeschel.

Meffert, H. and Kirchgeorg, M. (1995) Green Marketing. In: (ed.) M.J. Baker. *Companion Encyclopedia of Marketing.* London/New York: Routledge.

Palmer, K. *et al.* (1995) Tightening Environmental Standards: The Benefit-Cost or the No-Cost Paradigm? *Journal of Economic Perspectives* **9**(4), 119–132.

Piasecki, B. and Asmus, P. (1990) *In Search of Environmental Excellence—Moving beyond Blame.* New York: Simon & Schuster Inc.

Porter, M.E. (1985) *Competitive Advantage: Creating and Sustaining Superior Performance.* New York: The Free Press.

Rappaport, A. and Flaherty, M.F. (1992) *Corporate Responses to Environmental Challenges.* New York: Quorum Books.

Schmidheiny, S. (1992) *Changing Course: A Global Business Perspective on Development and the Environment.* Cambridge: MIT Press.

Schot, J. and Fischer, K. (1993) Conclusion. Research Needs and Policy Implications. In: (eds) K. Fischer and J. Schot. *Environmental Strategies for Industry.* Washington D.C.: Island Press.

Senghaas, D. (1994) *Wohin driftet die Welt?* Frankfurt: Suhrkamp.

Steger, U. (1993) *Umweltmangement.* Wiesbaden: Gabler.

Ullmann, A. (1982) *Industrie und Umweltschutz.* Frankfurt: Campus.

Vietor, R. and Reinhardt, F. (1995) *Business Management and the Natural Environment.* Cincinnati: Southwestern Publishing Company.

Wieselhuber, N. and Stadlbauer, W.J. (1992) Ökologie-Management als strategischer Unternehmensfaktor. Dr. Wieselhuber & Partner Unternehmensberatung, München.

Wiesenthal, H. (1990) Unsicherheit und Multiple-Self-Identität: Eine Spekulation über die Voraussetzungen strategischen Handelns. MPIFG Discussion Paper 90/2, Köln.

Wiesenthal, H. (1994) Lernchancen der Risikogesellschaft. *Leviathan* **22**, 135–159.

Wiesenthal, H. (1995) Konventionelles und unkonventionelles Organisationslernen: Literaturreport und Ergänzungsvorschlag. *Zeitschrift für Soziologie* **24**, 137–155.

Willke, H. (1989) *Systemtheorie entwickelter Gesellschaften*. Weinheim/München: Juventa.

Winter, G. (1993) *Das umweltbewußte Unternehmen*. München: Beck.

17. ANY GENERALIZABLE PATTERN OF SUCCESSFUL ENVIRONMENTAL MANAGEMENT?

Jobst Conrad

If one asks once again for essential factors explaining environmental management development and success, one may, referring to corresponding literature cited in Chapter 2, distinguish the following systematic theoretical-analytical perspectives, which focus – on the micro-, meso- or macro-level – on different explanatory dimensions of environmental management:

1. Ecological problem structure and related social objectives (concerning the type and severity of underlying environmental problems or the range of environmental goals: economic compensation, technical correction, technical prevention, sustainable ecological development);

2. Technical solutions of environmental problems (with regard to their feasibility and type: additive (end-of-pipe) technology, integrated technology, system technology; complexity and coupling modes of technologies applied);

3. Appropriate information (referring to environmental monitoring, information processing, information evaluation, measuring eco-performance, and the overall structure of the environmental information system);

4. Appropriate corporate organization and resources (regarding formal organization, informal organization, management systems and functions, degree of self-organization, available financial manpower, know-how resources; general inquiry and screening of company-internal influencing factors);

5. The economics of environmental management (concerning (changing) ecology–economy trade-offs, environmental protection where pollution prevention pays, environmentally oriented demand and ecological market chances and marketing, thereby induced ecological competition, corporate risk management and coping with (normal) accidents);

6. Company-external determinants (with regard to public pressure and environmental (pressure) groups, environmental policy and regulations, environmental conflict or cooperation with external actors, lack of social credibility and legitimacy, environmental ethics as cultural driving force);

7. System structures (in relation to the fundamental prerequisites of differentiated socio-functional systems dealt with by systems theory, structurally different development phases of (industrial) environmental protection and management, the time horizon of environmental management strategy);

8. Social actors (regarding the type of actors involved, the actor constellation and the respective pattern of problem perception as well as interest and power

distribution; and the explanation of actors' behaviour by rational choice and game theories);

9. Social learning and strategy (focusing on social processes and intentions such as (typical) innovation processes, organizational learning, corporate (environmental) strategy and its consistency, strategic environmental management portfolio (internal ecological strengths and weaknesses, external environmental chances and risks), multiple rationalities, actor identities and bifurcation points);

10. History and situation (referring to strict historical analysis and reconstruction or situational explanation, e.g. utilizing windows of opportunity or not).

Whereas each of these conceptual models addresses important and valuable characteristics of environmental management in a specific analytical perspective, the empirical findings of environmental management case studies demonstrate that it is rarely feasible to explain most of the variance of (successful) environmental management by influencing factors in only one of these dimensions. Instead, the crucial importance of the favourable interplay of influencing factors in different dimensions has to be emphasized. Thus, it is at best of limited use to discuss the relative strengths and weaknesses of the various theoretical-analytical approaches to environmental management listed, when we now try to answer the question about generalizable patterns of successful environmental management on the basis of the preceding comparative evaluations.

The spread and deepening of environmental management during the last decade indicates neither whether programmatic and formal-organizational anchoring of environmental protection has already led to substantive measures of environmental protection, nor which corporate strategies, organizational forms and modes of implementation of environmental management imply substantive successes beyond mere compliance with public regulations. In Chapters 13 and 14, the relatively general common findings of the case studies and of other investigations have been addressed. Empirically they pointed to quite similar features of successful environmental management on a general abstract level and to quite different features on more concrete levels of analysis.[1] One would thus expect a generalizable pattern of successful environmental management in abstract but not in concrete terms. If, on the one hand, the respective basic features of actor constellations, structural boundary conditions, and social networks correspond to typical diagnoses in innovation, organization, and social (politico-economic) theory and if, on the other

---

[1] These can be summarized by the following points, listed in Chapter 14:

1. Successful (innovative) environmental management depends on the positive interaction of company-internal and -external determinants leading to the technical, organizational, socio-economic and cultural entrenchment of environmental management efforts, thus generating a self-supporting interaction dynamics towards substantive environmental protection.

2. Counteractive qualifying factors typically play some rôle but are never strong enough to balance out or even supersede supportive qualifying factors, and thus block successful environmental protection efforts. So these factors are typically in a minority position able to generate time delay and to prevent environmental management systems comprising many companies which contribute to a whole value chain.

hand, there is no best way to successful environmental management, and similar firms act quite differently with respect to environmental protection (Cebon, 1993), then it is of theoretical and methodological interest to look for possible systematic reasons for these findings. They can be found in the positive interaction of the following explanatory dimensions:[2]

1. If one tries to explain the results of social processes in more complex and differentiated terms rather than in terms of only one or two qualifying factors, one usually still assumes diverse probabilistic social rules and regularities. Their significance and weighting may, however, vary strongly from case to case, so that at best case-specific findings can be still expected on a high level of abstraction because of the high complexity of explanation. For instance, both public scandals as well as innovative, ecology-minded top management can induce successful environmental management, but these conditions need not be simultaneously fulfilled in any given case.

   At best, the statement that corporate efforts towards improved and/or system-atic environmental protection are mostly triggered by, broadly speaking, nega-tively perceived events may be generalizable within limits. These events can be due to direct environment-related factors, such as foreseeable violations of stand-ards, changes in demand preferences due to growing customer environmental awareness, or political and public pressure because of inadmissible environmental pollution, or to pressure for action to be taken only indirectly relevant to environmental questions, such as the necessary utilization of economically significant cost-saving potentials or strategic corporate siting decisions; and they can be caused by factors internal as well as external to the company. By contrast, genuinely positive starting points for enhanced environmental management in the sense of, for example, a purely humanistic environmental commitment without any (economic) pressure for action can hardly be expected.

---

3. The specific configuration of substantive influencing factors explaining individual success stories is particular for each case, allowing little generalization.
4. Some common ground is given by the preponderance of economically profitable environ-mental management, by the limited number of individual and organization actors usually involved, and by the mostly minor significance of public environmental policy for innovative environmental management.
5. Furthermore, perceived pressure for change (for varying reasons), initial experiences of success, the availability of resources, individuals who champion environmental management ideas within the organization, strategic action and cooperation of the actors, frequently seem to be necessary conditions for successful environmental management.
6. The joint impact of ecology push-and-pull factors on the degree of a company's (successful) environmental management efforts appears to be higher than expected from the mere addition of both types of influencing factors (Ostmeier, 1990), confirming the importance of the positive interaction among influencing factors.
7. Thus, successful environmental management requires rather comprehensive, systematic, resource consuming and long-term, concerted corporate action addressing the various levels, functions, and groups of a company and taking into account the whole product life-cycle (Fischer/Schot, 1993; Freimann/Hildebrandt, 1995; Hildebrandt et al., 1995; Huber, 1991; Kirchgeorg, 1990; Rappaport/ Flaherty, 1992; Steger, 1992, 1993; Wieselhuber/Stadlbauer, 1992).

[2] The following discussion is taken essentially from Conrad, 1995.

2. If one interprets social processes with reference to different dimensions and domains of the extended sociosphere, it is no surprise that successful management relies on roughly the following preconditions: it has to lead to (ecologically) favourable effects; appropriate technical means and manpower have to be available for its implementation; it has to be economically viable or, even better, profitable; it relies on corresponding programme and implementation decisions and regulations; it has to be reproducible in cognitive information terms as well as socioculturally anchored in the zeitgeist; it requires the appropriate disposition and commitment on the part of the actors involved; and the regulative characteristics of sociofunctional subsystems and institutionalized structures as well as the lifeworlds of individual actors and primary social systems, shaping their pattern of perception and behaviour, must not impede it. What remains to be explained is whether successful environmental management occurs without one or even several of these preconditions. It appears to be a valid general assumption, however, that successful environmental management is highly probable under favourable preconditions in all (14) domains of the extended sociosphere.

It has been empirically demonstrated that since the 1990s favourable preconditions have indeed existed in the semiotic-symbolic dimension of the sociosphere, but much less so in its other dimensions. It is therefore small wonder that there has hitherto been much more talk about, than substantial action towards environmental management (Freimann, 1994).

3. With regard to different classes of determinants of successful environmental management, as discussed in the preceding chapters,[3] it is again plausible that positive interaction among such types of determinant favouring environmental management makes success highly probable, whereas in case of predominantly counteractive and counterproductive qualifying factors successful environmental management becomes improbable and more dependent on situational chances. However, specific individual qualifiers within certain classes of determinant, such as company-internal and external situation-structural conditions and facts, need not by any means be always in parallel. Thus, if positive interaction among the majority of qualifying factors rather than among a specific selection of factors is responsible for the technical, organizational, socio-economic, sociopsychological and sociocultural entrenchment of patterns of orientation and behaviour, which is important for successful environmental management, this makes it clear why generalizable empirical findings can at best be expected on an adequately high level of abstraction.

4. In so far as comprehensive environmental management is connected not only with trade-offs with other corporate interests, but also necessarily has to take into account trade-offs among different ecological concerns (e.g. using plastics instead of steel in NedCar automobile manufacturing), which can only be weighed

---

[3] As a reminder: substantive (ecological) problem structure, techno-economic context, wider social context, economic boundary conditions, psychological context, situational structure, historiographic events, and personality structure of key individuals.

normatively and seldom unequivocally offset,[4] no general normative statements about environmental management strategies and measures to be preferred can be made on a concrete substantive level. Instead, success frequently depends merely on the situational utilization of options that present themselves within a given context, which, besides offering a window of opportunity, need not – in the sense of scientific objectivity – be the ecologically most favourable ones by any means.

5. In a double sense, the time dimension plays an important role with respect to generalizable empirical findings on successful environmental management. On the one hand, in the majority of case studies more far-reaching environmental management efforts addressing not just one specific environmental problem have occurred only since the early 1990s. They are therefore ecological pioneers in a restricted sense only because an environmentally oriented corporate management is no longer a novelty but is already more or less state of the art, at least on the level of corporate culture. Successful environmental management *cum grano salis* has occurred since 1990 in an environmentally more aware and friendly atmosphere and under economically more favourable competitive conditions than in 1980, let alone 1970.

On the other hand, the preconditions for, as well as the pattern of motivations and interests behind successful environmental management change in the course of its institutionalization. A successful launch and initial success demand other management qualities than well-established, routinized environmental management in everyday corporate practice. And the first environmental protection measures often lead to greater economic savings as well as environmental improvements than subsequent measures. Therefore, rather easily attainable initial successes may well contribute to developing a motivational self-dynamics in favour of further environmental management efforts.

Finally, a corporate product policy aiming at ecological quality and rationality for its (sustainable) products, which usually develops at a later stage, requires more comprehensive environmental management than that mainly addressing the efficient in-house use of input resources and operational environmental protection. Thus, different forms of corporate environmental management can be expected according to a company's given evolutionary stage.[5]

These not only historically changing contextual conditions but also requirements for a substantively broadening environmental management that modify in the course of development imply that the empirically determinable conditions for successful environmental management vary over time and can therefore hardly be generalized on a concrete level. Thus, environmental management as an object of empirical research largely constitutes a moving target, where causal relationships are in a permanent state of change.

6. If one furthermore understands the upcoming process of ecological modernization as also connected with sociostructural and sociocultural breaches, which in their societal and economic dynamics are necessarily accompanied by intensified

---

[4] This is well known from environmental impact assessments.
[5] Ciba represents a pioneer in this respect.

phenomena of crisis and innovation, social conflict, and a corresponding renewal of capital stock, then it is known from formal theories of dynamic systems that the development path of such (social) transformation processes cannot be predicted (cf. Mayntz, 1996) and therefore cannot be controlled by deliberate organizational development and corporate strategy. Thus, generalizable findings on successful environmental management can hardly be expected in this perspective.

7. (Individual) actors and therefore their motivations and interests play a decisive role in successful (innovative) environmental management in particular. The possibility of coupling environmental concerns with parallel (individual) interests and motives frequently is therefore of essential importance. However, these motivations and interests can significantly differ in substance from case to case.[6] Thus only the statement about the enormous importance of viable couplings of interests and motives appears generalizable.

8. In so far as industrial corporations as primarily economic actors can in the final resort survive only if they dispose in the long run of sufficient economic resources, the various competitive forces, described by Porter (1985, 1990),[7] represent decisive factors in maintaining a company's position in the market, apart from utilizing situational chance and from politically determined framework conditions relating to the sociopolitical order. These competitive forces have – in analogy to the generalized profit-motive – to be conceptualized theoretically as essential economic framework conditions of corporate behaviour which cannot of themselves explain a company's decision to establish or to forgo systematic environmental management. Consequently, the preference for cost-saving over cost-demanding environmental management measures, and the relative lack of innovative, ecologically oriented corporate strategies become quite evident, but less so the occurrence of successful (innovative and costly) environmental management.[8]

9. Finally, the requirements of environmental management differ for different components of the value chain as well as for different functional domains in a company, such as primary production versus distribution, or research and development versus marketing, so that at best abstract socio-organisational and socio-cultural conditions can be deduced as general conditions of successful environmental management in these different spheres.

In sum, in spite of the theoretically feasible possibility of genuine longitudinal case studies and of systematic comparison of success and failure cases of

---

[6] In the ABC-Coating and Glasuld case studies, the coupling of environmental and health concerns played an important role. The managing director of Amecke Fruchtsaft realized savings in input resources in connection with a corporate strategy of returnable packaged fruit beverages, economic savings, and gaining personal profile via environmental protection.

[7] Rivalry among existing competitors, bargaining power of suppliers, bargaining power of buyers, threat of new entrants, threat of substitute products or services.

[8] As corresponding examples of lacking demand indicated in the Hertie, Kunert, and Ciba case studies, environmental management strategies have to take all competitive forces into account.

environmental management, the following factual findings together make it highly plausible that empirical studies of (successful) environmental management do not yield substantive generalizable results except on a high level of abstraction or in the form of checklists reflecting everyday knowledge (cf. Winter, 1993):

1. Serious methodological limitations to simple (unilinear) models of explanation;
2. The essential importance of interaction among different qualifying factors and of the interaction of different dimensions of the sociosphere;
3. Vague trade-offs between different company goals;
4. Contextual conditions and scope of environmental management changing over time;
5. Uncertain development paths of adaptation processes to breaches in socio-economic structures;
6. Fluctuating couplings of interests and motives;
7. Indeterminate, though restricting impacts of competitive forces on (innovative) environmental management strategies;
8. And the differences of environmental management requirements for different components of the value chain as well as for different functional domains of a company.

We may hence conclude by summarizing the major general reasons for growing corporate efforts in (systematic) environmental management by:

1. Reaction to cost pressure to remain competitive;
2. Political loyalty to the rule of law; and
3. Efforts towards social integration, towards embedding the company in its societal environment (Dyllick, 1989).

As Huber *et al.* put it in Chapter 12: 'If...industry is open today to ecological matters, it is not due to a Rousseau-like, romantic, and vital love of Nature. A much more plausible reason is the utilitarian attitude of homo oeconomicus, supplemented by the insight that the ecological expansion of the production model [with the internalization of externalized environmental costs] lies in the enlightened self-interest of the economic participants ... One can permanently build on environmental trade only if the positive sentiments for Brother Tree and Mother Earth stand up to the sharp wind of competition and the cool climate of rational calculation.'

LITERATURE

Cebon, P.B. (1993) The Myth of Best Practices: The Context Dependence of Two High-performing Waste Reduction Programs. In: (eds) K. Fischer and J. Schot, *Environmental Strategies for Industry.* Washington D.C.: Island Press.

Conrad, J. (1995) Erfolgreiches Umweltmanagement im Vergleich: generalisierbare empirische Befunde? In: (eds) J. Freimann and E. Hildebrandt, *Praxis der betrieblichen Umweltpolitik.* Wiesbaden: Gabler.

Dyllick, Th. (1989) *Management von Umweltbeziehungen. Öffentliche Auseinandersetzungen als Herausforderung*. Wiesbaden: Gabler.

Fischer, K. and Schot, J. (eds) (1993) *Environmental Strategies for Industry*. Washington D.C.: Island Press.

Freimann, J. (1994) Vom Nutzen des nüchternen Hinsehens. Umweltmanagement in der Abwicklung? In: (eds) E. Schmidt and S. Spelthahn, *Umweltpolitik in der Defensive – Umweltschutz trotz ökonomischer Krise*. Frankfurt: Fischer.

Freimann, J. and Hildebrandt, E. (eds) (1995) *Praxis der betrieblichen Umweltpolitik*. Wiesbaden: Gabler.

Hildebrandt, E. *et al.* (1995) Industrielle Beziehungen und ökologische Unternehmenspolitik. Final Report, Volume 1 and 2, Science Center, Berlin.

Huber, J. (1991) *Unternehmen Umwelt*. Frankfurt: Fischer.

Kirchgeorg, M. (1990) *Ökologieorientiertes Unternehmensverhalten*. Wiesbaden: Gabler.

Mayntz, R. (1996) Gesellschaftliche Umbrüche als Testfall soziologischer Theorie. In: (ed.) L. Clausen, *Gesellschaften im Umbruch*. Frankfurt: Campus.

Ostmeier, H. (1990) *Ökologieorientierte Produktinnovationen*. Frankfurt: Peter Lang.

Porter, M.E. (1985) *Competitive Advantage: Creating and Sustaining Superior Performance*. New York: The Free Press.

Porter, M.E. (1990) *The Competitive Advantage of Nations*. London: MacMillan.

Rappaport, A. and Flaherty, M.F. (1992) *Corporate Responses to Environmental Challenges*. New York: Quorum Books.

Steger, U. (ed.) (1992) *Handbuch des Umweltmanagements*. München: Beck.

Steger, U. (1993) *Umweltmanagement*. Wiesbaden: Gabler.

Wieselhuber, N. and Stadlbauer, W.J. (1992) Ökologie-Management als strategischer Erfolgsfaktor. Dr. Wieselhuber & Partner Unternehmensberatung, München.

Winter, G. (1993) *Das umweltbewuß*te Unternehmen. München: Beck.

18. SIGNIFICANCE OF, AND IMPLICATIONS AND RECOMMENDATIONS FOR ENVIRONMENTAL POLICY

Jobst Conrad

This final chapter discusses environmental management from the perspective of (public) environmental policy[1] in three dimensions:

1. Analysing the significance and (specific) influence of environmental policy for environmental management activities and evolution;
2. Deriving recommendations on what environmental policy, particularly on EU level, should do to enlarge and improve environmental management;
3. Assessing the actual capacity of environmental policy to follow these recommendations in view of its internal and external restrictions and of its actual mode of proceeding, taking up corresponding considerations in Chapter 2.

1 MUTUAL INTERPLAY OF ENVIRONMENTAL MANAGEMENT AND ENVIRONMENTAL POLICY

As stated in Chapters 13 to 15 the importance of environmental policy for environmental management depends on national culture, the innovation-diffusion dimension, the (situational) specificities of individual cases, including the willingness and interest of the company, the size of the company, the degree of advanced development in environmental policy, the pioneer character of environmental management, the generality (and spread) of environmental management activities and instruments addressed by environmental policy (individual cases versus cases of general concern; eco-auditing versus specific items), and its expected (future) development as a significant context for corporate strategy.

As explained in Chapter 13, for systematic reasons environmental policy regulations cannot be expected to play an important (favourable) role for successful (innovative) environmental management. Instead, environmental policy can be expected to play the role of an important contextual determinant by, on the one hand, facilitating or impeding the success of (corporate) environmental management efforts and, on the other, by contributing or not to the diffusion of environmental improvements throughout the economy and social institutions. Particularly relevant in this respect are the politically determined (ecologically

---

[1] As mentioned earlier, environmental policy in this book usually means public environmental policy, if not stated otherwise, as opposed for instance to corporate environmental policy.

stringent) legal boundary framework conditions, supporting or complicating innovative environmental management, thus contributing or not to the societal push-and-pull dynamics described above in Chapter 15, necessary for the social entrenchment of environmental management, and – in order to allow for the political enforceability of corresponding environmental policy regulations and programmes – therefore necessary for the relative consistency and viability of environmental policy measures, which prevent their fragmentation or undermining by other policy segments and programmes. So environmental policy in favour of corporate environmental management for structural reasons has primarily to start on a meta-level of providing appropriate boundary conditions (regulation of self-regulation; cf. Huber, 1991; Ladeur, 1987). As pointed out in Chapters 2 and 15, this amounts to a push-and-pull strategy of environmental policy with an increasingly cooperative posture besides a command-and-control orientation with corresponding organizational principles for action programmes, administration and enforcement. Environmental policy of this type would be concerned to establish a constellation and combination favourable to environmental interests of weltanschauung, lifestyle, law, politics, economics, and technology, which evoke and promote the paratactic cooperation among social actors, structures and functional systems; to set appropriate spatial, temporal, and functional priorities; and concentrate on strategic points of departure favouring a broadly effective and long-term ecologization dynamic (e.g., raising energy prices or imposing a liability to take back materials) (see Huber, 1991; Jänicke, 1993; Knoepfel, 1993).[2]

This environmental policy perspective is based on the recognition that – in modern industrialized societies in particular – the successful treatment and solution of a social problem depends on the mutual capacity for interlinkage, interplay and reinforcement of the various dimensions of the extended sociosphere as well as the various actor strategies and behaviours. Without such favourable push-and-pull dynamics even optimal task orientation and fulfilment by one social actor or subsystem disposing of sufficient resources will not lead to the solution of a general social problem, such as ecological sustainability. Thus, on the one hand, environmental policy is but one component in the total interaction of determinants like public pressure, legal prescriptions, cost pressure, changes in perception, corporate commitment, reorganization and self-development efforts, leading to the diffusion of improved environmental management. So it has to place its hope on synergy effects and cannot rely on its own impacts alone. However, on the other hand without appropriate legal and economic framework conditions, to be fixed by environmental policy, industrial corporations have insufficient options for effective environmental management that does not undermine their competitiveness and profitability (cf. Huber, 1991; Mazur *et al.*, 1995). Public and private environmental action may thus not conflict; instead, each social actor has to play his part in the overall interplay of actor behaviour and system dynamics, if effective environmental protection or even more sustainable ecological

---

[2] Certainly, such a long-term, strategic environmental policy presupposed by these conclusions cannot simply be assumed in everyday political practice.

development is to be achieved. These general social goals, which have undoubtedly gained considerable and widespread recognition in the course of recent decades, can only be operationalized by negative demarcation of ecologically incompatible actions and structures, but rarely, however, by positive specification (cf. van den Daele, 1993).

2   THE TECHNICAL, SOCIAL AND TEMPORAL CONTEXT OF ENVIRONMENTAL POLICY RECOMMENDATIONS

Before listing in somewhat more detail corresponding environmental policy recommendations, it is important to give some indication in an analytically distinct manner of the technical, social and time context of environmental policy.

In technical terms, different environmental policy strategies for dealing with environmental problems (environmental protection, risk minimization, or sustainable development; cure by compensation of environmental damages or by additive environmental technology, or prevention by environmentally compatible technology or by environmentally favourable structural change) reflect different substantive environmental policy objectives. These are also reflected in the specific combination of sufficiency, efficiency and consistency strategies chosen in tackling sustainable development (cf. Huber, 1995a). Furthermore, environmental policy should be, but sometimes is not, aware of the physical limits to environmental improvements which have certain policy implications, for instance the acceptance of some remaining environmental impacts of producing socially desired products versus the imposition of (disproportionately) high costs for additional, only marginal environmental improvements, or the favourableness of new integrated production systems over further improvements in end-of-pipe technologies within a given system of production.

In social terms, the feasibility of (specified) environmental policy strategies and programmes depends on the various (legal, cultural, informational, material, manpower etc.) resources and skills of the actors pursuing or addressed by such strategies and programmes. The environmental policy options available to a local water authority clearly differ from those at the disposition of a federal environmental agency in charge of promoting environmental management in industry, and these differ again from those available to water utilities. Different actors hence have differential environmental policy capacities. Environmental policy at the EU level (cf. Héritier et al., 1994; Hey/Brendle, 1994; Jachtenfuchs/Strübel, 1992; Johnson/Corcelle, 1989; Liefferink et al., 1993), mainly developed within the European Commission and by DG XI, clearly reflects these features in its emphasis on formal regulatory activities. 'The weak points of the EC environmental policy have always been the low level at which quality requirements are set, the slow pace at which EC rules are negotiated and decided upon, and the great delay with which EC measures are implemented by member states. It often seems as if member states are able to pursue their own policies irrespective of EC-acts. The EC lacks enforcement measures to pressure defaulting states to comply. The Single European Act does not offer hope that this picture will be changed through binding legal steps. It expresses the general willingness of the member states to continue the existing environmental policy.'[3] (Koppen, 1988: 65ff.)

Favourable conditions for effective environmental policy (implementation) are, in particular, sufficient monitoring with clearly specified criteria and methods of measurement, well defined administrative norms and standards, organization, financing, instruments and procedures, separation of monitoring and administrative action, separation of programme formulation and implementation, intrapolicy cooperation transcending ecological media, interpolicy cooperation networking environmental authorities with other environmentally relevant authorities, regionalized enforcement, personnel education (Knoepfel, 1993). It is then essential to differentiate between external and 'internal' environmental policy addressees, i.e. industrial corporations or the consumer, and public organizations themselves.[4] Environmentally relevant internal public administration activities, which are not considered further in this chapter, are ecologically perhaps quite as important as (external) enforcement activities since they also relate to regional planning, the organization of transportation systems, energy supply, water and clean-air policy. Typically, the extent and success of environmental management within public institutions differ strongly from institution to institution depending on political undercurrents and majorities and depend at least as much on dynamic entrepreneurial individuals as in industrial corporations (Stitzel, 1992).

In the time dimension, both the time horizon of environmental policy and the actual point in time are significant for environmental policy recommendations. Furthermore, in the case of irreversible environmental degradation, the possibility of retarded environmental policy action coming too late has to be taken into account in environmental policy recommendations. The discrepancy between the required long-term time horizon for tackling and solving environmental problems (especially creeping disasters, cf. Böhret, 1990) and actual short-term political time perspectives (in modern industrial societies) among others due to legislative periods is well known. On the other hand, environmental protection and environmental policy have developed considerably in most western industrialized countries over recent decades (see Chapter 2; Hajer, 1995; Jänicke, 1996; Jänicke/Weidner, 1997; Mez/Jänicke, 1996; Weidner, 1996; Wolf, 1992) so that both the type of environmental problem and the type of regulatory regime now differ strongly from those of about 20 years ago. Environmental policy recommendations to improve (industrial) environmental management have to take this into account. Since almost all larger industrial companies have meanwhile become aware of evident opportunities for making pollution prevention pay, and none of them is eager to be prosecuted for violating

---

[3] 'It is difficult to overstate the significance of this structural difference between regulatory policies and policies involving the direct expenditure of public funds. The distinction is particularly important for the analysis of Community policy-making, since not only the economic, but also the political and administrative cost of enforcing EC regulations is borne by the Member States.' (Majone, 1989: 166)

[4] Although the improvement of environmental management by environmental policy has an easier course within public institutions which can also demonstrate the social (and economic) viability of corresponding environmental protection measures, experience shows that the substantive environmental policy success of in-house environmental management is typically no more likely than that of externally oriented management.

environmental laws or regulations, successful environmental management depends on how a company realizes these objectives effectively and efficiently.

Furthermore, the limits to bureaucratic environmental policy regulation appear to have been increasingly reached. However, precisely this experience prepares the ground for the spread of more advanced environmental policy approaches towards contextual control and integrated system technologies: regulation of self-regulation. Besides, despite quite limited implementation, (traditional) command-and-control policy has in the past not only been relatively influential and successful compared to persuasive and economic policy instruments, but also remains an important backing for the viability of these more reflexive environmental policy approaches. Past environmental policy experience has led to both rising demand for, and growing disillusion with environmental policy. In the context of the environmental debate and environmental protection, no withering away of the state is observed but rather an increasing politicization of production (cf. Conrad, 1990a, 1990b; Jänicke/ Weidner, 1995).

When now deriving environmental policy recommendations within this context, three levels of recommendation have to be distinguished, namely general environmental policy recommendations, environmental policy recommendations concerning environmental management in particular, and environmental policy recommendations specifically addressing the EU Commission. In any case, it has to be borne in mind that scientific policy recommendations can provide policy-makers with guidelines and unconventional perspectives but hardly with new concrete instructions on how to act in particular[5] (cf. Conrad, 1992). Furthermore, the limited generalizability of paths to successful environmental management, as discussed in the previous chapter, implies methodological limitations to general environmental policy recommendations. Finally, environmental policy recommendations relate to intentional policy with the objective of improving the quality of the environment. This supposition marks a specific analytical perspective but does not deny that actual policy rarely ever follows one main purpose.[6] It implies, however, that other

---

[5] Scepticism towards excessive concretization of social-science recommendations for action is appropriate if it is more than a question of conceptual models and initiatives. Where the knowledge needed to deal with practical problems is specific to a situation and not liable to generalization and abstraction, the experience, intuition and skill of the 'practician' and the prompt processing of information in everyday practice within the organization is generally superior to every academic data collection (Kaufmann, 1977: 51). All too often 'practical' suggestions from the social sciences may be not only irrelevant but also dysfunctional and counterproductive. It is all too easy to draw conclusions about the functional or causal conditions of practical proposals on the basis of their feasibility or even plausibility. Since the social sciences are only seldom in a position to communicate recipes, in other words products and techniques that would be feasible independently of handed-down theoretical and interpretation patterns in societal spheres of action, recommendations directed towards policy formation must exercise restraint in their claims to concretization where it is a matter of the systematization and generalization of reality description for actual conditions for action. 'Only where sociological insights are integrated in practical theories can they become practically relevant. In the process of scienticization, academic concepts and theorems change the perception and definition of practical problems and thus potential problem-solving behaviour.' (Kaufmann, 1977: 53)

[6] 'Politics and policy may only secondarily be concerned with problem-solving, policymaking entails a great deal more than decisions, policymaking is partly nonrational and also involves multiple rationales, and policymaking is concerned with a great deal more than substantive outcomes.' (Bartlett, 1994: 175)

sociostructural development trends or other policies without environmental objectives, which often have tremendous environmental effects (Jänicke *et al.*, 1992), may be important objects of environmental policy-making (e.g. via interpolicy cooperation) but not genuine addressees of the subsequent environmental policy recommendations.

3   GENERAL ENVIRONMENTAL POLICY RECOMMENDATIONS

With reference to six general principles of effective policy-making, namely multiple common action, subsidiarity, active democracy, policy monitoring and evaluation, networking, and capacity building (see Galtung, 1992; Jänicke, 1992), general environmental policy recommendations are concerned with:

1. Gaining political and technical in-house (modernization) capacity and competence in environmental policy programmes embedding environmental management in a sustainable economy perspective;

2. A (preventive) environmental policy orientation towards sustainable development instead of regulations favouring end-of-pipe technologies and a policy focusing mainly on existing pollution and contamination from past industrial activities because of scandal-driven political priorities;

3. Striving for the internal consistency and reliability of environmental policy, particularly in coordination with other, often opposing policies;

4. Conceiving industrial corporations as (in the last resort) cooperative actors with their own legitimate interests, when aiming at improving the environment and sustainable development;

5. Therefore favouring contextual policy steering by attempting to install ecologically beneficial legal and economic framework conditions.

4   ENVIRONMENTAL POLICY RECOMMENDATIONS IN FAVOUR OF ENVIRONMENTAL MANAGEMENT

Addressing (corporate) environmental management somewhat more specifically, environmental policy recommendations should, as stated above, relate essentially to providing adequate framework conditions that facilitate and do not hinder corporate environmental management. The following environmental policy recommendations are advanced in this respect:

1. There are no hard and fast rules on promoting environmental management apart from general regulative ideas aiming to improve or even create the context facilitating it. State environmental policy ought consequently to contribute to providing a favourable setting for innovative environmental management and not dampen it by bureaucratic rules or undermine its internal (ecological) coherence by apportioning it to a multiplicity of separate jurisdictions.

2. Nothing can therefore be said about *specific* policy instruments, institutions, and regulatory patterns that are of general, and not just individual, environmental policy advantage.

3. Precisely because of the important role played by individuals in small and medium-sized enterprises with often quite divergent corporate cultures and inplant flows of operations, this is indicative of the usefulness of regulators having decentralized scope for action and flexible negotiatory processes.

4. Moreover, this shows the great significance for environmental policy and enterprises alike of engaging good staff, and of providing good initial and further training to ensure that the personnel in positions of responsibility for policy and company dispose of the necessary judgment and competence in making decisions and taking action suitable to the given situation.

5. In general it is advisable for state environmental policy, even where it is preventive in orientation, to direct firstly its resources and activities less towards unproblematic instances of successful environmental management such as that described in the case studies, but rather mobilize its (relatively limited) possibilities for regulating and (partially) controlling ecologically serious, socially conflictual environmental problems that may require successful coordination across a range of areas. Secondly, it may contribute through direct or indirect measures to the rapid diffusion of successful patterns of innovative environmental management. And thirdly, as we have indicated, it ought to aim at creating a favourable general setting for preventive environmental protection at the meta-level, e.g. instilling a preference for waste avoidance rather than waste recycling (to be decided from case to case), and setting up the appropriate legal liability arrangements (limited absolute liability and reversal of the onus of proof), insisting on producer obligations to take back materials, internalizing environmental costs,[7] emphasizing the increased relevance of (obligatory) eco-audits, eco-labels and environmental impact assessments, which the various social actors with their diverging interests would have to take into account in safeguarding their different options for action because pollution prevention would really pay.

6. This is true both with regard to the promotion of a gradual ecologization process occurring anyway in society and industry in general, with industry making increased efforts towards more comprehensive environmental management, and especially to the maintenance of such an environmental policy course in times of economic crisis evidencing a trend towards a rollback in environmental protection (cf. Schmidt/Spelthahn, 1994).

In somewhat more concrete terms, such an environmental policy is oriented towards:

1. Helping to break up closed-shop actor constellations, which systematically block substantial participation and thereby the actual influence of environmentally

---

[7] 'The advantage of ecological taxes over administrative rules and tied special charges is precisely that they exert a *relatively non-specific*, general and constant pressure to improve environmental efficiency.' (Huber, 1991: 250)

oriented actors, and to generate actor networks capable of coordinating and organizing rather comprehensive (cradle-to-grave) environmental management systems;[8]

2. Providing infrastructural support in the form of valuable information, environmental education, environmental research and well-targeted subsidies for environmental management efforts of (small and medium-sized) companies (e.g. co-funding of ecologically valuable high-cost options);

3. Allowing for flexible decentralized organization of concrete environmental management measures and programmes;

4. But combining this cooperative approach to support the diffusion and improvement of (industrial) environmental management with rather stringent and enforceable environmental regulations as its necessary backbone;[9]

5. Supported by the strategic use of prospective intervention, which gives industrial corporations the time to adapt to clearly foreseeable environment-related rules and standards and to establish cooperative commitments;[10]

6. Pursuing strategies of coupling environmental-management-oriented policy measures with parallel interests of the social actors involved, particularly keeping stringent workplace regulations, because considerable innovation and change is driven by worker health and safety concerns (EHS-programmes);

7. Making firms accountable to the public to ensure high behaviour transparency;[11]

8. Cooperating with environmental management pioneers as useful partners in developing new environmentally-more-compatible production systems and products, exploring actual barriers to, and possibilities for, the implementation of

---

[8] 'Short-term planning, the distortions of end-of-pipe incentives, and tendencies of firms to favour incremental improvements over radical changes require new kinds of partnerships between industry, government, labor, academia and the community to develop a sustainable economy... The Dutch program (Sustainable Technological Development) seeks to engage industry, government and other stakeholders in a communication and learning process that will launch those technological developments that lead to a sustainable society. The program can be seen as a practical experiment to influence existing networks and to create new ones.' (Ashford/Meima, 1994: 19, 20)

[9] 'The central task of government action could be described as the deepening and broadening of existing pressures. Deepening of pressures implies that regulatory pressure should be tightened in two ways. First, the use of different instruments and actions should be seen as complementary. Stringent and certain regulation is needed and should not be replaced by voluntary action. Voluntary action is necessary but will work only when it has the force of law behind it. Voluntary action should take the form of agreements between industry and government and the appointment of regional prevention teams to stimulate, supervise, and advise companies... Broadening of pressures implies that government action should be explicitly focused to elicit pressures from other and new sources, such as investors, insurers, user firms, workers, unions, and the public.' (Schot/Fischer, 1993: 371, 372)

[10] 'Industry's ability to do long-term product planning is dependent on government establishing clear quality and environmental performance requirements that will exist in the distant future.' (Ashford/Meima, 1994: 20)

[11] 'Second, the existing regulatory trend to force firms to become more open should be intensified. Firms should be forced to publish waste and emission figures, environmental plans and targets for the future, and results of environmental audits. Thus, the formulation and implementation of prevention plans should be made compulsory.' (Schot/Fischer, 1993: 372)

cleaner production methods in reality, and winning allies for advanced environmental regulation, though limiting their influence on standard-setting to avoid barriers to entry and implicit protectionism;

9. Dealing with, and learning from, the failures of environmental management, especially by comparing them with structurally related success stories;

10. And contributing to the diffusion of successful environmental management through public rewards, which help to propagate wider knowledge and recognition of it and support image-conscious firms that appreciate public recognition.

If environmental policy follows its own administrative and political requirements as listed above it will probably contribute most to improving environmental management and protection in society.[12] Such an environmental policy aims at:

1. Restructuring administrative arrangements internally to enhance long-term environmental policy in the direction of ecological sustainability;

2. Enlarging its own capacity, including the reduction of counterproductive administrative capacities;

3. The consistency, linkability and social enforceability of environmental regulation and measures;

4. And interpolicy cooperation to raise environmental concern in other policy fields and to prevent significant negative environmental impacts of other policies.

Among other things, this would imply turning away from the currently (explicitly or implicitly) predominant policy orientation towards end-of-pipe technology, bureaucratic regulation, unclear goal priorities, cumbersome public conflict management, and setting eco-performance conditions for public (investment) subsidies.

Corresponding innovation-friendly environmental regulation then follows the following principles (Porter/van der Linde, 1995: 124):

- Focus on outcomes, not technologies.
- Enact strict rather than lax regulation.
- Regulate as close to the end-user as practical, while encouraging up-stream solutions.
- Employ phase-in periods.
- Use market incentives.
- Harmonize or converge regulations in associated fields.
- Develop regulations in sync with other countries or slightly ahead of them.
- Make the regulatory process more stable and predictable.
- Require industry participation in setting standards from the beginning.

---

[12] In addition, such an environmental policy should be able to improve environmental management within public (political) institutions themselves, where it formally has easier access and can demonstrate the social (and economic) viability of corresponding environmental protection measures.

- Develop strong technical capabilities among regulators.
- Minimize the time and resources consumed in the regulatory process itself.

5   ENVIRONMENTAL POLICY RECOMMENDATIONS FOR THE EU-COMMISSION

EU environmental policy is particularly dependent on such a communicative policy orientation indicated above because of its even more limited (in comparison to national environmental policies) political resources and power to make its many directives actually work. So its main efforts to improve environmental management should be oriented towards communication, coordination, networking, besides strengthening uniform environmental rules and standards, where involving firms with advanced environmental management seems favourable. This demands competence in intrapolicy cooperation (communication with and coordination of various national environmental policies as well as among different industries in their environmental management practices) and in interpolicy cooperation (internalization of environmental concerns in other policies, especially economic, regional and agricultural policy).

With respect to eastern European countries, the current strategy to provide targeted subsidies and to encourage collaboration and joint ventures with western institutions in the EU has to be emphasized, allowing a transfer of knowledge and resources promoting future self-organized environmental management in eastern European companies.

6   EVALUATING THE ENVIRONMENTAL POLICY RECOMMENDATIONS

The above environmental policy recommendations are not specified with respect to the technical, social and temporal context called for in this chapter. Except for few general recommendations referring to the EU-Commission, no public actors have been mentioned as their addressees. No specified substantive goals have been listed underlying these recommendations. Nor has their time horizon been specified. The reason for these deficits lies in their level of abstraction as general orientations, which refer to the current situation of recognized, but not yet widely practised, (corporate) environmental management. These guidelines for environmental policy can thus be pursued by practically all the diverse (public) environmental policy actors on different levels of environmental protection and with any medium or long-term time horizon. More specific recommendations would have to take their respective technical, social and temporal context into account. Some can be found in the detailed case studies carried out in this project.

The above environmental policy recommendations can be derived quite plausibly or even stringently from the assumed model of multi-layered complex interplay among different determinants of environmental management and policy, be it on the level of social actors, of endogenous resources and capacities, of sociocultural and socioeconomic conditions, or of power and interest constellations. However, as for environmental management, the question is less what are substantially appropriate environmental policy orientations and strategies, which in essence have been well-known for years in science and often in politics, but rather how to achieve them under the actual social conditions of policy-making. Again, the answer is that these strategic environmental policy orientations may increasingly serve as mind-framing cognitive guidelines that cannot simply be enforced, but only aspired to in everyday politics. Promising concrete policy measures to comply with them will depend on the case-specific framework conditions, actor constellation, situational pattern and opportunity structure.

Analytically one may distinguish between structural and situational restrictions to successful environmental policy. Past and present environmental policy capacity thus appears to be great enough to induce and enforce improved environmental protection by end-of-pipe technology but not by changing civilizational patterns that are obviously deeply entrenched from the sociocultural point of view, such as a sociostructurally-embedded strong mobility orientation and the favoured private transportation (cf. Jänicke, 1996; Wolf, 1992). Its capacity is systemically insufficient to achieve more comprehensive environmental policy objectives, such as an ecologically sustainable society, even under conditions of eco-dictatorship. Therefore, even if systematically obvious solutions to the environmental problematique, such as ecological productivity, immaterialization of production, sufficiency (besides efficiency) goals through limitation of (material) volumes are thoroughly discussed in academia, environmental policy can only propagate but not substantially pursue them. It is possible, however, to enlarge gradually environmental policy capacity and overcome situational policy restrictions by following the above environmental policy recommendations. Although comparative analyses of successful (national) environmental policies suffer from considerable methodological and data problems (Jänicke, 1996), some conclusions about achievable conditions of success are feasible and supported by sophisticated policy analysis (see Hajer, 1995; Héritier, 1993; Huber, 1991, 1995b; Jänicke, 1990, 1993, 1996; Jänicke/Weidner, 1995, 1997; Vig/Kraft, 1994; Wallace, 1995; Weidner, 1996).

Thus, steps and measures to improve environmental policy (capacity and success) that are in principle socially and politically viable can be indicated today. However, their actual (political) realization clearly cannot be guaranteed by appropriate (scientific) policy recommendations. The above environmental policy recommendations are – as part of a secular social (learning) process fostered by crisis incidents, a gradually growing social appreciation of environmental protection and the self-dynamics of environmental policy (see Chapter 2) – considered to be at least partly viable in the longer run although they are rather demanding and therefore not very likely to be actually realized.

LITERATURE

Ashford, N. and Meima, R. (1994) Designing the Sustainable Enterprise. Summary Report Second International Conference. The Greening of Industry Network. ERP Environment, Wetherby.

Bartlett, R.V. (1994) Evaluating Environmental Policy Success and Failure. In: (eds) N.J. Vig and Kraft, M., *Environmental Policy in the 1990s*. Washington D.C.: CQ-Press.

Böhret, C. (1990) *Folgen. Entwurf für eine aktive Politik gegen schleichende Katastrophen.* Opladen: Leske + Budrich.

Conrad, J. (1990a) Do Public Policy and Regulation Still Matter for Environmental Protection in Agriculture? EUI Working Paper EPU 90/6, Florence.

Conrad, J. (1990b) Technological Protest in West Germany: Signs of a Politization of Production? *Industrial Crisis Quarterly* **4**, 175–191.

Conrad, J. (1992) *Nitratpolitik im internationalen Vergleich.* Berlin: edition sigma.

Daele, W. van den (1993) Sozialverträglichkeit und Umweltverträglichkeit. Inhaltliche Mindeststandards und Verfahren bei der Beurteilung neuer Technik. *Politische Vierteljahresschrift* **34**, 219–248.

Galtung, J. (1992) Prinzipien des ökologischen Überlebens. In: (eds) G. Altner *et al. Jahrbuch Ökologie 1992.* München: Beck.

Hajer, M. (1995) *The Politics of Environmental Discourse. Ecological Modernization and the Policy Process.* Oxford: Clarendon Press.

Héritier, A. (ed.) (1993) Policy-Analyse. Kritik und Neuorientierung. *Politische Vierteljahresschrift*, Special Issue 24. Opladen: Westdeutscher Verlag.

Héritier, A. *et al.* (1994) *Die Veränderung von Staatlichkeit in Europa.* Opladen: Leske + Budrich.

Hey, Ch. and Brendle, U. (1994) *Umweltverbände und EG.* Opladen: Westdeutscher Verlag.

Huber, J. (1991) *Unternehmen.Umwelt.* Frankfurt: Fischer.

Huber, J. (1995a) Nachhaltige Entwicklung durch Suffizienz, Effizienz und Konsistenz. In: (eds) P. Fritz *et al. Nachhaltigkeit in naturwissenschaftlicher und sozialwissenschaftlicher Perspektive.* Stuttgart: Wissenschaftliche Verlagsgesellschaft.

Huber, J. (1995b) *Nachhaltige Entwicklung. Strategien für eine ökologische und soziale Erdpolitik.* Berlin: edition sigma.

Jachtenfuchs, M. and Strübel, M. (eds) (1992) *Environmental Policy in Europe.* Baden-Baden: Nomos.

Jänicke, M. (1990) Erfolgsbedingungen von Umweltpolitik im internationalen Vergleich. *Zeitschrift für Umweltpolitik und Umweltrecht* **13**, 213–232.

Jänicke, M. (1992) Umweltpolitik 2000. Erfordernisse einer langfristigen Strategie. In: (eds) G. Altner *et al. Jahrbuch Ökologie 1992.* München: Beck.

Jänicke, M. (1993) Über ökologische und politische Modernisierungen. *Zeitschrift für Umweltpolitik und Umweltrecht* **16**, 159–175.

Jänicke, M. (ed.) (1996) *Umweltpolitik der Industrieländer. Entwicklung—Bilanz—Erfolgsbedingungen.* Berlin: edition sigma.

Jänicke, M. *et al.* (1992) *Umweltentlastung durch industriellen Strukturwandel?* Berlin: edition sigma.

Jänicke, M. and Weidner, H. (eds) (1995) *Successful Environmental Policy. A Critical Evaluation of 24 Cases.* Berlin: edition sigma.

Jänicke, M. and Weidner, H. (eds) (1997) *National Environmental Policies: A Comparative Study of Capacity Building.* Berlin: Springer.

Johnson, S. and Corcelle, G. (1989) *The Environmental Policy of the European Communities.* London: Graham and Trotman.

Kaufmann, F.X. (1977) Sozialpolitisches Erkenntnisinteress und Soziologie. In: (eds) Ch. von Ferber and F.X. Kaufmann, Soziologie und Sozialpolitik. *Kölner Zeitschrift für Soziologie und Sozialpsychologie*, Special Issue 19, 35–75.

Knoepfel, P. (1993) Bedingungen einer wirksamen Umsetzung umweltpolitischer Programme—Erfahrungen aus westeuropäischen Staaten. Cahiers de l'IDEAP 108, Lausanne.

Koppen, I. (1988) The European Community's Environmental Policy From the Summit in Paris, 1972, to the Single European Act, 1987. EUI Working Paper 88/328, Florence.

Ladeur, K.-H. (1987) Jenseits von Regulierung und Ökonomisierung der Umwelt: Bearbeitung von Ungewißheit durch (selbst-) organisierte Lernfähigkeit – eine Skizze. *Zeitschrift für Umweltpolitik und Umweltrecht* **10**, 1–22.

Liefferink, J.D. *et al.* (eds) (1993) *European Integration and Environmental Policy.* London: Belhaven Press.

Majone, G. (1989) Regulating Europe: Problems and Prospects. In: (eds) Th. Ellwein *et al. Jahrbuch zur Staats- und Verwaltungswissenschaft 3.* Baden-Baden: Nomos.

Mazur, K.P. *et al.* (1995) *Energieverbrauch: Kostenwahrheit ohne Staat?* Stuttgart: Schaeffer-Poeschel.

Mez, L. and Jänicke, M. (eds) (1996) *Sektorale Umweltpolitik.* Berlin: edition sigma.   •

Porter, M.E. and van der Linde, C. (1995) Green and Competitive: Ending the Stalemate, *Harvard Business Review* September–October 1995: 120–134.

Schmidt, E. and Spelthahn, S. (eds) (1994) *Umweltpolitik in der Defensive—Umweltschutz trotz ökonomischer Krise*. Frankfurt: Fischer.

Schot, J. and Fischer, K. (1993) Conclusion: Research Needs and Policy Recommendations. In: (eds) K. Fischer and J. Schot, *Environmental Strategies for Industry*. Washington D.C.: Island Press.

Stitzel, M. (1992) Das Umweltmanagement in der öffentlichen Verwaltung. In: (ed.) U. Steger, *Handbuch des Umweltmanagements*. München: Beck.

Vig, N.J. and Kraft, M.E. (eds) (1994) *Environmental Policy in the 1990s. Toward a New Agenda*. Washington D.C.: CQ Press.

Wallace, D. (1995) *Environmental Policy and Industrial Innovation. Strategies in Europe, the USA and Japan*. London: Earthscan.

Weidner, H. (1996) *Basiselemente eiuner erfolgreichen Umweltpolitik. Eine Analyse und Evaluation der Instrumente der japanischen Umweltpolitik*. Berlin: edition sigma.

Wolf, R. (1992) Sozialer Wandel und Umweltschutz. Ein Typologisierungsversuch. *Soziale Welt* **43**, 351–376.

SUBJECT INDEX